A Short Course on Relativistic Heavy Ion Collisions

A Short Course on Relativistic Heavy Ion Collisions

A K Chaudhuri

Variable Energy Cyclotron Centre, Kolkata, India and Homi Bhabha National Institute, Kolkata, India

IOP Publishing, Bristol, UK

ISBN 978-0-750-31060-4 (ebook)
ISBN 978-0-750-31061-1 (print)

DOI 10.1088/978-0-750-31060-4

Version: 20140901

IOP Expanding Physics
ISSN 2053-2563 (on-line)
ISSN 2054-7315 (print)

British Library Cataloguing-in-Publication Data: A catalogue record for this book is available from the British Library.

Published by IOP Publishing, wholly owned by The Institute of Physics, London

IOP Publishing, Temple Circus, Temple Way, Bristol BS1 6HG, UK

US Office: IOP Publishing, Inc., 190 North Independence Mall West, Suite 601, Philadelphia, PA 19106, USA

In memory of my mother, the late Tarubala Chaudhuri.

Contents

Preface

Relativistic heavy ion collision is a fascinating field of research. In recent years, the field has seen an unprecedented level of progress. A new state of matter, deconfined quark–gluon plasma (QGP), was predicted. An accelerator was built to detect this new state of matter. Experiments were performed and the discovery of the 'most perfect fluid' was made. Conclusive identification of the most perfect fluid state with the deconfined state has yet to be achieved. One of the impediments towards such identification is the fundamental property of the strong interaction, the 'color confinement', i.e. the constituents of the theory, the 'colored' quarks and gluons, are confined within a hadron. Any information about the deconfined state must be amassed from the color-neutral hadrons. And yet the process by which colored building blocks convert into a color singlet state is not properly understood. This necessitates model building. To young researchers, the field poses a problem in that it is multi-disciplinary, requiring knowledge of thermodynamics, statistical physics, kinetic theory, group theory, quantum chromodynamics (QCD), etc. The complexity of heavy ion collisions has necessarily led to a proliferation of models, e.g. the thermal model, blast wave model hydrodynamic model and models based on transport equations, etc, the physics of which need to be understood.

The Cyclotron Centre at Kolkata is actively engaged in theoretical and experimental research in high-energy nuclear physics. As a part of the training offered to our students, both in theory and experiment, we offer a course on relativistic heavy ion collisions. While teaching the course, I noted that there was a need for a book in which students could find the minimum knowledge to pursue a career in this field. I was then contacted by Dr John Navas of Institute of Physics Publishing, UK, who had come across our course material (which had been made publicly available for the benefit of our students) and suggested to me that I convert it into a book on the subject. I accepted the job. It was not easy; the initial plan was to write a book of approximately 200 pages. I selected a few topics that I felt were relevant to be discussed in the book, and ended up with a book that was closer to 300 pages in length.

The book is divided into nine chapters and an appendix. Each chapter is divided into sections and subsections. The first chapter is introductory, and discusses the subject of the deconfined state or QGP. A brief history of nuclear physics is given. The conceptual basis for the deconfinement of quarks and gluons is also given.

The second chapter deals with relativistic kinematics. Very fundamental concepts such as the space–time continuum, Lorentz transformation, center-of-mass frame and phase space are introduced. A short description of Dalitz decay is also included. I may mention here that there is a very good book on kinematics by Rolf Hagedorn, and my exposition on kinematics can in no way surpass his work.

The third chapter is entitled 'The Glauber model'. The Glauber model is routinely used in relativistic heavy ion collisions to characterize the collisions as a function of collision centrality. The fundamentals of the model are discussed briefly. Glauber's solution of quantum scattering from composite objects is given. The concepts of collision centrality and measurement by binary collisions, participant numbers, wounded nucleon numbers and so on are introduced.

I have entitled the fourth chapter 'Classical theories for macroscopic systems'. It is divided into three sections: (i) thermodynamics, (ii) statistical physics and (iii) kinetic theory. The discussions are not exhaustive. In thermodynamics, after stating the thermodynamics laws, I introduce the concept of thermodynamic potential and derive the Maxwell's equations. The physical meanings of different thermodynamic potentials, e.g. enthalpy, Gibbs potential, etc, are given. In statistical physics, after introducing the important concept of ensemble, three types of ensembles—microcanonical, canonical and grand canonical—are distinguished. Explicit expressions for various thermodynamic quantities are obtained from the partition function. The last topic discussed in this chapter is the kinetic theory. After giving some basic definitions and their relations with physical quantities, I derive the relativistic kinetic (transport) equation, with and without collision, with external force, for mixtures. The equilibrium distribution function is obtained from entropy maximization. Lastly, conservation equations are derived from the transport equation.

The fifth chapter deals with QCD. QCD is a gauge theory with SU(3) as the gauge group. It is a complex theory. Following a brief discussion on gauge theory, the QCD Lagrangian is studied. QCD is renormalizable and the physics of renormalization is given. The perturbative and non-perturbative approaches of QCD are distinguished. The important 'factorization theorem' is stated and concepts such as parton distribution function and fragmentation function are introduced. In the non-perturbative QCD, only the lattice QCD is discussed. Gauge and fermion action on a lattice is derived. The problems associated with fermion action are stated. The concepts of the Wilson loop, Polyakov loop, Z(3) symmetry, etc, are introduced. Some selected results of lattice simulations are discussed.

Chapter 6 discusses phenomenological equations of state for the two phases of QCD matter: QGP, the high temperature, high density phase, and hadronic resonance gas, the low temperature, low density phase. In the phenomenological approach, following the MIT bag model, a bag pressure is introduced to the QGP equation of state. To appreciate the bag pressure, a brief description of the MIT bag model is given. A van der Waals treatment of finite volume correction is also discussed here. I also discuss the historically important statistical bootstrap model.

Chapter 7 discusses some selected models for heavy ion collisions. The vastly successful statistical/thermal model, where particles are evaporated from a thermal source, the blast wave model where the thermal source rapidly explodes to freeze-out, and the hydrodynamic model where the fireball expands following hydrodynamic equations, are discussed in detail. The very useful Cooper–Frye prescription for the invariant particle distribution is also derived in this chapter. The chapter contains brief discussions on the transport approach based models: BUU/VUU and the quantum molecular dynamics based UrQMD.

Chapter 8 deals with two-particle correlation. 'Amplitude interferometry' and 'intensity interferometry' are distinguished. The quantum mechanical reason for intensity interferometry is discussed. The expression for the two-particle correlation function is derived. Gaussian parameterization for the correlation function, Bertsch–Pratt parameterization and Yano–Koonin–Podgoreskií parameterization are given. Finally, a few experimental results are discussed.

Since quarks and gluons are not observable, identifying QGP in a collision is non-trivial. Chapter 9 briefly discusses selected probes for QGP detection, J/ψ suppression, strangeness enhancement, electromagnetic possibilities (i.e. photons and dileptons), and jet quenching.

I hope that this book will be useful to students and young researchers in the field of relativistic heavy ion collisions. I will appreciate any comments and criticism of the material presented in this publication. Please email them directly to me at either akc@vecc.gov.in or asiskumarchaudhuri@gmail.com.

I wish to take this opportunity to thank my students and colleagues who have helped me in writing this book. I would particularly like to mention Dr Partha Barat, D N Basu, Jane Alam, Victor Roy, Partha Bhaduri and Zuber Ahmed, who read some portions of the book. I also thank the various authors who have kindly permitted me to use selected graphics from their publications for illustration purposes in this book. Lastly, I thank my wife, Suparna, and my son, Turja. Their unstinting support and patience made it possible for me to labor for eight months to prepare the book.

IOP Publishing

A Short Course on Relativistic Heavy Ion Collisions

A K Chaudhuri

Chapter 1

Introduction to the deconfined state

1.1 Introduction

Surprisingly, our diverse Universe consists of a handful of 'elementary' or 'fundamental'[1] particles. In figure 1.1, we have listed the currently known elementary particles. These elementary particles can be classified as (i) matter particles, the half integer spin fermions, and (ii) mediator particles, the integral spin bosons. There are two types of matter particles, called quarks and leptons. As indicated in figure 1.1, there are only six quarks, named up (u), down (d), strange (s), charm (c), bottom or beauty (b) and top or truth (t). Likewise, there are six leptons, electron (e), muon (μ), tau (τ), and their associated neutrinos, electron neutrino (ν_e), muon neutrino (ν_μ) and tau neutrino (ν_τ). Note that the matter particles are further classified into three generations. Each generation contains two quarks and two leptons, i.e. the first generation matter particles are the two quarks (u, d) and two leptons (e, ν_e), the second generation matter particles are the two quarks (c, s) and two leptons (μ, ν_μ), and the third generation matter particles are the two quarks (b, t) and two leptons (τ, ν_τ). The properties of the particles, with the exception of mass, remain unchanged between the generations. The mass increases over successive generations. For each of these matter particles, there is an antiparticle, e.g. a positron (e^+) is the antiparticle of an electron (e^-). The handful of fundamental particles can interact in only four definite manners: (i) strong interaction, (ii) electromagnetic (EM) interaction, (iii) weak interaction and (iv) gravitational interaction. In table 1.1, we have listed the mediators of the interactions, and also shown the relative strength of the interactions. All these particles interact gravitationally.

The study of the strong interaction is generally called nuclear physics. Historically, nuclear physics started with Rutherford's discovery of the nucleus in his

[1] By elementary or fundamental particles we mean that they are simple and structureless, not made of anything smaller.

doi:10.1088/978-0-750-31060-4ch1

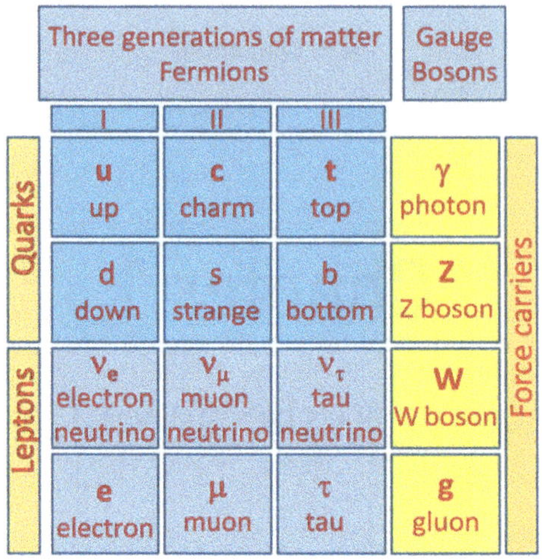

Figure 1.1. The three generations of matter particles in the standard model. The mediator particles (gauge bosons) are also shown.

Table 1.1. The four fundamental forces, their relative strengths and their mediators are listed.

Interaction	Theory	Mediators	Relative strength	Interaction range (m)
strong	QCD	gluon	10^{38}	10^{-15}
electromagnetic	QED	photon	10^{36}	infinity
weak	electroweak	W, Z	10^{25}	10^{-18}
gravitational	general relativity	graviton	1	infinity

celebrated gold foil experiment in 1909[2]. At that time, the prevailing model for atom was J J Thompson's plum pudding model. Massive positively charged particles were assumed to be distributed uniformly over the atomic volume, with electrons (the plum) dispersed throughout the positive mass (the pudding), to maintain charge neutrality of the atom. In this model, alpha particles would be expected to pass through the gold foil, with little deviation in their paths. In Rutherford's experiment, quite a few alpha particles were found to be deflected by large angles. The result of the gold foil experiment was so bizarre at that time that Rutherford said, 'It was almost as incredible as if you fired a 15-inch shell at a piece of tissue paper and it came back and hit you'. The experimental result can only be explained if positive charges are concentrated within a small volume. Rutherford used the term 'nucleus' to describe the concentration of positive charges at the centre of the atom. Incidentally, this term was first coined by Robert Brown, the botanist, in 1831, to describe cell structure (alternatively, by Michael Faraday in 1844), from the Latin word *nux* which means 'nut'. Rutherford also realized the necessity for a third

[2] To be specific, two of Rutherford's scholars, Hans Geiger and Ernest Marsden, conducted the experiment and wrote the paper. The paper was communicated to the Royal Society by Lord Rutherford.

atomic particle, the neutron, for the stability of the nucleus. The concept of the atomic nucleus was completed when James Chadwick, Rutherford's collaborator, discovered the neutron in 1932. Indeed, one can say that proper nuclear physics started in 1932 after the discovery of the neutron.

For a long time the atomic nucleus was assumed to be composed of protons (a term possibly coined by Rutherford for the hydrogen nucleus) and neutrons and they were to interact strongly. In the meantime there was much progress in the understanding of EM interaction. It was recognized that EM interaction arises due to the exchange of photons between two charged particles. In analogy to EM interaction, in 1934 Hideki Yukawa put forward the hypothesis that strong interaction between nucleons originated from the exchange of mesons. He also predicted the mass of a meson to be 200 times that of an electron. At that time mesons were not known. He made this bold conjecture to obtain a theory analogous to EM interaction. Yukawa was just 27 years old.

In 1937 pions were discovered and in 1949 Yukawa was awarded the Nobel Prize in Physics. However, in later years, with the advent of particle accelerators, experimentalists discovered hundreds of particles (mesons and baryons), many of which can be thought of as mediators of the strong interaction. People then tried to characterize these particles and study their internal symmetry[3]. We will not go into detail, but suffice to say that Murray Gell-Mann and George Zweig (1964) found that all of these particles, including protons and neutrons, consist of only a few building blocks which are called quarks. Quarks are fractionally charged particles. Since fractionally charged particles are not observed experimentally, it was theorized that they are confined inside the hadrons. Murray Gell-Mann picked the word 'quark' from the sentence 'Three quarks for Muster Mark' in the James Joyce novel *Finnegans Wake*. The simplest version of the quark model faces a problem. Some baryons, e.g. Ω^- or Δ^{++}, would then be composed of identical quarks and violate Pauli's exclusion principle. To eliminate the contradiction, the concept of color was introduced. Color is a new quantum number. Only three colors are required to be hypothesized.

Murray Gell-Mann was born in September 1929. When he postulated quarks, he was 35 years old. He won the Nobel prize in 1969. One can borrow the words of G H Hardy (known for discovering Ramanujan, the Indian mathematical wizard) and say, 'creative physics is a young man's game'. Take, for example, Sir Isaac Newton, who at the age of 23–24 gave us the law of gravitation and discovered fluxions (calculus), or Albert Einstein, who discovered relativity at the age of 25–26. Wolfgang Pauli formulated his exclusion principle when he was 25 years old.

Traditionally, nuclear physics is the study of nuclear matter at zero temperature and at densities of the order of the atomic nuclei, nucleon density $\rho \sim 0.17\ \text{fm}^{-3}$ or energy density $\epsilon \sim 0.16\ \text{GeV fm}^{-3}$. The advent of accelerators has extended the study to hundreds of MeV in temperature and to energy densities several orders of

[3] Internal symmetry refers to the fact that one generally finds a family of particles called multiplets, all with the same or nearly the same mass. Each multiplet can be considered as a realization of some internal symmetry.

magnitude higher. At such high density/temperature, individual hadrons lose their identity and the matter is best described in terms of its constituents, e.g. quarks and gluons, commonly called quark–gluon plasma (QGP). Historically, T D Lee, in collaboration with G C Wick, first speculated about an abnormal nuclear state in which nucleon mass is zero or near zero in an extended volume and non-zero outside the volume [1, 2]. They also suggested that an effective way to search for these new objects was through high-energy heavy ion collisions. Around the same time, Collins and Perry [3] conjectured that super-dense matter, found in neutron stars, exploding supernovae, the early Universe, etc, consists of quarks rather than of hadrons. They argued as follows: a neutron has a radius of ~0.5–1.0 fm and a density of $\sim 8 \times 10^{14}$ gcm^{-3}. The central density of a neutron star is $\sim 10^{16} - 10^{17}$ gcm^{-3}. It is then expected that the neutrons overlap and their individuality is lost. Collins and Perry then suggested that the matter at the high densities of a neutron star is a quark soup. In 1979, Susskind [4] studied quark confinement at finite temperature on a lattice. He found that confinement is strictly a low-temperature phenomenon. A transition to a plasma-like phase occurs at high temperatures. In this plasma-like phase, free gluons Debye screen the quarks. In 1980 Shuryak [5] wrote the first review paper on quantum chromodynamic (QCD) thermodynamics. In 1983, J D Bjorken [6] developed the 1+1 dimension hydrodynamic model for the space–time evolution of matter created in relativistic heavy ion collisions. Incidentally, the hydrodynamic solution now known as the Bjorken scaling solution was earlier obtained by Chiu, Sudarshan and Wang [7]. In 1984, the US government sanctioned plans to build the Relativistic Heavy Ion Collider (RHIC) to expressly search for QGP. Construction work started in 1991 and the first physics result was published in 2001. To date, the four experimental groups, (i) STAR (**S**olenoid **T**racker **A**t **R**HIC), (ii) PHENIX (**P**ioneering **H**igh **E**nergy **N**uclear **I**on e**X**periment) (iii) BRAHMS (**B**road **R**ange **HA**dron **M**agnetic **S**pectrometers) and (iv) PHOBOS[4], at RHIC have published more than 400 scientific papers in peer reviewed journals. Approximately 40% of the papers were published in *Physical Review Letters*, which is known to publish only 'important' physics results.

In this short course, we will try to discuss some aspects of matter at such high density and temperature. For a general introduction to the subject, see [8–11]. More detailed information can be obtained in the white papers published by the four experimental groups at RHIC [12–15] and also from the proceedings of the Quark Matter conference series [16].

1.2 Conceptual basis for QGP formation

For composite hadrons, with finite spatial extension, the concept of hadronic matter appears to lose its meaning at sufficiently high density. Once we have a system of

---

[4] Unlike the other three experiments, PHOBOS is not an acronym. Experimentalists from MIT first proposed an experiment called MARS (**M**odular **A**rray for **R**HIC **S**pectra) which was rejected for being too expensive. A much scaled down project was accepted. Professor John Negele suggested that it be called PHOBOS, after the largest moon of Mars.

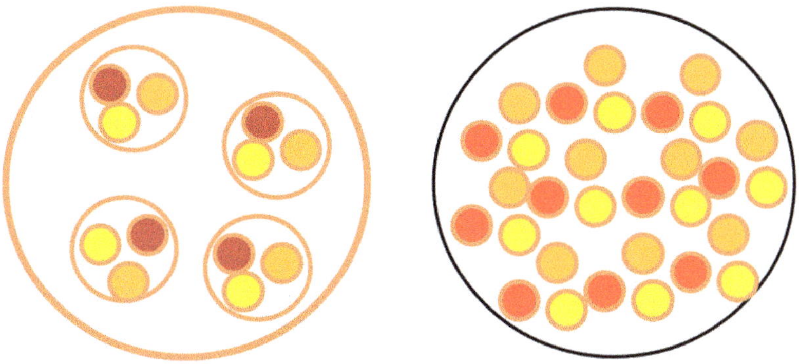

Figure 1.2. The left panel shows a nucleus at normal density. The right panel shows the same at high density.

mutually interpenetrating hadrons, each quark will find in its vicinity a number of quarks, at a distance less than the hadron radius. The situation is shown schematically in figure 1.2. At low density, a particular quark in a hadron knows its partner quarks. However, at high density, when the hadrons start to interpenetrate each other, a particular quark will not be able to identify the quark which was its partner at lower density. Indeed, a similar consideration led Collins and Perry [3] to suggest that the matter in the neutron star will behave as a quark soup. Similar phenomena can happen at high temperature. Hagedorn [17] studied the thermodynamics of interacting hadron gas where resonance formation is the basis of interaction. He developed the so-called 'statistical bootstrap model'. Later, in section 6.6, the model will be discussed in more detail. It was shown that the density of states increases exponentially with mass and there is a temperature T_H beyond which the partition function[5] diverges. The temperature is known as the Hagedorn temperature and Hagedorn interpreted the temperature as the limiting temperature beyond which hadronic matter cannot exist. Cabibbo and Parisi [18] gave an alternative interpretation to the exponentially increasing mass spectrum. They argued that the exponentially increasing spectrum is associated with any system that is undergoing second order phase transition. The Hagedorn temperature is the critical temperature, associated with the transition. We can visualize it as follows: with increasing temperature, more and more resonances will be formed from thermal fluctuations. Resonance production will be dominated by low mass resonances, as the large mass resonance production will be suppressed by the Boltzmann factor $e^{-m_H/T}$. A large number of resonances will make the system dense again and hadrons can start to interpenetrate each other.

The system in which hadrons interpenetrate is best considered as quark matter, rather than being made of hadrons. It is customary to call the quark matter QGP. We define QGP[6] as a thermalized, or near thermalized, state of quarks and gluons, where quarks and gluons are free to move over a nuclear volume rather than a

[5] The partition function is a concept in statistical physics and describes the properties of a system at equilibrium. It will be discussed in a later chapter.

[6] The *Oxford English Dictionary* defines QGP as a hypothetical, highly energized form of matter that contains unbound quarks and gluons, believed to have been present ten millionths of a second after the Big Bang.

nucleonic volume. Model calculations indicate that beyond a critical energy density $\epsilon_{cr} \sim 1$ GeV fm^{-3}, or temperature $T_{cr} \sim 200$ MeV, matter can exist only as QGP.

QGP is the deconfined state of strongly interacting matter. Since at low density or low temperature quarks are confined within the hadrons, and at high density or high temperature quarks are deconfined, one can talk about a confinement–deconfinement phase transition. This will be discussed later, but it turns out that for a baryon-free matter, the confinement–deconfinement transition is not a phase transition in the thermodynamic sense[7]; rather, it is a smooth cross-over, from confinement to deconfinement or vice versa. The mechanism of deconfinement is provided by the screening of the color charges [4]. It is analogous to the Mott transition in atomic physics [19]. In dense matter, the long range Coulomb potential, which binds ions and electrons into electrically neutral atoms, is partially screened due to the presence of other charges and the potential become much more short range,

$$V(r) = \frac{e_0^2}{r} \rightarrow \frac{e_0^2}{r} \times \exp\left(-\frac{r}{r_\mathrm{D}}\right). \tag{1.1}$$

Here r is the distance of the probe from the test charge, e_0. r_D is the Debye screening radius and is inversely proportional to density,

$$r_\mathrm{D} \sim n^{-1/3}. \tag{1.2}$$

At sufficiently high density, r_D can be smaller than the atomic radius. A given electron can no longer feel the binding force of its ion; alternatively, at such density, the Coulomb potential can no longer bind the electron and ion into a neutral atom. The insulating matter becomes a conducting matter. This is the Mott transition. We expect deconfinement to be the QCD analog of the Mott transition. Due to screening of the color potential, quarks cannot be bound into a hadron. Now one may wonder about the very different nature of QCD and quantum electrodynamic (QED) forces. Interaction potential in QED and QCD can be expressed as

$$QED\colon\ V(r) \sim -\frac{e^2}{r} \tag{1.3}$$

$$QCD\colon\ V(r) \sim -\frac{\alpha}{r} + \sigma r. \tag{1.4}$$

In equation (1.4), $\alpha = \frac{g^2}{4\pi}$ is the strong coupling constant and the second term σr accounts for the QCD confinement, σ being the string constant. While in QED the potential decreases continuously with increasing distance, in QCD the potential increases with increasing distance. However, screening is a phenomenon that occurs at high density, or at short distance. The difference between QED and QCD at large distances is then of no consequence. Moreover, due to asymptotic freedom (to be

[7] In thermodynamic phase transition, free energy or its derivative at some order has singularity at the transition point.

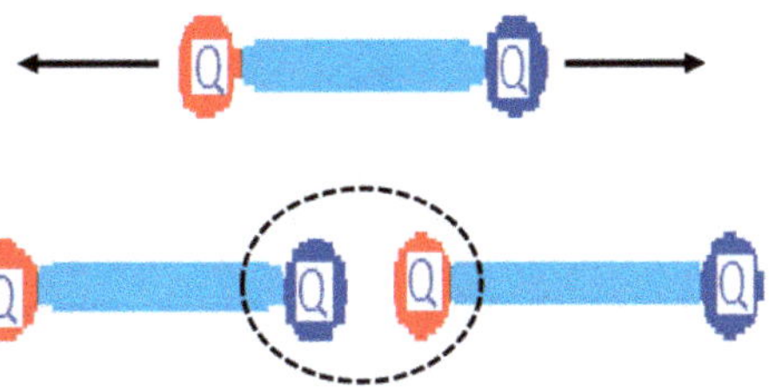

Figure 1.3. Schematic representation of string fragmentation. When two quarks are pulled apart, at some stage the string breaks to produce two pairs of quarks.

discussed later), in QCD the interaction strength decreases at short distances, thereby enhancing the deconfinement.

It may be noted that in an insulating solid, at $T > 0$, conductivity is not exactly zero, but it is exponentially small

$$\sigma_E \sim e^{-\Delta E/T} \tag{1.5}$$

where ΔE is the ionization potential. Above the Mott transition temperature, σ_E is non-zero because Debye screening has globally dissolved Coulomb binding between the ion and electrons, but below the Mott transition temperature ionization can produce locally free electrons, making σ_E small but non-zero. A corresponding phenomenon in QCD is string fragmentation. This is shown schematically in figure 1.3. Consider a hadron made up of two quarks. The two quarks interact via exchange of gluons. Now consider pulling the quarks apart. The confining potential will rise with the distance of separation and a flux tube will grow between the quarks until it reaches the value m_H, the lowest $q\bar{q}$ state. At this point, an additional hadron will form, whose anti-quark neutralizes the quark we were trying to separate. This is the mechanism of quark fragmentation.

1.3 Why is it important to study QGP?

QGP must have existed in the very early Universe. In the Big Bang theory[8], our Universe has undergone several stages of evolution since its birth some 14 billion years ago. In figure 1.4 , different stages of the evolution of the Universe, in the Big Bang model, are shown. They can be broken down as follows.

(i) The earliest time scale one can talk about is Planck time[9], $t_P = \sqrt{\frac{\hbar G}{c^5}} \approx 5.39 \times 10^{-44}$ s. At the Planck time scale, the temperature of the Universe is

[8] In 1929, Edwin Hubble discovered what is now known as the expanding Universe: that all the galaxies outside our own Milky Way are moving away from us at a speed proportional to their distance from us. It is easy to understand that if we go back in time the Universe becomes smaller and smaller. If we continue to travel backward, we come to a time (now known to be about 14 billion years ago) when the entire Universe was contained in a single point in space. The Universe must have been born in this single violent event which came to be known as the 'Big Bang'.

[9] Planck time is a unique combination of the gravitational constant (G), speed of light (c) and quantum constant ($\hbar$) that produces a constant with the unit of time (see appendix A.2).

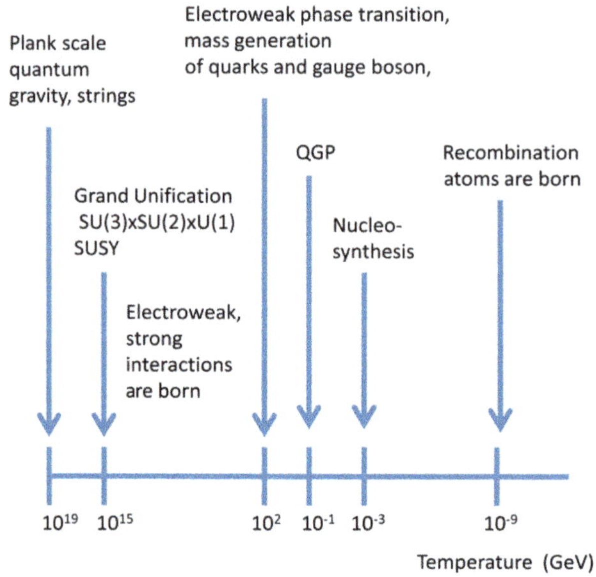

Figure 1.4. Schematic representation of the temporal evolution of the Universe in the Big Bang theory.

of the order of $T \sim 10^{19}$ GeV, which is the Planck scale temperature. Despite an enormous effort by string theorists, little is understood about this era. It is believed that at this stage quantum gravity was important and all the four forces were unified.

(ii) As the Universe continues to expand it cools. We have a better understanding of the later stage of evolution, when the temperature of the Universe was, say, around $T \sim 10^{16}$ GeV. This is in the grand unification scale, where gravity is separated and only the strong, EM and weak interactions are unified. The Universe at this scale may also be supersymmetric[10].

(iii) As the Universe further expands and cools, the strong and electroweak interactions are separated. At much lower temperatures of $T \sim 100$ GeV, electroweak symmetry breaking takes place. Baryon asymmetry may be produced here. The Universe exists as QGP, a deconfined state of quarks and gluons.

(iv) Somewhere around $T \sim 100$ MeV, deconfinement–confinement transition occurs, and hadrons are formed. RHIC at Brookhaven National Laboratory and the Large Hadron Collider (LHC) at CERN were designed to study matter around this temperature.

(v) At a temperature of $T \sim 1$ MeV, nucleosynthesis starts and light elements are formed. This temperature range has been well studied in nuclear physics

[10] Supersymmetry is a new kind of symmetry, where each particle boson (fermion) has a superpartner fermion (boson). For perfect supersymmetry, superpartners must have the same mass. It is evident that supersymmetry is not currently observed in nature. Up to now, experiments have failed to detect any superpartner.

experiments. For example, nuclear collisions at the Variable Energy Cyclotron Centre, Kolkata, produce matter around this temperature.

(vi) At a temperature of $T \sim 1$ eV, the Universe changes from being an ionized gas to a gas of neutral atoms, and structures begin to form.

QGP may also exist at the core of a neutron star. Neutron stars are remnants of the gravitational collapse of massive stars. They are small objects, with a radius of ~10 km, but very dense, with a central density of ~10 times the normal nuclear matter density. At such high density, hadrons lose their identity and matter is likely to be in the form of QGP. One important difference between the QGP in the early Universe and that in neutron stars is the temperature. While in the early Universe QGP temperature was $T \sim 100$ MeV, at the core of a neutron star the QGP is cold, $T \sim 0$ MeV. Hot and dense matter with an energy density exceeding 1 GeV fm^{-3} may also occur in supernova explosions, collisions between neutron stars or between black holes.

Bibliography

[1] Lee T D and Wick G C 1974 *Phys. Rev.* D **9** 2291

[2] Lee T D 1975 *Rev. Mod. Phys.* **47** 267

[3] Collins J C and Perry M J 1975 *Phys. Rev. Lett.* **34** 1353

[4] Susskind L 1979 *Phys. Rev.* D **20** 2610

[5] Shuryak E V 1980 *Phys. Reports* **61** 71

[6] Bjorken J D 1983 *Phys. Rev.* D **27** 140

[7] Chiu C B, Sudarshan E C G and Wang K-h 1975 *Phys. Rev.* D **12** 902

[8] Wong C-Y (ed) *Introduction to High Energy Heavy Ion Collisions* (Singapore: World Scientific)

[9] Csernai L P and Vogt R (ed) *Introduction to Relativistic Heavy Ion Collisions* (Amsterdam: Elsevier)

[10] Vogt R 2013 *Ultrarelativistic Heavy-ion Collisions* (Amsterdam: Elsevier)

[11] Florkowski W (ed) *Phenomenology of Ultra-relativistic Heavy-ion Collisions* (Singapore: World Scientific)

[12] Arsene I *et al* (BRAHMS Collaboration) 2005 *Nucl. Phys.* A **757** 1

[13] Back B B *et al* (PHOBOS Collaboration) 2005 *Nucl. Phys.* A **757** 28

[14] Adcox K *et al* (PHENIX Collaboration) 2005 *Nucl. Phys.* A **757** at press (arXiv: nucl-ex/0410003)

[15] Adams J *et al* (STAR Collaboration) 2005 *Nucl. Phys.* A **757** at press (arXiv: nucl-ex/0501009)

[16] Proceedings of Quark Matter conferences

[17] Hagedorn R 1965 *Nuovo. Cim. Suppl.* **3** 147
Hagedorn R 1968 *Nuovo. Cim.* **56A** 1027

[18] Cabibbo N and Parisi G 1975 *Phys. Lett.* B **59** 67

[19] Mott N 1968 *Rev. Mod. Phys.* **40** 677

Chapter 2

Kinematics of heavy ion collisions

2.1 Introduction

Our knowledge of the Universe is gained through experiments. The horizons of the human mind and of science are increased by solving puzzles posed by newer and newer experiments. It is thus appropriate that we start our discussion with relativistic kinematics for heavy ion collisions. Kinematics is the branch of dynamics that deals with the motion of bodies, without consideration of the forces acting on the bodies. It is very relevant for experimentalists. Readers may consult [1–3] for additional information on kinematics.

Throughout the note, we have used natural units; see appendix A.1 for details of what this means. We have also used Einstein's summation convention, by which repeated indices are summed over (unless otherwise stated). Thus,

$$\partial_\mu J^\mu \equiv \partial_t J^t + \partial_x J^x + \partial_y J^y + \partial_z J^z. \tag{2.1}$$

2.2 Space–time continuum

All of us have some notion of space and time. In Galilean mechanics, three-dimensional space and one-dimensional time are distinctly different entities. Soon after Einstein's discovery of relativity, Hermann Minkowski developed a new scheme for thinking about space and time that emphasized its geometric qualities. In relativistic mechanics, space and time are mixed mathematically. Minkowski then reasoned that they are but one entity which he called the space–time continuum. Continuum because, according to our experience, there is no void in space or in time. In a public lecture on relativity, he said:

> The views of space and time which we wish to lay before you have sprung from the soil of experimental physics, and therein lies their strength. They are radical. Henceforth, space by itself, and time by itself, are doomed to fade away into mere shadows, and only a kind of union of the two will preserve an independent reality.

doi:10.1088/978-0-750-31060-4ch2

The space–time continuum is also called Minkowski space. It is a metric space[1]. The metric is

$$\eta^{\mu\nu} = g^{\mu\nu} = g_{\mu\nu} = \begin{pmatrix} 1 & 0 & 0 & 1 \\ 0 & -1 & 0 & 0 \\ 0 & 0 & -1 & 0 \\ 0 & 0 & 0 & -1 \end{pmatrix}. \tag{2.2}$$

It is also written more briefly as

$$g^{\mu\nu} = \mathrm{diag}\left(1, -1, -1, 1\right). \tag{2.3}$$

For a metric as in equation (2.2), the space–time is flat. Each point in the space–time continuum can be designated by four real numbers, three for the spatial directions and one for the time, which together form a four-vector. Two types of four-vectors are defined. Contravariant four-vector x^μ is

$$x^\mu \equiv \left(x^0, x^1, x^2, x^3\right) = (t, x, y, z). \tag{2.4}$$

Using the metric $g^{\mu\nu}$ one also defines a covariant four-vector,

$$x_\mu \equiv g_{\mu\nu} x^\nu = \left(x_0, -x_1, -x_2, -x_3\right) = (t, -x, -y, -z). \tag{2.5}$$

The scalar product of two four-vectors is denoted as

$$a \cdot b = a^\mu b_\mu = a_0 b_0 - a^1 b^1 - a^2 b^2 - a^3 b^3 = a_0 b_0 - \mathbf{a} \cdot \mathbf{b} \tag{2.6}$$

where the bold face quantities denote vectors in three dimensions.

One immediately finds the invariant square of a four-vector,

$$x^2 = x \cdot x = x^\mu x_\mu = x_0^2 - \sum_{i=1}^{3} x^{i2} = t^2 - x^2 - y^2 - z^2. \tag{2.7}$$

The distance between two space–time points, $x_1^\mu = (t_1, x_1, y_1, z_1)$ and $x_2^\mu = (t_2, x_2, y_2, z_2)$ is calculated as

$$\begin{aligned} d\tau^2 &= (t_1 - t_2)^2 - (x_1 - x_2)^2 - (y_1 - y_2)^2 - (z_1 - z_2)^2 \\ &= dt^2 - dx^2 - dy^2 - dz^2 \end{aligned} \tag{2.8}$$

which is invariant under the Lorentz transformation. $\tau = \sqrt{t^2 - x^2 - y^2 - z^2}$ is called proper time.

In figure 2.1, we have depicted the space–time continuum in two dimensions, t and z. The region of space–time for which $\tau^2 > 0$ is called the time-like region, while $\tau^2 < 0$ is called the space-like region. The $t = z$ line is called light-like (only light or

[1] A metric space is a set X that has the notion of distance $d(x, y)$ between every pair of points $x, y \in X$.

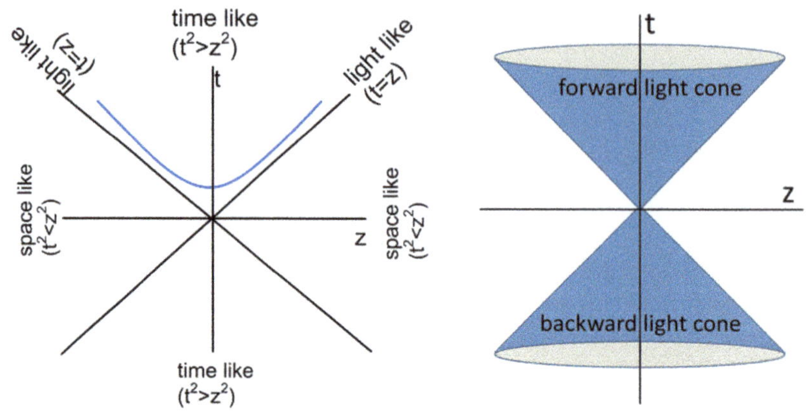

Figure 2.1. Left panel: schematic diagram of the space–time continuum in two dimensions. The time-like and space-like regions of the space–time continuum are shown. Also shown are the light-like lines. The blue curve is a constant proper time curve. Right panel: schematic picture of a forward and backward light cone. In relativistic heavy ion collisions, nuclei collide at $t = x = y = z = 0$. Only the forward light cone is accessible to the various particles that are produced after the collision.

massless particles can travel along this line). The space-like region is inaccessible to a physical particle, since it needs to travel faster than light. For a massive particle, with speed $v < c$, the only accessible region is the time-like region. One understands that in 3+1 dimensions, the light-like surfaces will form a cone as shown in figure 2.1. Only the forward light-cone region is accessible to physical particles.

2.3 Lorentz transformation

Einstein's theory of special relativity rests on the equivalence principle: *physical processes are the same in all the inertial*[2] *frames.*

Measurements made in one inertial frame are related to the measurements in another inertial frame by a transformation, called the Lorentz transformation. In relativistic nucleus–nucleus collisions it is convenient to use kinematic variables which take a simple form under Lorentz transformation. For completeness, we briefly discuss Lorentz transformation.

If x^μ is the coordinate in one frame of reference, then in any other frame of reference the coordinate x'^μ must satisfy

$$g_{\mu\nu} \mathrm{d}x'^\mu \mathrm{d}x'^\nu = g_{\mu\nu} \mathrm{d}x^\mu \mathrm{d}x^\nu \tag{2.9}$$

or equivalently

$$g_{\mu\nu} \frac{\mathrm{d}x'^\mu}{\mathrm{d}x^\rho} \frac{\mathrm{d}x'^\nu}{\mathrm{d}x^\sigma} = g_{\rho\sigma}. \tag{2.10}$$

[2] Inertial frames move with *constant* velocity with respect to each other.

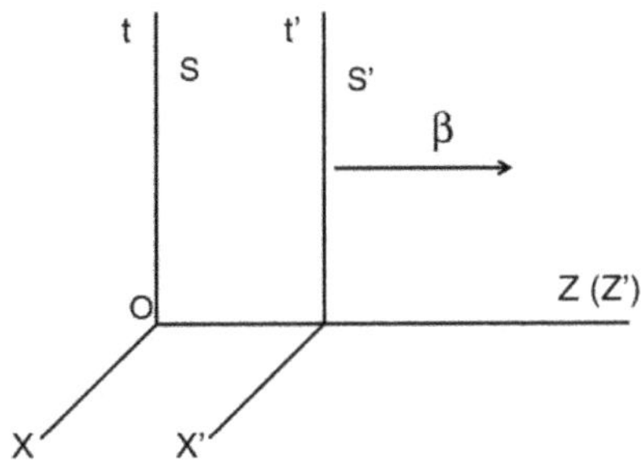

Figure 2.2. Schematic diagram of two inertial frames S and S', S' moving with a longitudinal velocity β.

The transformation has the special property that the speed of light is the same in the two frames of reference: a light wave travels at the speed $|\mathrm{d}\mathbf{x}/\mathrm{d}t| = 1$. The transformation $x^\mu \to x'^\mu = \Lambda^\mu_\nu x^\nu + a^\mu$, a^μ being an arbitrary constant, satisfying equation (2.10), i.e.

$$g_{\mu\nu}\Lambda^\mu_\rho \Lambda^\nu_\sigma = g_{\rho\sigma}, \tag{2.11}$$

is called a Poincaré transformation. Lorentz transformation is the special case of Poincaré transformation when $a^\mu = 0$. The matrix Λ^μ_ν forms a group called the Lorentz group.

A general Lorentz transformation consists of rotation and translation. Lorentz transformation without rotation is called Lorentz boost. As an example, consider the Lorenz boost along the z direction by velocity β (see figure 2.2). The transformation leaves the transverse coordinates x and y unchanged. Only t and z are changed. The transformation of t and z can be written as

$$\begin{pmatrix} t' \\ z' \end{pmatrix} = \begin{pmatrix} \gamma & -\beta\gamma \\ -\beta\gamma & \gamma \end{pmatrix} \begin{pmatrix} t \\ z \end{pmatrix} \tag{2.12}$$

where $\gamma = 1/\sqrt{1-\beta^2}$ is the Lorentz factor.

Inverse Lorentz transformation can be easily obtained by changing the direction of the velocity, $\beta \to -\beta$,

$$\begin{pmatrix} t \\ z \end{pmatrix} = \begin{pmatrix} \gamma & -(-\beta)\gamma \\ -(-\beta)\gamma & \gamma \end{pmatrix} \begin{pmatrix} t' \\ z' \end{pmatrix}. \tag{2.13}$$

2.4 Space–time rapidity

In equation (2.12) we replace the velocity β by a new variable η_s,

$$\beta = \tanh(\eta_s). \tag{2.14}$$

It is easy to see that the Lorentz factor is $\gamma = \cosh(\eta_s)$. The variable η_s is called space–time rapidity (the subscript s is to distinguish space–time rapidity from pseudo-rapidity of particles, to be defined later). In terms of space–time rapidity, the Lorentz transformation can be written as

$$\begin{pmatrix} t' \\ z' \end{pmatrix} = \begin{pmatrix} \cosh(\eta_s) & -\sinh(\eta_s) \\ -\sinh(\eta_s) & \cosh(\eta_s) \end{pmatrix} \begin{pmatrix} t \\ z \end{pmatrix}. \tag{2.15}$$

Note the similarity of the transformation matrix with the rotation matrix[3]. Transformations given by equation (2.15) are called hyperbolic rotation. Lorentz boost along the z direction can be thought of as a hyperbolic rotation by rapidity η_s.

Rapidity variables have the interesting property that they are additive under Lorentz boost. Successive Lorentz boost with rapidity η_{s1} and η_{s2} is equivalent to a boost with rapidity $\eta_{s1} + \eta_{s2}$. This can be shown by direct computation,

$$\begin{aligned} \begin{pmatrix} t' \\ z' \end{pmatrix} &= \begin{pmatrix} \cosh(\eta_{s2}) & -\sinh(\eta_{s2}) \\ -\sinh(\eta_{s2}) & \cosh(\eta_{s2}) \end{pmatrix} \begin{pmatrix} \cosh(\eta_{s1}) & -\sinh(\eta_{s1}) \\ -\sinh(\eta_{s1}) & \cosh(\eta_{s1}) \end{pmatrix} \begin{pmatrix} t \\ z \end{pmatrix} \\ &= \begin{pmatrix} \cosh(\eta_{s2} + \eta_{s1}) & -\sinh(\eta_{s2} + \eta_{s1}) \\ -\sinh(\eta_{s2} + \eta_{s1}) & \cosh(\eta_{s2} + \eta_{s1}) \end{pmatrix} \begin{pmatrix} t \\ z \end{pmatrix}. \end{aligned} \tag{2.16}$$

The additive property of rapidity is very useful in transforming quantities from one frame to another.

We have discussed Lorentz transformation with respect to the space–time co-ordinates. However, the transformation is generic and any of the four-vector is transformed similarly. Consider, for example, that in a frame S, a particle has energy E and three-momentum $\mathbf{p} = (p_x, p_y, p_z)$. In four-vector notation, $p^\mu = p = (E, p_x, p_y, p_z)$. If observed from a frame S', moving along the z direction with a velocity β, E and p_z are transformed as

$$\begin{pmatrix} E' \\ p_z' \end{pmatrix} = \begin{pmatrix} \gamma & -\beta\gamma \\ -\beta\gamma & \gamma \end{pmatrix} \begin{pmatrix} E \\ p_z \end{pmatrix}. \tag{2.17}$$

The inverse transformation is

$$\begin{pmatrix} E \\ p_z \end{pmatrix} = \begin{pmatrix} \gamma & \beta\gamma \\ \beta\gamma & \gamma \end{pmatrix} \begin{pmatrix} E' \\ p_z' \end{pmatrix}. \tag{2.18}$$

---

[3] A two-dimensional rotation matrix, where the axes are rotated by an angle θ, has the form,

$$R(\theta) = \begin{pmatrix} \cos(\theta) & -\sin(\theta) \\ \sin(\theta) & \cos(\theta) \end{pmatrix}$$

The following relation between the Lorentz factor and particle energy is useful,

$$\gamma = \frac{1}{\sqrt{1-\beta^2}} = \frac{1}{\sqrt{1-(\mathbf{p}/E)^2}} = \frac{E}{M}. \tag{2.19}$$

2.5 Transverse mass, rapidity and pseudo-rapidity

In relativity, the four-momentum of a particle is written as $p^\mu = (E, \mathbf{p}) = (E, p_x, p_y, p_z)$. However, it is more convenient to use variables, transverse mass m_T and rapidity Y. The transverse mass m_T and rapidity Y of a particle are defined as

$$\text{Transverse mass: } m_T = \sqrt{m^2 + p_T^2} \tag{2.20}$$

$$\begin{aligned}\text{Rapidity: } Y &= \frac{1}{2}\ln\frac{E+p_z}{E-p_z} = \frac{1}{2}\ln\frac{1+p_z/E}{1-p_z/E} \\ &= \tanh^{-1}\left(\frac{p_z}{E}\right) = \tanh^{-1}(\beta_L).\end{aligned} \tag{2.21}$$

Unlike the transverse momentum which is a vector quantity, transverse mass is scalar; consequently, it is invariant under a Lorentz transformation. Rapidity is not invariant under Lorentz transformation but it has a simple transformation law under Lorentz transform. The reader may notice the similarity between the space–time rapidity in equation (2.14) and particle rapidity defined above. As with spatial rapidity η_s, particle rapidity is additive under a longitudinal boost. A particle with rapidity Y in a given inertial frame has rapidity $Y + \mathrm{d}Y$ in a frame which moves relative to the first frame with rapidity $\mathrm{d}Y$ in the $-z$ direction. One can see this from the addition formula of relativistic velocity β_1 and β_2. The resultant velocity,

$$\beta = \frac{\beta_1 + \beta_2}{1 + \beta_1\beta_2}, \tag{2.22}$$

is also the addition formula for hyperbolic tangents,

$$\tanh(Y_1 + Y_2) = \frac{\tanh(Y_1) + \tanh(Y_2)}{1 + \tanh(Y_1)\tanh(Y_2)}. \tag{2.23}$$

In terms of transverse mass m_T and rapidity Y, particle four-momentum can be written as

$$p = p^\mu = (E, p_x, p_y, p_z) = (m_T \cosh Y, p_x, p_y, m_T \sinh Y). \tag{2.24}$$

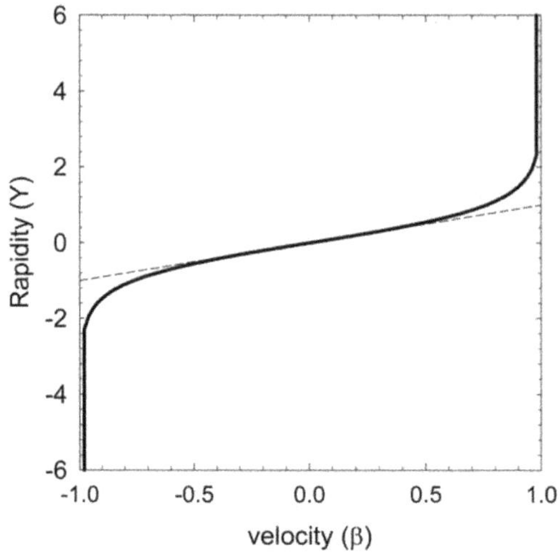

Figure 2.3. Rapidity $Y = \frac{1}{2}\ln\frac{1+\beta}{1-\beta}$ is plotted against the velocity β. The dashed line corresponds to $Y = \beta$.

Rapidity is the relativistic analog of non-relativistic velocity. In figure 2.3, rapidity is plotted against the velocity. For low velocity $\beta \leqslant 0.5$, Y is identical to β. Only with a large velocity limit is Y substantially different from β. This can also be shown explicitly. Equation (2.21), in the non-relativistic limit $p \ll m$, can be written as

$$Y = \frac{1}{2}\ln\frac{\sqrt{p^2 + m^2} + mv_z}{\sqrt{p^2 + m^2} - mv_z} = \frac{1}{2}\ln\frac{m + mv_z}{m - mv_z}$$
$$= \frac{1}{2}\left[\ln(1 + v_z) - \ln(1 - v_z)\right] \approx v_z. \tag{2.25}$$

Figure 2.3 also demonstrates the utility of rapidity in characterizing a particle. In relativistic collisions, the particles produced can have a very large velocity $\beta \approx 1$. Near $\beta \approx 1$, the rapidity variable is very sensitive to small changes in β.

In high-energy nuclear scattering, there is a natural choice of axis, the collision axis. Traditionally, it is taken along the z axis, and is usually called the beam axis. For a particle emitted at an angle θ with respect to the beam axis, the rapidity variable is

$$Y = \frac{1}{2}\ln\frac{E + p_z}{E - p_z} = \frac{1}{2}\ln\frac{\sqrt{m^2 + p^2} + p\cos\theta}{\sqrt{m^2 + p^2} - p\cos\theta}.$$

At very high energy, $p \gg m$, the mass can be neglected,

$$Y = \frac{1}{2}\ln\frac{p + p\cos\theta}{p - p\cos\theta}$$
$$= -\ln\tan\theta/2 \equiv \eta. \tag{2.26}$$

η is called pseudo-rapidity. Only the angle θ determines the pseudo-rapidity. It is a convenient parameter for experimentalists when details of the particle, e.g. mass, momentum, etc, are not known, and only the angle of emission is known (for example in emulsion experiments).

2.6 Light-cone coordinates

For a particle with four-momentum $p(p_0, \mathbf{p}_T, p_z)$, forward and backward light-cone variables are defined as

$$p_+ = p_0 + p_z \tag{2.27}$$

$$p_- = p_0 - p_z. \tag{2.28}$$

It is apparent that for a particle traveling along the beam axis, forward light-cone momentum is higher than for a particle traveling opposite to the beam axis. An important property of the light cone is that in case of a boost, light-cone momentum is multiplied by a constant factor. It can be seen as follows, writing the momentum in terms of the rapidity variable, $p^\mu = (m_T \cosh y, p_x, p_y, m_T \sinh y)$,

$$p_+ = m_T e^y$$
$$p_- = m_T e^{-y}.$$

Under a Lorentz boost by Δy, they will be transformed as,

$$p_+' = m_T e^{(y+\Delta y)} = p_+ e^{\Delta y}$$
$$p_-' = m_T e^{-(y+\Delta y)} = p_- e^{-\Delta y}.$$

Consider two four-vectors in light-cone coordinates,

$$p^\mu = \left(p_+, p^1, p^2, p_-\right) = \left(p^0 + p^3, p^1, p^2, p^0 - p^3\right) \tag{2.29}$$

$$q^\mu = \left(q_+, q^1, q^2, q_-\right) = \left(q^0 + q^3, q^1, q^2, q^0 - q^3\right). \tag{2.30}$$

The Lorentz invariant scalar product has the form

$$p \cdot q = p_+ q_- + p_- q_+ - p^1 q^1 - p^2 q^2. \tag{2.31}$$

One naturally wonders about the motivation for defining such a coordinate system, which depends on the choice of z-axis. One motivation is obviously the nice transformation property under the Lorentz boost. However, that is not all; there is another, more compelling reason. In high-energy nuclear scattering, there is a natural choice of axis, the collision axis. Traditionally, it is taken along the z-axis. When a vector is highly boosted along the z-axis, in light-cone coordinates, the large and small components of the momentum are clearly indicated.

2.7 Invariant distribution

Let us show that $\frac{d^3p}{E}$ is Lorentz invariant. The differential of the Lorentz boost in the longitudinal direction is

$$\mathrm{d}p_z^* = \gamma(\mathrm{d}p_z - \beta\mathrm{d}E) = \gamma\left(\mathrm{d}p_z - \beta\frac{p_z\mathrm{d}p_z}{E}\right),$$

$$= \frac{\mathrm{d}p_z}{E}\gamma(E - \beta p_z) = \frac{\mathrm{d}p_z}{E}E^* \tag{2.32}$$

where we have used $E^2 = m^2 + p_\mathrm{T}^2 + p_z^2 \Rightarrow E\mathrm{d}E = p_z\mathrm{d}p_z$. Then $\mathrm{d}p_z/E$ is Lorentz invariant. Since p_T is Lorentz invariant, d^3p/E is also Lorentz invariant.

The Lorentz invariant differential yield is

$$E\frac{\mathrm{d}^3N}{\mathrm{d}^3p} = E\frac{\mathrm{d}^3N}{\mathrm{d}^2p_\mathrm{T}\mathrm{d}p_z} = \frac{\mathrm{d}^3N}{\mathrm{d}^2p_\mathrm{T}\mathrm{d}y} \tag{2.33}$$

where the relation $\mathrm{d}p_z/E = \mathrm{d}y$ is used. Sometimes experimental results are given in terms of pseudo-rapidity. The transformation from (y, p_T) to (η, p_T) can be obtained as follows: write the rapidity and pseudo-rapidity variables as

$$y = \frac{1}{2}\ln\frac{E + p_\mathrm{L}}{E - p_\mathrm{L}} = \frac{1}{2}\ln\frac{\sqrt{m^2 + p_\mathrm{T}^2 + p_\mathrm{L}^2} + p_\mathrm{L}}{\sqrt{m^2 + p_\mathrm{T}^2 + p_\mathrm{L}^2} - p_\mathrm{L}} \tag{2.34}$$

$$\eta = \frac{1}{2}\ln\frac{p + p_\mathrm{L}}{p - p_\mathrm{L}} = \frac{1}{2}\ln\frac{\sqrt{p_\mathrm{T}^2 + p_\mathrm{L}^2} + p_\mathrm{L}}{\sqrt{p_\mathrm{T}^2 + p_\mathrm{L}^2} - p_\mathrm{L}}. \tag{2.35}$$

Assuming p_T is fixed, the rapidity density can be written as

$$\frac{\mathrm{d}N}{\mathrm{d}\eta} = \frac{\mathrm{d}N}{\mathrm{d}y}\frac{\mathrm{d}y}{\mathrm{d}\eta}. \tag{2.36}$$

A little algebra will show that

$$\frac{\mathrm{d}y}{\mathrm{d}\eta} = \frac{\mathrm{d}y/\mathrm{d}p_z}{\mathrm{d}\eta/\mathrm{d}p_z} = \frac{p}{E} = \sqrt{1 - \frac{m^2}{m_\mathrm{T}^2\cosh^2 y}}, \tag{2.37}$$

and rapidity and pseudo-rapidity densities are related as

$$\frac{\mathrm{d}N}{\mathrm{d}\eta\mathrm{d}p_\mathrm{T}} = \sqrt{1 - \frac{m^2}{m_\mathrm{T}^2\cosh^2 y}}\frac{\mathrm{d}N}{\mathrm{d}y\mathrm{d}p_\mathrm{T}}. \tag{2.38}$$

2.8 Luminosity

Let us first understand the concept of the cross section. In most of the experiments, a beam of projectiles scatters off a target at rest. For a sufficiently thin target, the flux of projectiles inside the target is constant. The rate of any interaction W_i will be proportional to the flux J of the incoming projectiles times the total number of target particles,

$$W_i \propto NJ. \tag{2.39}$$

The constant of proportionality is called the cross section for the process. It has the dimension of area (the dimension of flux J being number per unit area per unit time). Empirically, the cross section is the effective area presented to the beam by a target scatterer.

The cross section is a useful concept and it is desirable to use it even in collider experiments, where unlike the experiments described earlier, two beams of particles collide. The interaction rate will still be proportional to the cross section and on a number of machine dependent parameters, e.g. the number of particles per bunch, their spatial distribution and the frequency of the collision of those bunches. All the machine dependent parameters are lumped together into a parameter called luminosity, $\mathcal{L}$, and the interaction rate in a collider is written as

$$W = \mathcal{L}\sigma. \tag{2.40}$$

From (2.40), one can read the dimension of luminosity, $[\mathrm{L}^{-2}\mathrm{T}^{-1}]$. Time integrated luminosity times the cross section then gives the expected number of counts in any given process,

$$N_{\text{count}} = \sigma \int \mathcal{L} \mathrm{d}t. \tag{2.41}$$

Luminosity is a machine dependent parameter and it is necessary to know the parameters of the collider, e.g. number of bunches, particle density of each bunch and frequency of revolution. Let the machine collide two beams, each containing n bunches. Let each bunch contain N_1 and N_2 particles respectively. The revolution frequency is f, and the luminosity can be calculated as

$$\mathcal{L} = fn\frac{N_1 N_2}{A} \tag{2.42}$$

where A is the cross-sectional area of the beams.

2.9 Center-of-mass and laboratory system

Our knowledge about nature is gained mainly through the collision process, by which one body is made to collide with another. Traditionally, in laboratory experiments, one body is kept fixed when the other body hits it. The stationary body is called the target and the moving body the projectile. The frame of reference of such a system is called the laboratory frame (or lab frame for short). Theoretical calculations, however, are generally simpler in a frame of reference where the center-of-mass (cm) is at rest, or the total momentum is zero. It is called the centre of mass frame (or cm frame for short).

Let us consider the collision of two particles of mass m_1 and m_2, in the lab and cm frame (see figure 2.4). In the lab frame, m_2 is at rest and the four-momenta of the two particles are $p_1 = (E_1, \mathbf{p}_1)$ and $p_2 = (E_2, \mathbf{p}_2) = (m_2, \mathbf{0})$. In the cm frame, the total three-momentum is zero and the four-momenta can be written as $p_1^* = (E_1^*, \mathbf{p}_{cm})$ and $p_2^* = (E_2^*, -\mathbf{p}_{cm})$ (we designate the cm coordinates by a superscript *). Energy momentum is conserved in the collision. In four-vector notation, the conservation equation is

$$p_1 + p_2 = p_1^* + p_2^* \tag{2.43}$$

$$\text{in the cm frame: } E_{cm}^2 = \left(p_1^* + p_2^*\right)^2 = \left(E_1^* + E_2^*\right)^2 - (\mathbf{p}_{cm} - \mathbf{p}_{cm})^2 = \left(E_1^* + E_2^*\right)^2 \tag{2.44}$$

$$\begin{aligned}\text{in the lab frame: } E_{cm}^2 = (p_1 + p_2)^2 &= \left(E_1 + m_2\right)^2 - |\mathbf{p}_1|^2 \\ &= \left(E_1^2 - |\mathbf{p}_1|^2\right) + m_2^2 + 2m_2E_1 \\ &= m_1^2 + m_2^2 + 2m_2E_1.\end{aligned} \tag{2.45}$$

Hence, in a collision of a particle of mass m_1 and energy E_1 with a particle of mass m_2 at rest, the cm energy[4] is

$$E_{cm} = \sqrt{s} = \sqrt{m_1^2 + m_2^2 + 2m_2E_1} = \sqrt{m_1^2 + m_2^2 + 2m_2E_{\text{projectile}}}\,. \tag{2.46}$$

The magnitude of the cm momentum $\mathbf{p}_{cm}$ can be easily obtained from equation (2.44)

$$s = E_{cm}^2 = \left(E_1^* + E_2^*\right)^2 = \left[\sqrt{m_1^2 + |\mathbf{p}_{cm}|^2} + \sqrt{m_2^2 + |\mathbf{p}_{cm}|^2}\right]^2. \tag{2.47}$$

The equation can be solved to yield the cm momentum,

$$\begin{aligned}|\mathbf{p}_{cm}|^2 &= \frac{\left(s - m_1^2 - m_2^2\right)^2 - 4m_1^2m_2^2}{4s} \\ &= \frac{\left[s - (m_1 + m_2)^2\right]\left[s - (m_1 - m_2)^2\right]}{4s} \\ &= \frac{1}{4s}\lambda\left(s, m_1^2, m_2^2\right)\end{aligned} \tag{2.48}$$

[4] It is a common practice in high-energy experiments to denote cm energy by $\sqrt{s}$. Both E_{cm} and $\sqrt{s}$ are used here.

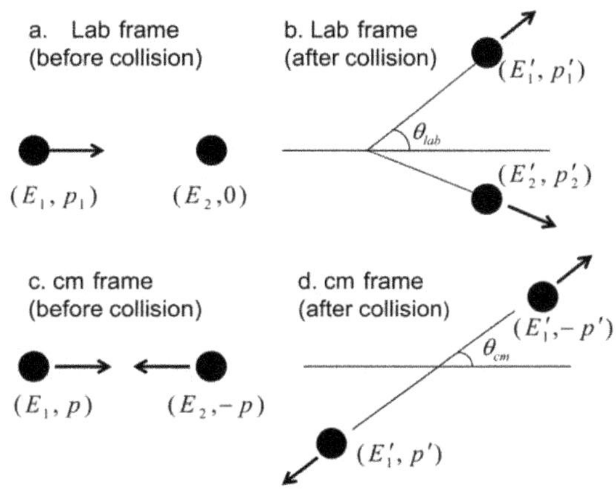

Figure 2.4. Pictorial diagram of two-body collision, 1 + 2 → 3 + 4, in the laboratory frame (a and b) and in the cm frame (c and d).

where $\lambda(x, y, z) = x^2 + y^2 + z^2 - 2xy - 2yz - 2zx$ is the Källén function (used frequently in relativistic kinematical relations). The cm momentum, in Lorentz invariant quantities, is then obtained as

$$|\mathbf{p}_{cm}| = \frac{1}{2\sqrt{s}}[\lambda(s, m_1, m_2)]^{1/2}. \tag{2.49}$$

We can also express lab momentum in terms of Lorentz invariants. From equation (2.46), $E_1 = \frac{s - m_1^2 - m_2^2}{2m_2}$ and

$$\begin{aligned} |\mathbf{p}_1|^2 &= E_1^2 - m_1^2 = \left(\frac{s - m_1^2 - m_2^2}{2m_2}\right)^2 - m_1^2 \\ &= \frac{\left(s - m_1^2 - m_2^2\right)^2 - 4m_1^2 m_2^2}{4m_2^2} = |\mathbf{p}_{cm}|^2 \frac{s}{m_2^2} \end{aligned} \tag{2.50}$$

or

$$|\mathbf{p}_1| = |\mathbf{p}_{cm}| \frac{\sqrt{s}}{m_2}. \tag{2.51}$$

The velocity of the cm in the lab frame is

$$\beta_{cm} = \frac{\mathbf{p}_1 + \mathbf{p}_2}{E_1 + E_2} = \frac{\mathbf{p}_1}{E_1 + m_2}, \tag{2.52}$$

and the Lorentz factor

$$\gamma_{cm} = \frac{1}{\sqrt{1 - \beta_{cm}^2}} = \frac{E_1 + m_2}{E_{cm}} = \frac{E_1 + m_2}{\sqrt{s}}. \tag{2.53}$$

Noting that the Lorentz factor $\gamma_{cm} = \cosh(y_{cm})$, the cm rapidity is then expressed as

$$y_{cm} = \cosh^{-1}\left(\frac{E_1 + m_2}{\sqrt{s}}\right). \tag{2.54}$$

2.9.1 Relation between y_{cm} and y

From equation (2.18)

$$E = \gamma\left(E^* + \beta p_z^*\right) \tag{2.55}$$

$$p_z = \gamma\left(\beta E^* + p_z^*\right) \tag{2.56}$$

$$\begin{aligned} y &= \frac{1}{2}\ln\frac{E + p_z}{E - p_z} = \frac{1}{2}\ln\frac{\left(E^* + \beta p_z^*\right) + \left(\beta E^* + p_z^*\right)}{\left(E^* + \beta p_z^*\right) - \left(\beta E^* + p_z^*\right)} \\ &= \frac{1}{2}\ln\frac{\left(E^* + p_z^*\right)(1 + \beta)}{\left(E^* - p_z^*\right)(1 - \beta)} = \frac{1}{2}\ln\frac{E^* + p_z^*}{E^* - p_z^*} + \frac{1}{2}\ln\frac{1 + \beta}{1 - \beta} \\ &= y^* + y_{cm}. \end{aligned} \tag{2.57}$$

The rapidity of a particle in the lab system is equal to the rapidity of the particle in the cm system plus the rapidity of the centre of mass.

2.9.2 Transformation of angles

Let, in the cm system, the scattered particle of momenta p_{cm} make an angle θ_{cm} with the longitudinal axis z. The momentum can be decomposed into two parts,

$$\begin{aligned} p_{cm\parallel} &= p_{cm}\cos(\theta_{cm}) \\ p_{cm\perp} &= p_{cm}\sin(\theta_{cm}). \end{aligned}$$

In the lab frame they will be transformed as

$$p_{lab\parallel} = \gamma_{cm}\left(p_{cm\parallel} + \beta_{cm}E_{cm}\right) = \gamma_{cm}\left(p_{cm}\cos(\theta_{cm}) + \beta_{cm}E_{cm}\right) \tag{2.58}$$

$$p_{lab\perp} = p_{cm\perp} = p_{cm}\sin(\theta_{cm}). \tag{2.59}$$

Giving the desired relation between the angles in lab and cm frame

$$\tan(\theta_{lab}) = \frac{p_{lab\perp}}{p_{lab\parallel}} = \frac{p_{cm}\sin(\theta_{cm})}{\gamma_{cm}\left(p_{cm}\cos(\theta_{cm}) + \beta_{cm}E_{cm}\right)}. \tag{2.60}$$

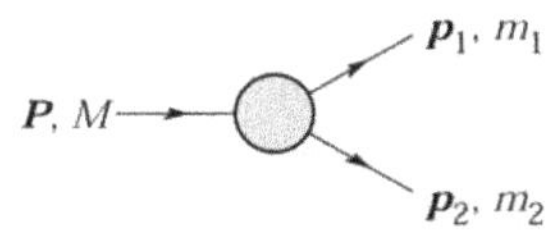

Figure 2.5. Pictorial diagram of two-body decay, where a mother particle of mass M and four-momentum P decays to daughters of mass m_1, m_2, with four-momentum p_1 and p_2, respectively.

2.10 Decay of particles

In relativistic energy heavy ion collisions, hundreds of particles are produced, many of which are decay product. It is then very important to understand the kinematics of particle decay. The decaying particle is generally called the mother particle and the decay products the daughter particles. Two- and three-particle decay are the predominant decay modes in heavy ion collisions. Four or more particle decay is also possible, but such decay modes are negligible in general and are not considered. In the following some details of the kinematics of two- and three-body decay are presented.

2.10.1 Two-body decay

Two-body decay is the simplest decay process. As depicted in figure 2.5, a mother particle of mass M at rest decays to two particles of mass m_1 and m_2 with four-momenta p_1 and p_2 respectively. The four-momenta of the particles can be chosen as

$$\text{Mother particle:} \quad P = (M, 0, 0, 0), \tag{2.61a}$$

$$\text{Daughter particle 1:} \quad p_1 = (E_1, 0, 0, p_z), \tag{2.61b}$$

$$\text{Daughter particle 2:} \quad p_2 = (E_2, 0, 0, -p_z). \tag{2.61c}$$

From conservation of four-momentum,

$$M = E_1 + E_2 = \sqrt{m_1^2 + p_z^2} + \sqrt{m_2 + p_z^2}. \tag{2.62}$$

Solving the above equation (see equations (2.47) and (2.48)), one obtains

$$p_z = \sqrt{\frac{\left(M^2 - m_1^2 - m_2^2\right)^2 - 4m_1^2 m_2^2}{4M^2}} = \frac{\left[\lambda\left(M^2, m_1^2, m_2^2\right)\right]^{1/2}}{2M}. \tag{2.63}$$

The energy of the daughter particles can also be determined. Note that the four-vector P (see equation (2.61a)) has the simplest structure and a product involving P will be simpler than any other product. We re-write the four-momentum conservation equation as

$$\begin{aligned} p_2^2 = m_2^2 &= \left(P - p_1\right)^2 = P^2 + p_1^2 - 2(P \cdot p) \\ &= M^2 + m_1^2 - 2ME_1 \\ \Rightarrow E_1 &= \frac{1}{2M}\left(M^2 + m_1^2 - m_2^2\right). \end{aligned} \tag{2.64}$$

Similarly we can obtain

$$E_2 = \frac{1}{2M}\left(M^2 + m_2^2 - m_1^2\right). \tag{2.65}$$

It may be mentioned here that there is no preferred direction for the daughter particles; the decay is isotropic. However, if the direction of one particle is chosen, the other will move in the opposite direction. In the mother particle rest frame, daughter particles move back-to-back. However, if the mother particle is not at rest, but decays during a flight, the daughter particles will not be back-to-back. In relativistic heavy ion experiments, decay in a flight is more important than decay at rest. For decay in a flight, the four-momenta of the particles can be written as

$$\begin{aligned} &\text{Mother particle:} && P = (E, 0, 0, p), \\ &\text{Daughter particle 1:} && p_1 = \left(E_1, \mathbf{p}_\mathrm{T}, p_{1z}\right), \\ &\text{Daughter particle 2:} && p_2 = \left(E_2, -\mathbf{p}_\mathrm{T}, p_{2z}\right). \end{aligned}$$

If viewed from a frame, moving with velocity $\beta = p/M$, the daughter particles will be boosted by the Lorentz factor $\gamma = 1/\sqrt{1 - \beta^2}$. The Lorentz boosted energy and momenta of the daughter particles can be written as,

$$\begin{aligned} E_1 &= \gamma\left(E_1^* + \beta p_{1z}^*\right) \\ p_{1z} &= \gamma\left(p_{1z}^* + \beta E_1^*\right) \\ p_\mathrm{T} &= p_\mathrm{T}^* \end{aligned}$$

and similarly for the daughter particle 2. The superscript '*' indicates the values in the cm system where the mother particle is at rest.

2.10.2 Three-body decay

In three-body or three-particle decay, the mother particle decays into three particles. An example of three-particle decay is shown in figure 2.6. A well known example of three-body decay is β-decay, where a neutron decays into a proton, an electron and an anti-neutrino, $n \rightarrow p + e^- + \bar{\nu}_e$. As is well known, the experimentally observed continuous β spectrum led Fermi to the prediction of the existence of the neutrino.

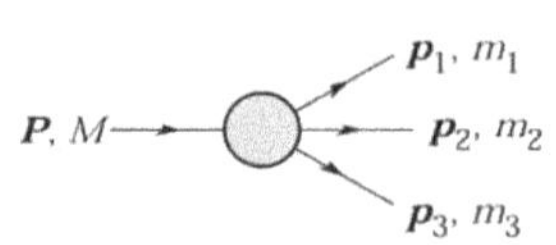

Figure 2.6. Pictorial diagram of three-body decay, where a mother particle of mass M and four-momentum P decays to daughters of mass m_1, m_2 and m_3, with four-momentum p_1, p_2 and p_3 respectively.

Analysis of three-body decay is greatly facilitated by what is now known as the Dalitz plot, after its inventor, Richard Dalitz[5]. A Dalitz plot is a way of assigning a point in phase space to every decay product according to its momenta. Let us denote the four-momentum by P, p_1, p_2 and p_3. The process involves four-particles and one can construct 16 invariants of the type, $A_i \cdot A_j$. Since $A_i \cdot A_j = A_j \cdot A_i$, only 10 of the 16 are independent. Four-momentum conservation,

$$P = p_1 + p_2 + p_3, \tag{2.66}$$

imposes four constraints to restrict the independent invariants to six. However, not all six are interesting, since there are four trivial invariants $A_i \cdot A_i = M_i^2$ which do not give any additional information. There are only two non-trivial invariants in terms of which the decay process can be specified. Generally one uses two of the following three invariants,

$$m_{23}^2 = (P - p_1)^2 = (p_2 + p_3)^2 \tag{2.67a}$$

$$m_{13}^2 = (P - p_2)^2 = (p_1 + p_3)^2 \tag{2.67b}$$

$$m_{12}^2 = (P - p_3)^2 = (p_1 + p_2)^2. \tag{2.67c}$$

The relevance of these invariants can be understood easily if the second parts of the definitions are considered. For example, $m_{23}^2 = (p_2 + p_3)^2$, i.e. m_{23} is the cm (or the invariant mass) of the subsystem consisting of the daughter particles 2 and 3. Similarly m_{13} is the cm (or the invariant mass) of the subsystem consisting of the daughter particles 1 and 3, and so on.

Let us evaluate m_{23}. In the rest frame of the mother particle,

$$m_{23}^2 = (P - p_1)^2 = P^2 + p_1^2 - 2P \cdot p_1 = M^2 + m_1^2 - 2ME_1.$$

For $E_1 \geqslant m_1$, the upper limit of $m_{23}^2 = m_{23_{\max}}^2 = (M - m_1)^2$. To obtain the lower limit, consider the alternative definition, $m_{23}^2 = (p_2 + p_3)^2$. Evaluate it in the rest frame of 2 and 3. Such a frame is called a Jackson frame and this particular frame can be designated as S23. In Jackson frame S23,

$$m_{23}^2 = (p_2 + p_3)^2 = (E_2^* + E_3^*)^2 \tag{2.68}$$

[5] In his obituary article on Richard Dalitz, Frank Close lamented that Dalitz was one of the great unsung scientists and that the geometrically based diagram he invented led to at least five Nobel prizes, and may lead to the discovery of Higgs [indeed, discovery of the Higgs boson was announced in July 2012], yet Dalitz himself was not honored with the prize, the ultimate scientific recognition.

where the superscript * denote the value in the frame S23. Since $E_3^* \geqslant m_3$, $E_2^* \geqslant m_2$, the lower limit of m_{23}^2 is $m_{23_{\min}}^2 \geqslant (m_2 + m_3)^2$. Similarly, the limits of the other invariants can be obtained,

$$m_{23}^2 \in \left[(m_2 + m_3)^2, (M - m_1)^2\right] \quad (2.69a)$$

$$m_{13}^2 \in \left[(m_1 + m_3)^2, (M - m_2)^2\right] \quad (2.69b)$$

$$m_{12}^2 \in \left[(m_1 + m_2)^2, (M - m_3)^2\right]. \quad (2.69c)$$

However, for a given $m_{23}^2 \in [(m_2 + m_3)^2, (M - m_1)^2]$ all of $m_{13}^2 \in [(m_1 + m_3)^2, (M - m_2)^2]$ is not kinematically accessible. The accessible limits can be obtained as follows.

In the Jackson frame S23, $p_2^* = -p_3^*$ and $p_1^* = P^*$.

$$m_{23}^2 = \left(P - p_1\right)^2 = \left(E^* - E_1^*\right)^2 = \left(\sqrt{M^2 + p_1^{*2}} + \sqrt{m_1^2 + p_1^{*2}}\right) \quad (2.70)$$

and solving for p_1^*,

$$p_1^{*2} = \frac{1}{4m_{23}^2}\lambda\left(m_{23}^2, M^2, m_1^2\right). \quad (2.71)$$

Similarly, from $m_{23}^2 = (p_2 + p_3)^2$, in the Jackson frame S23, one can obtain

$$p_2^{*2} = p_3^{*2} = \frac{1}{m_{23}^2}\lambda\left(m_{23}^2, m^2, m_3^2\right). \quad (2.72)$$

The invariant $m_{12}^2 = (p_1 + p_2)^2$ in the Jackson frame S23 can be calculated as

$$m_{13}^2 = (p_1 + p_3)^2 = m_1^2 + m_2^2 + 2\left(E_1^* E_3^* - p_1^* p_3^* \cos(\alpha)\right) \quad (2.73)$$

and for a fixed m_{23}^2, the dependence is only on α. Noting that $\cos(\alpha)$ varies between ± 1, the limits of m_{13}^2 are easily obtained,

$$m_{13\pm}^2 = m_1^2 + m_2^2 + 2\left(E_1^* E_3^* \pm p_1^* p_3^*\right). \quad (2.74)$$

One can rewrite $E_1^* = \frac{1}{2m_{23}}(M^2 - m_{23}^2 - m_1^2)$ and $E_2^* = \frac{1}{2m_{23}}(M^2 + m_3^2 - m_2^2)$ and get an explicit expression for $m_{13\pm}^2$,

$$m_{13\pm}^2 = m_1^2 + m_2^2 + \frac{1}{2m_{23}^2}\Big[\left(M^2 - m_{23}^2 - m_1^2\right)\left(m_{23}^2 - m_2^2 + m_3^2\right)$$
$$\pm \lambda^{1/2}\left(m_{23}^2, M^2, m_1^2\right)\lambda^{1/2}\left(m_{23}^2, m_2^2, m_3^2\right)\Big]. \quad (2.75)$$

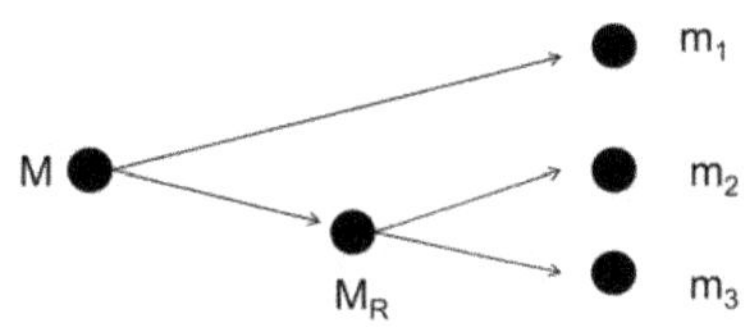

Figure 2.7. Pictorial diagram of three-body decay via a resonance formation.

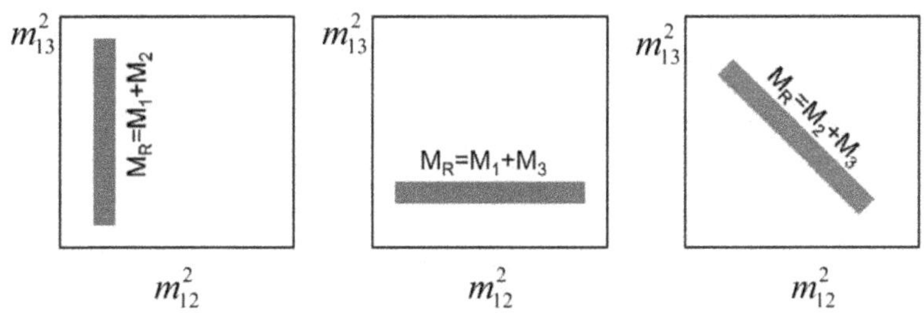

Figure 2.8. Pictorial diagram of three-body decay via a resonance formation.

Equation (2.75) defines the boundary of what is known as a Dalitz plot in the (m_{23}^2, m_{13}^2) plane. The main advantage of a Dalitz plot is its ability to identify short-lived resonances. Many times, the decay proceeds via resonance formation: in the first step, a mother particle decays to a daughter particle and a resonance M_R (see figure 2.7), which in the second step decays into two more daughter particles. Such a resonance can be easily identified in a Dalitz plot. This is because the three-body decay rate, averaged over the spin, can be written as

$$\Gamma = \frac{1}{8\pi}\frac{1}{32M^3}|\mathcal{M}|^2\, \mathrm{d}m_{12}^2\, \mathrm{d}m_{23}^2 \tag{2.76}$$

where $|\mathcal{M}|^2$ is the Lorentz invariant squared matrix element. If $|\mathcal{M}|^2$ is constant then the phase space will be uniformly populated. Any variation in phase space is due to a change in $|\mathcal{M}|$, which will change, for example, if a resonance is formed. If particles 1 and 2 are from the decay of a short-lived resonance, then the Dalitz plot will have a band of particles at $m_{12}^2 = M_R^2 = (m_1 + m_2)^2$. The width of the bands will correspond to the resonance decay width. Examples of bands for various resonance formations (1,2), (2,3) and (1,3) in Dalitz plots are shown in figure 2.8.

In an actual experiment, the phase space will not look as simple as shown in figure 2.8; rather, it will be filled uniformly/non-uniformly according to the physics of the decay process. As an example, in figure 2.9, the Dalitz plot for the decay $D_s^+ \to K^+K^-\pi^+$ is shown. There can be three possible pairings of the three particles, (i) (K^+K^-), (ii) $(K^-\pi^+)$ and (iii) $(K^+\pi^+)$.

In figure 2.9, each observed event is represented by a point in the $m_{K^+K^-}^2 - m_{K^-\pi^+}^2$ plane. One can see that the distribution of events is not uniform: they are band-like

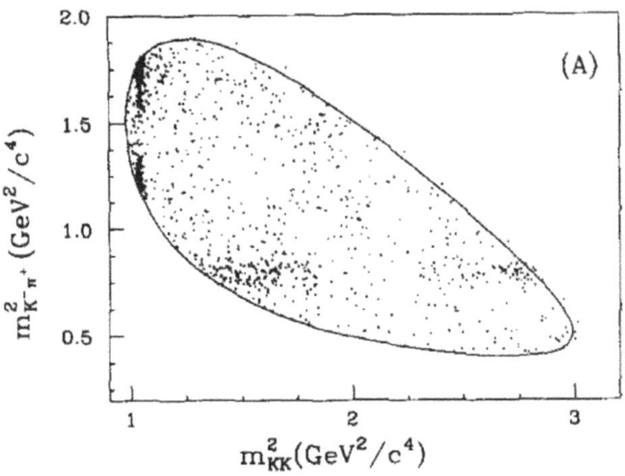

Figure 2.9. Dalitz decay of $D_s^+ \to K^+K^-\pi^+$. The figure is reproduced with permission from [4].

structures, one horizontal and one vertical. These bands, centered on specific masses, tell us that the final particles produced in the D_s^+ decay are not produced equally in all possible ways, rather they prefer to have particular mass combinations. As was discussed earlier, the band-like structures are the signature of resonances. In this particular example, the vertical band centered around $m^2_{K^+K-} \approx 1.0$ GeV^2 is the ϕ resonance from the (K^+K^-) combination, with mass $m_\phi = 1.02$ GeV. The horizontal band centered around $m^2_{K^-\pi^+} \approx 0.9$ GeV^2 is the signature of the resonance K^{*0} from the $(K^+\pi^+)$ combination with mass $m_{K_0^*} = 0.892$ GeV. From the band widths in the Dalitz plot, resonance lifetimes can also be obtained. It may be mentioned here that part of the band width is due to the finite resolution of the detectors.

One also notices that along the band, population is not uniform. For example, in the horizontal band shown in figure 2.9, two distinct lobes can be seen, one at $m^2_{K^+K^-} \approx 1.7$ GeV^2 and one at $m^2_{K^+K^-} \approx 3.1$ GeV^2. The lobes correspond to the spin of the resonance. Consider the spins of the particles: D_s^0, $K^\pm$ and π^- have zero spin angular momentum. The resonance $K^{*0}(892)$, however, is of spin 1. Then the decay $D_s^0 \to K^{*0}K^+ \to (K^-\pi^+)K^+$ proceeds through these steps: (i) a spin 0 particle decay to a spin 0 and spin 1 particle, (ii) the spin 1 decays into two spin 0 particles. The daughters of the spin 1 particle tend to be aligned or anti-aligned with the first spin 0 particle, giving rise to two lobe structures.

2.11 Mandelstam variables

In figure 2.10, a two-body collision process is shown. Two particles of momenta p_1 and p_2 and masses m_1 and m_2 scatter to particles of momenta p_3 and p_4 and masses m_3 and m_4. One can write the cross section for the process in terms of Lorentz invariant quantities. As was discussed earlier with reference to the three-body decay process, for a four-particle state, there can be 16 invariant quantities. Out of these 16,

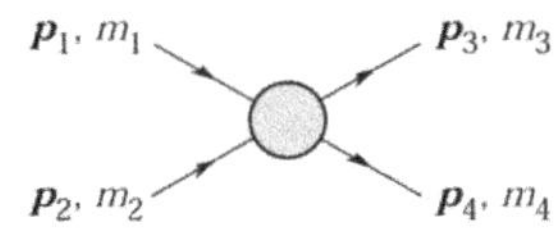

Figure 2.10. Pictorial diagram of the $A + B \rightarrow C + D$ process.

only six are independent. Of these six independent invariants, only two invariants are non-trivially independent, since the remaining four correspond to the mass of the particles and do not give any additional information.

Even though only two invariants suffice for the $1 + 2 \rightarrow 3 + 4$ processes, one generally defines three Lorentz invariants,

$$\begin{aligned} s &= (p_1 + p_2)^2 = (p_3 + p_4)^2 \\ &= m_1^2 + m_2^2 + 2E_1E_2 - 2\mathbf{p}_1 \cdot \mathbf{p}_2 \end{aligned} \tag{2.77}$$

$$\begin{aligned} t &= (p_1 - p_3)^2 = (p_2 - p_4)^2 \\ &= m_1^2 + m_3^2 - 2E_1E_3 + 2\mathbf{p}_1 \cdot \mathbf{p}_3 \end{aligned} \tag{2.78}$$

$$\begin{aligned} u &= (p_1 - p_4)^2 = (p_2 - p_3)^2 \\ &= m_1^2 + m_4^2 - 2E_1E_4 + 2\mathbf{p}_1 \cdot \mathbf{p}_4. \end{aligned} \tag{2.79}$$

They are called Lorentz-invariant Mandelstam variables or simply Mandelstam variables. Obviously, all of the three variables s, t and u are not independent. They are constrained by the relation

$$s + t + u = m_1^2 + m_2^2 + m_3^2 + m_4^2. \tag{2.80}$$

From the above identity, one also gets $0 \geqslant t, u \geqslant -s$.

Mandelstam variables are used extensively in calculating scattering amplitude, the complex function which completely defines the scattering process, such that they are frame independent. There are some specific significances of the variables s, t and u, e.g. s is the square of cm energy in the reaction $1 + 2 \rightarrow 3 + 4$, and t is the square four-momentum transfer in the reaction.

It is pertinent here to mention the crossing symmetry. The scattering amplitude for any process involving a particle with momentum p in the initial state is equal to the scattering amplitude for an otherwise identical process but with an antiparticle of momentum $k = -p$ in the final state.

$$\mathcal{M}\big(\phi_1(p_1)\phi_2(p_2)\ldots\big) = \mathcal{M}\big(\overline{\phi}_1(-p_1)\phi_2(p_2)\ldots\big) \tag{2.81}$$

where $\overline{\phi}$ denotes the antiparticle of ϕ.

In view of the crossing symmetry, s, t and u can be interpreted as follows:

s: square of cm energy when 1 and 2 (or 3 and 4) are incoming particles.
t: square of cm energy when 1 and $\overline{3}$ (or $\overline{2}$ and 4) are incoming particles.
u: square of cm energy when 1 and $\overline{4}$ (or $\overline{2}$ and 3) are incoming particles.

The three processes for which s, t and u are the square of cm energy are called the s-channel, t-channel and u-channel respectively.

Let us consider a specific reaction,

$$\begin{gathered} e^- + e^+ \rightarrow \mu^+ + \mu^- \\ (1) + (2) \rightarrow (3) + (4) \end{gathered} \tag{2.82}$$

$$\begin{aligned} s\text{-channel:}\ & e^- + e^+ \rightarrow \mu^+ + \mu^- \\ t\text{-channel:}\ & e^- + \mu^- \rightarrow e^- + \mu^- \\ u\text{-channel:}\ & e^- + \mu^+ \rightarrow \mu^+ + e^-. \end{aligned} \tag{2.83}$$

Thus the two reactions, (i) annihilation $e^- + e^+ \rightarrow \mu^+ + \mu^-$ and (ii) scattering $e^- + \mu^- \rightarrow e^- + \mu^-$, are related by the crossing symmetry.

2.12 Phase space

Theoretical evaluation of the cross section or decay rate for a process contains two distinct parts. One part is the calculation of the matrix elements, which contains dynamics of the specific process, while the other part is the calculation of the phase space integral. For example, the differential decay rate of a particle with four-momentum $P(E,M)$ into two daughter particles with four-momentum p_1 and p_2 is given by

$$d\Gamma = \frac{(2\pi)^4}{2E}|\mathcal{M}|^2 d^6\Phi(P: p_1, p_2). \tag{2.84}$$

The two-body phase space $d^6\Phi(P: p_1, p_2)$ can be explicitly written as

$$d^6\Phi(P: p_1, p_2) = \frac{d^3p_1}{(2\pi)^3 2E_1}\frac{d^3p_2}{(2\pi)^3 2E_2}\delta^4(P - p_1 - p_2). \tag{2.85}$$

The total decay rate can be obtained by integrating over the differential decay rate equation (2.84). For constant amplitude $|\mathcal{M}|$, one needs to evaluate integral $\int d^6\Phi(P: p_1, p_2)$.

$$I = \int d^6\Phi(P: p_1, p_2) = \int \frac{d^3p_1}{(2\pi)^3 2E_1}\frac{d^3p_2}{(2\pi)^3 2E_2}\delta^4(P - p_1 - p_2). \tag{2.86}$$

Let us evaluate it in the rest frame of the mother particle, $P = p_1 + p_2 = (M, 0, 0, 0)$. The delta function in equation (2.86) can be written as, $\delta^4(P - p_1 - p_2) = \delta(E - E_1 - E_2)\delta^3(\mathbf{p}_1 + \mathbf{p}_2)$ and integration over one of the momenta is trivial,

$$\begin{aligned} I &= \frac{1}{4(2\pi)^6}\int \frac{1}{E_1E_2} d^3p_1 \delta(M - E_1 - E_2) \\ &= \frac{1}{4(2\pi)^6}\int \frac{1}{E_1E_2} \mathbf{p}_1^2 d|\mathbf{p}_1| d\Omega \delta(M - E_1 - E_2) \end{aligned} \tag{2.87}$$

with $E_1 = \sqrt{\mathbf{p}_1^2 + m_1^2}$ and $E_2 = \sqrt{\mathbf{p}_1^2 + m_2^2}$. To evaluate the integral, we use the following identity,

$$\delta[f(p_1)] = \frac{1}{f'(p_{10})}\delta(p_1 - p_{10}) \tag{2.88}$$

where p_{10} is the root of $f(p_1) = 0$. Then,

$$f(p_1) = \sqrt{m_1^2 + p_1^2} + \sqrt{m_2^2 + p_1^2} - M \tag{2.89}$$

$$\text{and}\quad f'(p_1) = \frac{p_1}{\sqrt{m_1^2 + p_1^2}} + \frac{p_1}{\sqrt{m_2^2 + p_1^2}} = \frac{p_1}{E_1} + \frac{p_1}{E_2}. \tag{2.90}$$

The root p_{10} can be easily obtained (see (2.48))

$$p_{10} = p_{\rm cm} = \frac{1}{2M}\lambda^{1/2}\left(M^2, m_1^2, m_2^2\right) \tag{2.91}$$

and the desired integral is,

$$\begin{aligned} I &= \frac{1}{4(2\pi)^6}\int \frac{1}{E_1 E_2}\mathbf{p}^2 \mathrm{d}|\mathbf{p}|\mathrm{d}\Omega\delta\left(M - E_1 - E_2\right) \\ &= \frac{1}{4(2\pi)^6}\int \mathrm{d}\Omega\frac{p_{\rm cm}}{M}. \end{aligned} \tag{2.92}$$

2.12.1 *n*-body phase space

Lorentz invariant *n*-body phase space is written as

$$\mathrm{d}\Phi^n(p, p_1, p_2, \ldots p_n) = \prod_{i=1}^{n} \frac{\mathrm{d}^3 p_i}{(2\pi)^3 2E_i}\delta^3\left(\mathbf{p} - \sum_{i=1}^{n}\mathbf{p}_i\right)\delta\left(E - \sum_{i=1}^{n} E_i\right) \tag{2.93}$$

where we have designated the phase space by total momentum p and momenta of the *n*-particles. In the cm frame, $\mathbf{p} = 0$ and we rewrite the equation as

$$\begin{aligned} \mathrm{d}\phi^n\left(P, p_1, \ldots p_n\right) &= \prod_{i=1}^{n} \frac{\mathrm{d}^3 p_i}{(2\pi)^3 2E_i}\delta^3\left(\sum_{i=1}^{n}\mathbf{p}_i\right)\delta\left(E - \sum_{i=1}^{n} E_i\right) \\ &= \frac{\mathrm{d}^3 p_n}{(2\pi)^3 2E_n} \times \left\{\prod_{i=1}^{n-1} \frac{\mathrm{d}^3 p_i}{(2\pi)^3 2E_i}\delta^3\left((-\mathbf{p}_n) - \sum_{i=1}^{n-1}\mathbf{p}_i\right)\right. \\ &\quad \left. \times\, \delta\left((E - E_n) - \sum_{i=1}^{n-1} E_i\right)\right\}. \end{aligned} \tag{2.94}$$

The term under the curly bracket is the phase space factor for $(n - 1)$ particles with total momentum $-\mathbf{p}_n$ and energy $E - E_n$. Lorentz invariance allows this to be written as the phase space for a system of $(n - 1)$ particles with total momentum zero and energy ε,

$$\varepsilon = \sqrt{(E - E_n)^2 - p_n^2} = \sqrt{E^2 + m_n^2 - 2E\sqrt{m_n^2 + p_n^2}}\,. \tag{2.95}$$

We then obtain a recurrence relation, which can be used to calculate n-body phase space,

$$\mathrm{d}\Phi^n(P, p_1, \ldots p_n) = \frac{\mathrm{d}^3 p_n}{(2\pi)^3 2E_n} \times \mathrm{d}\Phi^{n-1}(\varepsilon, p_1, \ldots p_{n-1}). \tag{2.96}$$

Bibliography

[1] Hagedorn R 1963 *Relativistic Kinematics* (Reading, MA: Benjamin/Cummings)
[2] Byckling E and Kajantie K 1973 *Particle Kinematics* (New York: Wiley)
[3] Beringer J *et al* (Particle Data group) 2012 *Phys. Rev.* D **86** 010001
[4] Frabetti P L *et al* (E687 Collaboration) 1995 *Phys. Lett.* B **351** 591

IOP Publishing

A Short Course on Relativistic Heavy Ion Collisions

A K Chaudhuri

Chapter 3

The Glauber model

3.1 Introduction

A nucleus is an extended object. Accordingly, depending upon the impact parameter[1] of the collision, several types of collision can be defined, e.g. central collision, when two nuclei collide head on, or peripheral collision when only glancing interactions occur between the two nuclei. In high-energy collisions, geometry plays an important role and in the mid 1970s, a macroscopic model based on simple geometrical, kinematical and statistical assumptions was developed to explain the experimental data from intermediate energy (200 − 2000 MeV/nucleon) nuclear collisions. The model is called the nuclear fireball model or participant–spectator model [1–4]. A schematic diagram of a collision in the participant–spectator/fireball model is shown in figure 3.1. The model assumptions are briefly discussed here. The fireball model rests on three main assumptions.

(i) *Straight line geometry*. The projectile nucleus moves in a straight line even while interacting with the target nucleus. In this way, an overlap or interaction zone is defined for each impact parameter **b** (see figure 3.1). Nucleons in this zone are called the 'participants', while the remaining nucleons of the target and the projectile are called 'spectators'. Nucleons in the spectator region are more or less unaffected by the collision and continue their motion. Nucleons in the participant region interact and create a fireball. The baryon number and charge of the fireball are determined by the collision's geometry. The mass and velocity of the fireball are then determined uniquely by kinematics.

(ii) *Thermalization*. It is assumed that enough interactions occur during the initial formation and subsequent expansion of the fireball that thermal

[1] Impact parameter **b** is defined as the transverse distance between the centers of the colliding nuclei. Remember that it is a vector quantity.

doi:10.1088/978-0-750-31060-4ch3

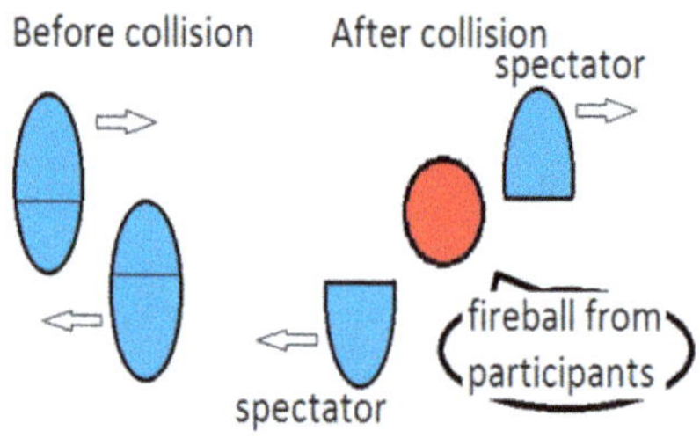

Figure 3.1. Schematic diagram showing a collision in the participant–spectator model. Participant and spectator regions are shown in red and blue, respectively.

equilibrium is established among all the hadron constituents, i.e., the total center-of-mass energy available is converted into random motion, consistent with the conservation laws.

(iii) *Decay of the fireball.* The fireball decays by statistical emission.

In intermediate energy range collisions, the fireball model could explain the energy spectra of proton and light nuclei. The model, with appropriate modification, has also found application in ultra-relativistic energy collisions. The concept that the nuclei maintain a straight line trajectory even while interacting has been widely used in ultra-relativistic energy to quantitatively distinguish between different collisions in terms of collision centrality. Collision centrality has become a key parameter in relativistic energy heavy ion collisions, because it is related directly to the initial overlap region of the colliding nuclei; overlap is maximum in central collisions and minimum in grazing collisions. The system created in a central collision can be qualitatively different from the system created in a peripheral collision. Different aspects of reaction dynamics can be understood if heavy ion collisions are studied as a function of impact parameter or equivalently in terms of the collision centrality.

The impact parameter of a collision cannot be measured experimentally. However, one can find one-to-one correspondence between the impact parameter of the collision and some experimental observable, e.g. charged particle multiplicity, transverse energy ($E_T = \sum_i E_i \sin\theta_i$), etc. For example, one can safely assume that multiplicity or transverse energy is a monotonic function of the impact parameter. High multiplicity or transverse energy events are from central collisions and low multiplicity or low transverse energy events are from peripheral collisions. One can then group the collisions according to multiplicity or transverse energy.

This can also be done quantitatively. Define a minimum bias collision where all possible collisions are allowed. In figure 3.2 a schematic picture of charged particle multiplicity (N_{ch}) in a minimum bias collision is shown. Minimum bias yield can be cut into successive intervals starting from the maximum value of multiplicity. The first 10% of the high N_{ch} events corresponds to the top 10%, or 0–10% collision centrality. Similarly, the first 20% of the high N_{ch} corresponds to 0–20% centrality. The overlap region between 0–10% and 0–20% corresponds to 10–20% centrality and so on. Similarly, centrality class can be defined by measuring the transverse energy.

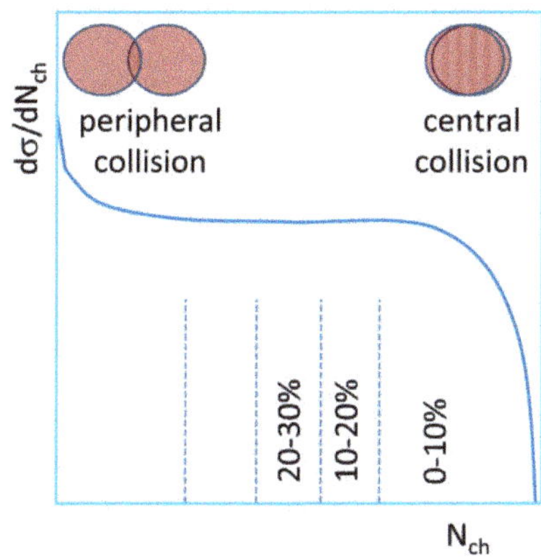

Figure 3.2. Schematic representation of multiplicity distribution in a minimum bias nucleus–nucleus collision.

Instead of impact parameter, one often defines centrality in terms of number of participating nucleons (the nucleons that undergo at least one inelastic collision) or in terms of binary nucleon collision number. These measures have a one-to-one relationship with impact parameter and can be calculated in a Glauber model.

The Glauber model is a highly successful model of high-energy collisions. In heavy ion collisions, in general, only a simple variant of the model is used. In order to appreciate the beauty of the model, we briefly describe the Glauber model of scattering, as envisaged by Glauber.

3.2 Glauber model of scattering

Nobel laureate Roy Jay Glauber is noted for his contributions to the field of quantum optics. In the 1950s he used quantum mechanical techniques to solve, analytically, the scattering problem for composite systems at high energy [5–9]. This is called the Glauber model or Glauber eikonal model. By high energy we mean the collisions in which the wavelength associated with the incident particle is much lower than the interaction range. Under that condition, the incident particle deviates little from its original path and the quantum mechanical scattering problem can be solved, allowing for small angle approximations and certain novel assumptions (to be described later). A simplified version of the model is widely used to classify heavy ion reactions in terms of collision centrality (to be explained in detail later). Let us briefly discuss the Glauber model of scattering.

In quantum mechanics, a scattering process is described by the solution of the Schrödinger equation,

$$\left[-\frac{\hbar^2}{2m}\nabla^2 + V(\mathbf{r}, t)\right]\psi'(\mathbf{r}, t) = -i\hbar\frac{\partial}{\partial t}\psi'(\mathbf{r}, t), \tag{3.1}$$

with the following boundary conditions: (i) the wave function must have an incident plane wave component, $e^{i\mathbf{k}\cdot\mathbf{r}} = e^{ikz}$, with energy $E = \frac{\hbar^2 k^2}{2m}$, moving in the positive z

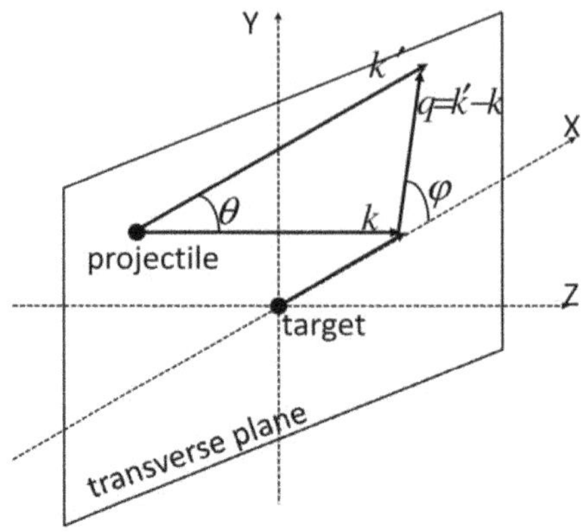

Figure 3.3. Schematic diagram of an elastic scattering process. A projectile with momentum **k** scatters off a target at rest. The momentum transferred in the process is $\mathbf{q} = \mathbf{k}' - \mathbf{k}$. The relevant angles are shown.

direction, and (ii) an outgoing spherical wave $\frac{e^{ikr}}{r}$. For time independent potential, $V(\mathbf{r}, t) = V(\mathbf{r})$, $\psi'(\mathbf{r}, t) = e^{-iEt/\hbar}\psi(\mathbf{r})$, the wave function $\psi(\mathbf{r})$ satisfies

$$\left[\nabla^2 + k^2\right]\psi(\mathbf{r}) = U(\mathbf{r})\psi(\mathbf{r}) \tag{3.2}$$

with $k^2 = \frac{2mE}{\hbar^2}$ and $U(\mathbf{r}) = \frac{2mV(\mathbf{r})}{\hbar^2}$. Dictated by the boundary condition, the wave function can be written as

$$\psi(\mathbf{r})_{r\to\infty} = e^{ikz} + f(\theta, \phi)\frac{e^{ikr}}{r} \tag{3.3}$$

where $f(\theta, \phi)$ is called the scattering amplitude. Here θ is the scattering angle measured with respect to the direction of the incident beam (assumed to be in the positive z-axis) and ϕ is the azimuthal scattering angle measured with respect to the positive x-axis (see figure 3.3). Determination of the scattering amplitude is the main concern in scattering theory. For the incident plane wave, the initial flux is just the velocity **v**. The flux scattered through solid angle $d\Omega$ is

$$|f(\theta, \phi)|^2\frac{1}{r^2}vr^2 d\Omega \tag{3.4}$$

so that the differential scattering cross section is

$$d\sigma = \frac{\text{flux through } d\Omega}{\text{incident flux}} = |f(\theta, \phi)|^2 d\Omega. \tag{3.5}$$

By the optical theorem[2], the forward ($\theta = 0$) scattering amplitude is related to the total cross section,

$$\sigma_{\text{tot}} = \sigma_{\text{el}} + \sigma_{\text{inel}} = \frac{4\pi}{k}\text{Im} f(\theta = 0). \tag{3.6}$$

[2] Optical theorem is a general law of scattering. In quantum theory of scattering it arises as a consequence of conservation of probability.

There are standard methods for solving for the scattering amplitude (see any standard textbook on quantum mechanics, e.g. [10, 11]). In partial wave analysis, for the azimuthally symmetric potential, the scattering amplitude $f(\theta, \phi)$ can be expanded as

$$f(\theta) = \sum_{l=0}^{l=\infty} \frac{1}{2ik}(2l+1)\left(e^{i2\delta_l} - 1\right)P_l(\cos(\theta)). \tag{3.7}$$

In equation (3.7), δ_l is the phase shift. Squaring equation (3.7), integrating over the angles and using the orthogonality condition for Legendre polynomials, one obtains for the scattering cross section,

$$\sigma_{\rm tot} = \frac{4\pi}{k^2} \sum_{l=1}^{\infty}(2l+1)\sin^2\delta_l. \tag{3.8}$$

Each partial wave then contributes to the scattering cross section. Low-energy collisions are dominated by s-wave scattering and one generally defines a scattering length $a = \frac{\sin\delta_0}{k}$ such that the dominant s-wave contribution is $\sigma_{l=0} = 4\pi a^2$.

In high-energy collisions, the wavelength associated with the incident particle is much smaller than the range of the interaction and scattering phase shifts at large l take on values different from zero. Furthermore, high-energy collisions are dominated by small angle collisions. For small angle and large l, the Legendre polynomials $P_l(\cos\theta)$ may be approximated as

$$P_l(\cos\theta) \approx \frac{1}{2\pi}\int_0^{2\pi} \exp\left\{i(2l+1)\sin\frac{\theta}{2}\cos(\alpha)\right\} d\alpha. \tag{3.9}$$

In high-energy collisions, the summation of l contains many terms and can be accurately replaced by an integral over l. Now, if we define an impact distance b with the relation, $kb = l + \frac{1}{2}$, the summation in equation (3.7) can be replaced as

$$\sum_{l=0}^{\infty} \to \int_0^{\infty} dl \to k\int_0^{\infty} db. \tag{3.10}$$

Introducing a function $\chi(b) = 2\delta_l$, the scattering amplitude may be rewritten as

$$f(\theta) = \frac{k}{2\pi i}\int_0^{\infty}\int_0^{2\pi} \left(e^{i\chi(b)} - 1\right)\exp\left\{2ikb\sin\frac{\theta}{2}\cos(\alpha)\right\} b\,db\,d\alpha. \tag{3.11}$$

The above equation can be written in a more convenient form in terms of the momentum transferred $\mathbf{q}$ in the collision. In spherical polar coordinates, the incident and final momenta of the particle can be written as

$$\mathbf{k} = (0, 0, k) \text{ and } \mathbf{k}' = k(\sin(\theta)\cos(\phi), \sin(\theta)\sin(\phi), \cos(\theta))$$

such that the transferred momentum, component-wise is

$$\mathbf{q} = \mathbf{k}' - \mathbf{k} = k\left(\sin(\theta)\cos(\phi), \sin(\theta)\sin(\phi), -\sin^2\left(\frac{\theta}{2}\right)\right). \tag{3.12}$$

In high-energy collisions, the scattering angle θ is small and the transferred momentum lies predominantly in the transverse plane. Taking the scalar product of $\mathbf{q}$ with the impact parameter vector which makes an angle α with positive x-axis $\mathbf{b} = (b\cos(\alpha), b\sin(\alpha))$, and making a small angle approximation,

$$\mathbf{q}\cdot\mathbf{b} = kb\sin(\theta)\cos(\phi-\alpha) \approx 2kb\sin\left(\frac{\theta}{2}\right)\cos(\phi-\alpha), \tag{3.13}$$

the scattering amplitude can be approximated as

$$f(\theta) = \frac{ik}{2\pi}\int e^{i\mathbf{q}\cdot\mathbf{b}}\left(1 - e^{i\chi(\mathbf{b})}\right)d^2b. \tag{3.14}$$

In equation (3.14), the integration is over the plane of impact parameter vectors. The relation between the phase shift function $\chi(\mathbf{b}) \equiv 2\delta_l$ and the potential is of the form

$$\chi(\mathbf{b}) = -\frac{1}{\hbar v}\int_{-\infty}^{\infty} V(\mathbf{b}, z)dz. \tag{3.15}$$

The conditions under which the Glauber approximation is valid were studied in detail. It turns out that the Glauber approximation is valid under the condition

$$\frac{V}{E} \ll 1 \quad \text{and} \quad \frac{1}{V/E} \ll ka \ll \frac{1}{(V/E)^2}. \tag{3.16}$$

Sometimes it is useful to use the 'Profile function' $\gamma(\mathbf{b})$,

$$\gamma(\mathbf{b}) = 1 - e^{i\chi(\mathbf{b})}, \tag{3.17}$$

to write the scattering amplitude as

$$f(\mathbf{q}) = \frac{ik}{2\pi}\int e^{i\mathbf{q}\cdot\mathbf{b}}\gamma(\mathbf{b})d^2b. \tag{3.18}$$

One notes that the profile function is the Fourier transform of the scattering amplitude,

$$\gamma(\mathbf{b}) = \frac{1}{2\pi ik}\int f(\mathbf{q})e^{-i\mathbf{q}\cdot\mathbf{b}}d^2q. \tag{3.19}$$

Using the optical theorem, from equation (3.14) one can immediately write for the total cross section,

$$\sigma_{\text{tot}} = \frac{4\pi}{k}\text{Im}f(\mathbf{k}=\mathbf{k}') = 2\text{Re}\int\left[1 - e^{i\chi(\mathbf{b})}\right]d^2b. \tag{3.20}$$

When the scattering is concentrated in the forward direction, the integration over the solid angle can be replaced by an integration over the plane, tangent to the sphere at $|\mathbf{k}|$,

$$d\Omega = \frac{d^2k}{k^2} \tag{3.21}$$

and the integrated elastic scattering cross section can be written as

$$\begin{aligned}\sigma_{\rm el} &= \int d\Omega |f(\mathbf{k}, \mathbf{k}')|^2 \\ &= \frac{1}{4\pi^2}\int d^2k d^2b d^2b' \left\{e^{i(k-k')\cdot b}\left[1 - e^{i\chi(b)}\right]\right\}\left\{e^{-i(k-k')\cdot b'}\left[1 - e^{i\chi(b')}\right]^*\right\} \\ &= \int d^2b |1 - e^{i\chi(b)}|^2 \end{aligned} \tag{3.22}$$

where we have used the two-dimensional δ-function,

$$\delta^{(2)}(\mathbf{b} - \mathbf{b}') = \frac{1}{(2\pi)^2}\int d^2k e^{i(k-k')\cdot(b-b')}. \tag{3.23}$$

Finally, the inelastic cross section can be obtained as

$$\sigma_{\rm in} = \sigma_{\rm tot} - \sigma_{\rm el} = \int d^2b\left(1 - \left|e^{i\chi(b)}\right|^2\right). \tag{3.24}$$

We have developed the Glauber model expression for elastic scattering amplitude with the assumption of azimuthal symmetry. However the expression is more general and also applicable in the case of azimuthally asymmetric potential (see [6]).

3.3 The Glauber model for a composite scatterer

The Glauber model views scattering of a composite projectile–target in terms of the individual interactions of the constituent nucleons. In figure 3.4, a nucleus–nucleus collision, as viewed in a Glauber model, is shown. It is assumed that at sufficiently high energy, nucleons carry enough momentum and are undeflected as the nuclei pass through each other. It is also assumed that the nucleons move independently in the nucleus and that the size is large compared to the nucleon–nucleon (NN) interaction range. The hypothesis of independent linear trajectories of nucleons makes it possible to obtain a simple analytical expression for nuclear cross section, number of binary collisions, participant nucleons, etc.

For scattering off a composite system like a nucleus, the major assumption Glauber made is that the total phase shift function is the sum of the phase shift function of each individual nucleon, considered separately. For example, in nucleon–nucleus collisions, if the target nucleus has A nucleons, at fixed positions

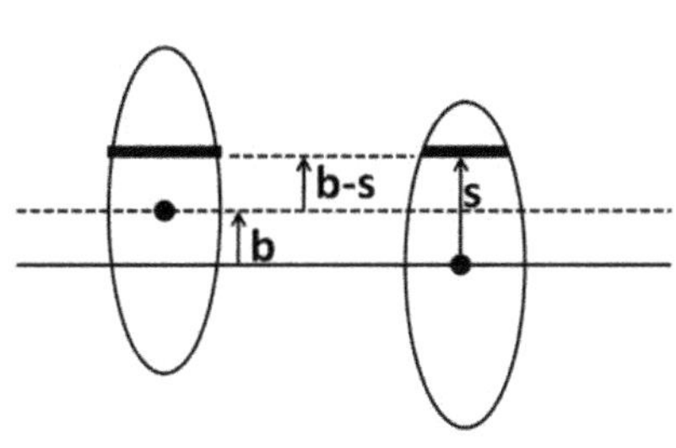

Figure 3.4. Nucleus–nucleus collisions as viewed in a Glauber model.

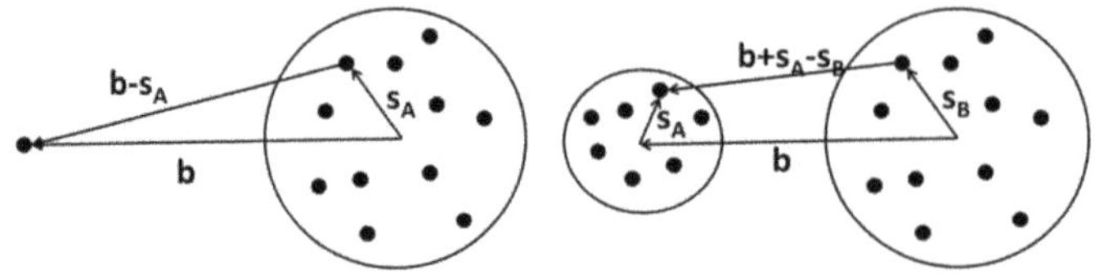

Figure 3.5. A schematic diagram of impact parameters **b** for nucleon–nucleus (left panel) and nucleus–nucleus (right panel) collisions, in the transverse plane.

$\mathbf{r}_1, \mathbf{r}_2, \ldots \mathbf{r}_A$, with components $\mathbf{s}_1, \mathbf{s}_2, \ldots \mathbf{s}_A$ in the plane perpendicular to the incident beam (see figure 3.5), the total phase shift function can be written as

$$\chi_{\text{tot}}(\mathbf{b}, \mathbf{s}_1, \mathbf{s}_2, \ldots \mathbf{s}_A) = \sum_{j=1}^{A} \chi_j\left(\mathbf{b} - \mathbf{s}_j\right), \tag{3.25}$$

or equivalently, the total profile function for the composite target is

$$\begin{aligned} \Gamma_{\text{tot}}(\mathbf{b}, \mathbf{s}_1, \mathbf{s}_2, \ldots \mathbf{s}_A) &= 1 - e^{i \sum_{j=1}^{A} \chi_j (\mathbf{b} - \mathbf{s}_j)} \\ &= 1 - \prod_{j=1}^{A} \left\{1 - \gamma_j\left(\mathbf{b} - \mathbf{s}_j\right)\right\}. \end{aligned} \tag{3.26}$$

Before we proceed further, we would like to stress that the assumption that the phase shift function of individual nucleons adds up (equation (3.25)) is a non-trivial assumption. We will not go into too much detail, but since the phase shifts produced by individual target nucleons are added up, not the scattering amplitudes, this is not a simple superposition; rather, it takes into account the effect of multiple scattering in all orders. It can be simply seen if the profile function is expanded,

$$\Gamma_{\text{tot}} = 1 - \left[1 - \sum_j \gamma_j + \sum_{ij} \gamma_i \gamma_j - \sum_{ijk} \gamma_i \gamma_j \gamma_k - \cdots\right]. \tag{3.27}$$

One can visualize the scattering process as follows: as the incident particle with momentum **k** scatters off the nucleus to momentum **k**′, simultaneously, the state of the target nucleus changes from $|i\rangle$ to $|f\rangle$. $\Gamma_{\text{tot}}(\mathbf{b}, \mathbf{s}_1, \mathbf{s}_2 \ldots \mathbf{s}_A)$ can be regarded as the operator which induces that change in the target nucleus. Introducing the configuration space wave function, the profile function averaged over nuclear volume is

$$\langle \Gamma_{\text{tot}}(\mathbf{b}, \mathbf{r}_1, \mathbf{r}_2, \ldots \mathbf{r}_A) \rangle = \int \psi_f^*(\mathbf{r}_1, \ldots \mathbf{r}_A) \left[1 - \prod_{j=1}^{A} \left\{1 - \gamma_j\left(\mathbf{b} - \mathbf{s}_j\right)\right\}\right] \psi_i(\mathbf{r}_1, \ldots \mathbf{r}_A) \, d^3 r_1 d^3 r_2 \ldots d^3 r_A. \tag{3.28}$$

For elastic scattering, initial and final states are the same. If the simplifying assumption is made that the ground-state wave function is a product wave function, $\psi_0 = \prod_j \phi(\mathbf{s}_j, z_j)$ with $|\phi|^2 = \frac{\rho(\mathbf{r})}{A}$, the profile function can be written as

$$\langle \Gamma_{\text{tot}}(\mathbf{b}, \mathbf{r}_1, \mathbf{r}_2 \ldots \mathbf{r}_A) \rangle = 1 - \int \prod_{j=1}^{A} |\phi_j(\mathbf{r}_j)|^2 \left(1 - \gamma_j(\mathbf{b} - \mathbf{s}_j)\right) \mathrm{d}^3 r_j$$
$$= 1 - \left(1 - \int \gamma(\mathbf{b} - \mathbf{s}) \frac{\rho(\mathbf{r})}{A} \mathrm{d}^3 r\right)^A \tag{3.29}$$

where in the last line we have dropped the subscript j as the profile function γ is the same for all the nucleons.

The elastic scattering amplitude is then given by

$$f_{\text{el}}(\mathbf{q}) = \frac{\mathrm{i}k}{2\pi} \int \mathrm{d}^2 b \mathrm{e}^{\mathrm{i}\mathbf{q}\cdot\mathbf{b}} \left\{ 1 - \left[1 - \int \gamma(\mathbf{b} - \mathbf{s}) \frac{\rho(\mathbf{r})}{A} \mathrm{d}^3 \mathbf{r} \right]^A \right\}. \tag{3.30}$$

If γ varies more rapidly than $\rho(r)$, one obtains

$$f_{\text{el}}(\mathbf{q}) = \frac{\mathrm{i}k}{2\pi} \int \mathrm{d}^2 b \mathrm{e}^{\mathrm{i}\mathbf{q}\cdot\mathbf{b}} \left\{ 1 - \left[1 + \frac{1}{A} \frac{2\mathrm{i}\pi}{k} f_{\text{NN}}(0) T(\mathbf{b}) \right]^A \right\} \tag{3.31}$$

where we have introduced the thickness function,

$$T(\mathbf{b}) = \int \rho(\mathbf{r}) \mathrm{d}z = \int \rho(\mathbf{b}, z) \mathrm{d}z \tag{3.32}$$

and used the identification (see equation (3.18)),

$$\frac{2\pi}{\mathrm{i}k} f_{\text{NN}}(0) = \int \gamma(\mathbf{b} - \mathbf{s}) \mathrm{d}^2 s. \tag{3.33}$$

For large A, using $\mathrm{e}^x = \lim_{n\to\infty}(1 + \frac{x}{n})^n$, the elastic scattering amplitude can be written as

$$f_{\text{el}}(\mathbf{q}) = \frac{\mathrm{i}k}{2\pi} \int \mathrm{d}^2 b \mathrm{e}^{\mathrm{i}\mathbf{q}\cdot\mathbf{b}} \left\{ 1 - \exp\left[\frac{2\pi \mathrm{i}}{k} f_{\text{NN}}(0) T(\mathbf{b}) \right] \right\}. \tag{3.34}$$

One can now define an effective phase shift function,

$$\chi_{\text{eff}}(\mathbf{b}) = \frac{2\pi}{k} f_{\text{NN}}(0) T(\mathbf{b}), \tag{3.35}$$

which contains only the geometry of the nucleus and forward scattering amplitude of NN scattering. In terms of the effective phase shift function, the form of the elastic scattering amplitude for proton-nucleus (pA) collisions is the same as in proton-proton (pp) collisions:

$$f_{\text{el}}(\mathbf{q}) = \frac{\mathrm{i}k}{2\pi} \int \mathrm{d}^2 b \mathrm{e}^{\mathrm{i}\mathbf{q}\cdot\mathbf{b}} \left\{ 1 - \mathrm{e}^{\mathrm{i}\chi_{\text{eff}}(\mathbf{b})} \right\}. \tag{3.36}$$

Using equations (3.20), (3.22) and (3.24), the total cross section (σ_{tot}) and integrated elastic scattering cross section (σ_{el}) for pA collisions can be written as

$$\sigma_{\text{tot}} = 2\text{Re} \int \mathrm{d}^2 b \left\{ 1 - \exp\left(\frac{\mathrm{i}2\pi T(\mathbf{b}) f_{\text{NN}}(0)}{k} \right) \right\} \tag{3.37}$$

$$\sigma_{\text{el}} = \int \mathrm{d}^2 b \left| \left\{ 1 - \exp\left(\frac{\mathrm{i}2\pi T(\mathbf{b}) f_{\text{NN}}(0)}{k} \right) \right\} \right|^2. \tag{3.38}$$

The integrated inelastic scattering cross section (σ_{in}) can also be calculated,

$$\begin{aligned} \sigma_{\text{in}} &= \int \mathrm{d}^2 b \{ 1 - \exp(-\sigma_{\text{tot}} T(\mathbf{b})) \} \\ &\approx \int \mathrm{d}^2 b \{ 1 - \exp(-\sigma_{\text{in}}^{\text{NN}} T(\mathbf{b})) \}. \end{aligned} \tag{3.39}$$

In obtaining equation (3.39) we have used the optical relation equation (3.6), $\sigma_{\text{tot}} = \frac{4\pi f(0)}{k}$ and used the fact that in NN collisions $\sigma_{\text{tot}} \approx \sigma_{\text{in}}^{\text{NN}}$.

We have briefly discussed the Glauber model for elastic scattering of pA collisions. The model can also be extended for nucleus–nucleus (AA) collisions, although the equations become more complicated. For example, the profile function 3.15 for pA collisions to be changed into AB collisions is as follows (see figure 3.5),

$$\Gamma\left(\mathbf{b}, \mathbf{s}_1^A, \mathbf{s}_2^A, \ldots \mathbf{s}_A^A, \mathbf{s}_1^B, \mathbf{s}_2^B, \ldots \mathbf{s}_B^B\right) = 1 - \prod_{i=1}^{A} \prod_{j=1}^{B} \left\{ 1 - \gamma\left(\mathbf{b} + \mathbf{s}_i^A - \mathbf{s}_j^B\right) \right\}. \tag{3.40}$$

In the above equations, the superscripts A and B denote that the nucleons are from projectile and target nuclei respectively. The nuclear volume averaged over the profile function can be written as

$$\begin{aligned} &\left\langle \Gamma\left(\mathbf{b}, \mathbf{s}_1^A, \mathbf{s}_2^A, \ldots \mathbf{s}_A^A, \mathbf{s}_1^B, \mathbf{s}_2^B, \ldots \mathbf{s}_B^B\right) \right\rangle \\ &= \left\langle \Psi_f \right| \left[1 - \prod_{i=1}^{A} \prod_{j=1}^{B} \left\{ 1 - \gamma\left(\mathbf{b} - \mathbf{s}_i^B - \mathbf{s}_j^B\right) \right\} \right] \left| \Psi_i \right\rangle \end{aligned} \tag{3.41}$$

with

$$\Psi = \prod_{i=1}^{A} \phi\left(\mathbf{r}_i^A\right) \prod_{j=1}^{B} \phi\left(\mathbf{r}_j^B\right). \tag{3.42}$$

In the end, one can obtain similar relations as in equations (3.37)–(3.39) for total, elastic and inelastic cross sections but with a modified thickness function,

$$T_{AB}(\mathbf{b}) = \int \mathrm{d}^2 s T_A(\mathbf{s}) T_B(\mathbf{b} - \mathbf{s}), \tag{3.43}$$

and the inelastic cross section can be written as

$$\sigma_{\rm in} = \int d^2b \left\{1 - \exp\left(-\sigma_{\rm in}^{\rm NN} T_{AB}(\mathbf{b})\right)\right\}. \tag{3.44}$$

Scattering amplitude for an inelastic process is more complicated and will not be discussed here. Interested readers may look into the excellent review by Glauber [5].

3.4 Nuclear density

In the Glauber model, the scattering cross section for nucleon–nucleus/nucleus–nucleus collisions is computed in terms of the nuclear one-body density $\rho(\mathbf{r})$ or more specifically the nuclear thickness function $T(\mathbf{b})$ and elementary NN inelastic cross section $\sigma_{\rm in}$.

A nucleus is made of protons and neutrons and one can define a proton, $\rho_{\rm p}(\mathbf{r})$, and neutron's, $\rho_{\rm n}(\mathbf{r})$, density functions, such that integrated over space they reproduce the atomic number (Z) and neutron number (N) of the nucleus. The two density distributions are not identical, especially for large mass nuclei when $N > Z$. In high-energy collisions, protons and neutrons are generally not distinguished and for an atomic mass A nucleus, nuclear density function is defined as

$$\int \rho(\mathbf{r}) d^3r = A. \tag{3.45}$$

The nuclear one-body density function $\rho(\mathbf{r})d^3r$ then gives the average number of nucleons within a small volume d^3r.

Density distributions can be obtained from self-consistent theories such as the Hartree–Fock/Hartree–Fock–Bogolyubov theories. For nuclei with $A \geqslant 12$, a convenient parametric form for the density function is

$$\rho(r) = \rho_0 \frac{1 + w(r/R)^2}{1 + \exp\left(\frac{r - R}{a}\right)} \tag{3.46}$$

where ρ_0 is the nucleon density, R the radius and a the skin thickness. w measures the deviation from a spherical shape. In table 3.1, these parameters are listed for selected nuclei.

Table 3.1. Parameters R, a and w of a three parameter Fermi distribution for selected nuclei are given.

Nucleus	R (fm)	a (fm)	w (fm)
^{16}O	2.608	0.513	−0.51
^{62}Cu	4.2	0.596	0.0
^{197}Au	6.38	0.535	0.0
^{208}Pb	6.62	0.594	0.0
^{238}U	6.81	0.6	0.0

For a large number of nuclei the parameters can be approximated as

$$R = 1.12A^{1/3} - 0.86A^{-1/3}(\text{fm}) \tag{3.47a}$$

$$a = 0.54(\text{fm}) \tag{3.47b}$$

$$w = 0. \tag{3.47c}$$

If in equation (3.46), we replace $\rho_0 \to \frac{\rho_0}{A}$, $\rho(r)$ normalized to unity can be interpreted as the probability of finding a given nucleon at a position $r(=x, y, z)$. Then,

$$T_A(\mathbf{b}) = \int \mathrm{d}z\rho_A(\mathbf{b}, z) \tag{3.48}$$

is the probability that a given nucleon in the nucleus is at a transverse distance **b**.

3.5 Binary collision numbers

In equation (3.24), we have obtained the expression for the total inelastic scattering cross section for NN collisions,

$$\sigma_{\text{in}} = \int \mathrm{d}^2b\left(1 - \left|e^{i\chi(\mathbf{b})}\right|^2\right). \tag{3.49}$$

We can introduce a probability function $p(\mathbf{b})$ such that two nucleons at impact parameter **b** interact inelastically,

$$\begin{aligned} p(\mathbf{b}) &= \left(1 - \left|e^{i\chi(\mathbf{b})}\right|^2\right)\sigma_{\text{in}} \\ &= t(\mathbf{b})\sigma_{\text{in}} \end{aligned} \tag{3.50}$$

where, in the last line, we have introduced the NN thickness function $t(\mathbf{b}) = 1 - |e^{i\chi(\mathbf{b})}|^2$. The thickness function has to be normalized to unity to fulfil the requirement that $\int p(\mathbf{b})\mathrm{d}^2b = \sigma_{\text{in}}$.

Let us now consider pA collision, a schematic picture of which is shown in the left panel of figure 3.5. Consider a particular nucleon in the target at $\mathbf{r}_A = (\mathbf{s}_A, z_A)$, i.e. transverse position $\mathbf{s}_A$. For an impact parameter **b** collision, the transverse distance between the projectile nucleon and the specific nucleon is $\mathbf{b} - \mathbf{s}_A$. According to equation (3.50), they will interact inelastically with the probability $t(\mathbf{b} - \mathbf{s}_A)\sigma_{\text{in}}$. Considering that the specific nucleon can be any one of the target nucleons, the probability that the incident nucleon at impact parameter **b** interacts inelastically with a target nucleon is

$$p(\mathbf{b}) = \int \mathrm{d}^3r_A\rho(\mathbf{r}_A)t(\mathbf{b} - \mathbf{s}_A)\sigma_{\text{in}} = T_A(\mathbf{b})\sigma_{\text{in}}, \tag{3.51}$$

where we have introduced the nucleon–nucleus thickness function,

$$T_A(\mathbf{b}) = \int \mathrm{d}^3r_A\rho(\mathbf{r}_A)t(\mathbf{b} - \mathbf{s}_A) = \int \mathrm{d}^2s_A\mathrm{d}z_A\rho(\mathbf{r})t(\mathbf{b} - \mathbf{s}_A). \tag{3.52}$$

The scale of variation of the NN thickness function ($t(\mathbf{b})$) is much smaller than the scale of variation of the nuclear density. One can approximate $t(\mathbf{b} - \mathbf{s}_A) \approx \delta^{(2)}(\mathbf{b} - \mathbf{s}_A)$ and the nucleon–nucleus thickness function reduces to

$$T_A(\mathbf{b}) = \int \mathrm{d}z_A \rho(\mathbf{b}, z_A) \tag{3.53}$$

with the normalization condition,

$$\int \mathrm{d}^2 b T_A(\mathbf{b}) = 1. \tag{3.54}$$

Similarly, in impact parameter $\mathbf{b}$ nucleus–nucleus collisions, one target nucleon at transverse position $\mathbf{s}_A$ and one projectile nucleon at transverse position $\mathbf{s}_B$ will interact inelastically with probability $t(\mathbf{b} - \mathbf{s}_A - \mathbf{s}_B)\sigma_{\text{in}}$. The volume averaged probability that one nucleon from the projectile will interact inelastically with one target nucleon can be written as

$$p(\mathbf{b}) = \int \mathrm{d}^3 r_A \mathrm{d}^3 r_B \rho(\mathbf{r}_A)\rho(\mathbf{r}_B) t(\mathbf{b} - \mathbf{s}_A - \mathbf{s}_B)\sigma_{\text{in}} = T_{AB}(\mathbf{b})\sigma_{\text{in}}, \tag{3.55}$$

with

$$T_{AB}(\mathbf{b}) = \int \mathrm{d}^2 s T_A(\mathbf{s}) T_B(\mathbf{s} - \mathbf{b}) = \int \mathrm{d}^2 s T_A\left(\mathbf{s} + \frac{\mathbf{b}}{2}\right) T_B\left(\mathbf{s} - \frac{\mathbf{b}}{2}\right)$$
$$\text{in the limit: } t(\mathbf{b}) \to \delta^{(2)}(\mathbf{b}). \tag{3.56}$$

$p(\mathbf{b}) = T_{AB}(\mathbf{b})\sigma_{\text{in}}$ is the probability that in an impact parameter $\mathbf{b}$ collision, a projectile nucleon interacts once with a target nucleon. The probability that they did not interact is then $(1 - T_A(\mathbf{b})\sigma_{\text{in}})$. If the target and projectile nucleus have A and B nucleons respectively, the probability of having n interactions is the binomial distribution,

$$P(n, \mathbf{b}) = \binom{AB}{n} \left[\sigma_{\text{in}} T_{AB}(\mathbf{b})\right]^n \left[1 - \sigma_{\text{in}} T_{AB}(\mathbf{b})\right]^{A-n}. \tag{3.57}$$

Each interaction is a binary collision and the average number of binary collisions can now be calculated as

$$\langle n(\mathbf{b}) \rangle = \sum_{n=1}^{AB} n P(n, \mathbf{b}) = AB T_{AB}(\mathbf{b})\sigma_{\text{in}}. \tag{3.58}$$

3.6 The inelastic cross section

The total probability of an interaction between A and B is

$$\frac{\mathrm{d}\sigma_{\text{in}}}{\mathrm{d}^2 b} = \sum_{n=1}^{AB} P(n, \mathbf{b}) = 1 - \left[1 - \sigma_{\text{NN}} T_{AB}(\mathbf{b})\right]^{AB}. \tag{3.59}$$

The total inelastic cross section is

$$\sigma_{\text{in}} = \int \mathrm{d}^2b \left(1 - \left[1 - \sigma_{\text{in}} T_{AB}(\mathbf{b})\right]^{AB}\right)$$
$$\approx \int \mathrm{d}^2b (1 - \exp(AB\sigma_{\text{in}} T_{AB}(\mathbf{b})). \tag{3.60}$$

3.7 Participant number

Using the Glauber model, we have calculated the average number of binary collisions. This has one-to-one correspondence with the impact parameter and can be used to characterize the collision centrality. However, it is preferable to characterize the collision centrality using the 'participant number' or 'wounded nucleon number'[3]. One experimental observation [16] is that in 200–2000 MeV A^{-1} proton induced collisions, the ratios of average multiplicities in pA collisions and in pp collisions are approximately a constant,

$$\frac{\overline{n}_{\text{pA}}}{\overline{n}_{\text{pp}}} = \frac{1}{2}[\overline{\nu} + 1], \tag{3.61}$$

where we have written the constant as $\frac{1}{2}(\overline{\nu} + 1)$. The result is explained by the wounded nucleon model, where it is assumed that the particle production is an incoherent superposition of particle production from wounded nucleons (those nucleons which have interacted inelastically at least once). If the incident nucleon is counted as a wounded nucleon, $\overline{w} = 1 + \overline{\nu}$ can be regarded as the total number of wounded nucleons in pA collisions. It appears that the incident nucleon contribution to the total multiplicity is the same as the contribution of each wounded nucleon in the target and equals approximately $\frac{1}{2}\overline{n}_{\text{pp}}$. There is no difference whether a nucleon is wounded once or several times. One may consider the physical process as follows: a soft collision excites the nucleon and particles are produced from the decay of the excited nucleon. Now if a nucleon is hit twice, the second hit only changes the excited state of the nucleon, not leading to any additional decay product. Thus whether a nucleon is wounded once or n times, the total number of particles from its decay remains the same.

Bialas, Bleszynski and Czyz [17] proposed to extend the above relation to nucleus–nucleus collisions. Average multiplicity in a collision of two nuclei with the mass numbers A and B is

$$\overline{n}_{AB} = \frac{1}{2}[\overline{\nu}_A + \overline{\nu}_B]\overline{n}_{\text{pp}} = \frac{1}{2}\overline{w}_{AB}\overline{n}_{\text{pp}} \tag{3.62}$$

where $\overline{\nu}_{A(B)}$ is the average number of wounded nucleons in the nucleus $A(B)$. The wounded nucleon model then predicts that $\overline{n}_{AB}$ per wounded nucleon is a constant.

The number of wounded nucleons or participant numbers can be calculated using the Glauber model. As mentioned previously, in AB collisions, the number of

[3] There is a subtle difference between number of wounded nucleons and number of participants. Those nucleons that interact only inelastically are called wounded nucleons. Participant numbers also include those nucleons interacting elastically.

wounded nucleons is the sum of the wounded nucleons in A and B and it suffices to calculate for one. Consider an AB collision at impact parameter $\mathbf{b}$. Let us denote the transverse positions of the nucleons in A as $\mathbf{s}_1^A, \mathbf{s}_2^A, \ldots, \mathbf{s}_A^A$ and similarly for the nucleons in B. Consider a nucleon i in B. The probability that the ith nucleon in B is wounded, i.e. it has collided with any one of the projectile nuclei, is then

$$p\left(\mathbf{s}_i^B;\, A;\, \mathbf{s}_1^A, \ldots \mathbf{s}_A^A,\, \mathbf{b}\right) = 1 - \prod_{j=1}^{A}\left[1 - t\left(\mathbf{b} + \mathbf{s}_i^B - \mathbf{s}_j^A\right)\sigma_{\text{in}}\right]. \tag{3.63}$$

After integrating over different configurations of the nucleons in B,

$$\begin{aligned}\bar{p}\left(A;\, \mathbf{s}_1^A, \ldots \mathbf{s}_A^A,\, \mathbf{b}\right) &= \int \mathrm{d}^2 s^B p\left(\mathbf{s}^B;\, A;\, \mathbf{s}_1^A, \ldots \mathbf{s}_A^A\right) T_B\left(\mathbf{s}^B\right) \\ &= \int \mathrm{d}^2 s^B T_B\left(\mathbf{b} - \mathbf{s}^B\right)\left(1 - \prod_{j=1}^{A}\left[1 - t\left(\mathbf{s}^B - \mathbf{s}_j^A\right)\sigma_{\text{in}}\right]\right).\end{aligned} \tag{3.64}$$

The probability of having w_B wounded nucleons is the binomial distribution,

$$\begin{aligned}P\left(w_B;\, B;\, A;\, \mathbf{s}_1^A, \ldots \mathbf{s}_A^A,\, \mathbf{b}\right) = &\binom{B}{w_B}\left[\bar{p}\left(A;\, \mathbf{s}_1^A, \ldots \mathbf{s}_A^A\right)\right]^{w_B} \\ &\times\left[1 - \bar{p}\left(A;\, \mathbf{s}_1^A, \ldots \mathbf{s}_A^A\right)\right]^{B - w_B}.\end{aligned} \tag{3.65}$$

The average number of wounded nucleons in B is

$$\left\langle w_B\left(B;\, A;\, \mathbf{s}_1^A, \ldots \mathbf{s}_A^A,\, \mathbf{b}\right)\right\rangle = B\bar{p}\left(A;\, \mathbf{s}_1^A, \ldots \mathbf{s}_A^A,\, \mathbf{b}\right). \tag{3.66}$$

Now the configuration of nucleons in nucleus A needs to be averaged over and we obtain

$$\begin{aligned}\langle w_B(B;\, A)\rangle = &B\int \mathrm{d}^2\mathbf{s}_1^A T_A(\mathbf{s}_1)\int \mathrm{d}^2\mathbf{s}_2^A T_A(\mathbf{s}_2)\ldots\int \mathrm{d}^2\mathbf{s}_A^A T_A\left(s_A^A\right) \\ &\times\int \mathrm{d}^2 s^B T_B\left(\mathbf{b} - \mathbf{s}^B\right)\left(1 - \prod_{j=1}^{A}\left[1 - t\left(\mathbf{s}^B - \mathbf{s}_j^A\right)\sigma_{\text{in}}\right]\right) \\ = &B\int \mathrm{d}^2 s^B T_B\left(\mathbf{b} - \mathbf{s}^B\right)\left(1 - \left[1 - \int \mathrm{d}^2 s^A T_A\left(\mathbf{s}^A\right) t\left(\mathbf{s}^B - \mathbf{s}^A\right)\sigma_{\text{in}}\right]^A\right).\end{aligned}$$

In the limit, $t(\mathbf{b}) \to \delta^{(2)}(\mathbf{b})$,

$$\langle w_B(B;\, A)\rangle = B\int \mathrm{d}^2 s T_B(\mathbf{b} - \mathbf{s})\left(1 - \left[1 - T_A(\mathbf{s})\sigma_{\text{in}}\right]^A\right).$$

Similarly, one can obtain the wounded nucleon numbers in nucleus A, and the total number of wounded nucleons in an impact parameter $\mathbf{b}$ collision is

$$\begin{aligned}\langle w\rangle(\mathbf{b}) = &A\int \mathrm{d}^2 s T_A(\mathbf{s})\left(1 - \left[1 - \sigma_{\text{in}} T_B(\mathbf{b} - \mathbf{s})\right]^B\right) \\ &+ B\int \mathrm{d}^2 s T_B(\mathbf{b} - \mathbf{s})\left(1 - \left[1 - \sigma_{\text{in}} T_A(\mathbf{s})\right]^A\right).\end{aligned} \tag{3.67}$$

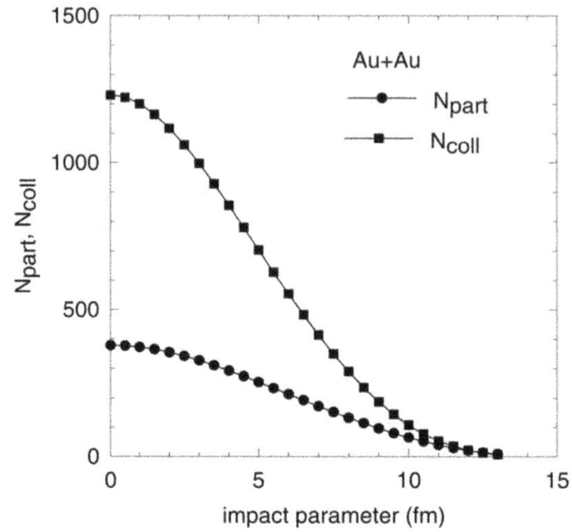

Figure 3.6. An optical Glauber model calculation for the impact parameter dependence of number of participant nucleons (N_{part}) and number of binary collisions (N_{coll}) in a $\sqrt{s}_{\text{NN}} = 200\,\text{GeV}$ Au+Au collision. The inelastic NN cross section is $\sigma_{\text{NN}} = 42$ mb.

Glauber model calculation of the binary collision number or participant number is energy dependent through the inelastic NN cross section σ_{in}. It is common to find $\sigma_{\text{NN}} \approx 30$ mb at the Super Proton Synchrotron (SPS; $\sqrt{s}_{\text{NN}} \approx 20\,\text{GeV}$), 40 mb at the RHIC ($\sqrt{s}_{\text{NN}} \approx 200\,\text{GeV}$) and 70 mb at the LHC ($\sqrt{s}_{\text{NN}} \geqslant 1000\,\text{GeV}$). For demonstration purposes, in figure 3.6 we have shown a Glauber model calculation for N_{part} and N_{coll} as a function of impact parameter in a $\sqrt{s}_{\text{NN}} = 200\,\text{GeV}$ Au+Au collision. One can see that there is a one-to-one correspondence between impact parameter b and participant number or collision number.

3.8 The Monte-Carlo Glauber model

In the Monte-Carlo (MC)-Glauber model, individual nucleons are stochastically distributed event-by-event and collision properties are calculated by averaging over many events. The optical Glauber model and the MC-Glauber model give very close results for average quantities like binary collision number or participant number. However, in quantities where fluctuations are important, e.g. participant eccentricity, the results are different.

MC-Glauber model calculations proceed as follows.

(i) An impact parameter b for the collision is selected randomly from the distribution,

$$\frac{\mathrm{d}N}{\mathrm{d}b} \propto b, \tag{3.68}$$

where N is the number of events.

(ii) The nucleons in the colliding nuclei are distributed randomly following the probability distribution $\rho(r)$. To be more specific, generate three random numbers, R_1, R_2 and R_3. In spherical polar coordinates, the volume element is given by $r^2 dr \sin(\theta)d\theta d\phi = r^2 dr d\cos(\theta)d\phi$. R_2 and R_3 are uniform random numbers varying between $[-1, +1]$ and $[0, 2\pi]$. R_1 is a random number following probability distribution $\rho(r)$ and weighted by r^2.
(iii) The centers of the two colliding nuclei are shifted to $(-b/2, 0, 0)$ and $(b/2, 0, 0)$ respectively.
(vi) Assuming the nuclei are moving in a straight line, two nuclei are collided.
(v) If the transverse separation between two colliding nucleons is less than the 'ball diameter' $D = \sqrt{\sigma_{\rm NN}/\pi}$, they are tagged as having interacted, and a register keeping the coordinates of the colliding nucleons is updated. More details about the model can be found in [13, 14].

3.9 The two-component model for multiplicity

In ultra-relativistic energy collisions, the final state contains large numbers of particles. For example, in $\sqrt{s}_{\rm NN} = 200$ GeV Au+Au collisions, in a central 0–5% collision, rapidity density, i.e. charged particle multiplicity in central rapidity, $\left.\frac{dN_{\rm ch}}{d\eta}\right|_{\eta=0} \approx 700$. Over the whole rapidity range charged particle multiplicity is ~ 4600 [15]. The majority, $\sim 85\%$, of the particles are pions. Considering that for every pair of charged pions there is a neutral pion, in a central $\sqrt{s}_{\rm NN} = 200$ GeV Au+Au collision more than 6000 particles are created. Most of the particles are low $p_{\rm T}$ particles, with a small (less than 5%) contribution from high $p_{\rm T}$ particles.

The modern theory of strong interaction is quantum chromodynamics (QCD). Later, we shall discuss QCD in some detail. In QCD one distinguishes between two types of processes: (i) soft processes, involving low momentum transfer of $q \leqslant$ 1–2 GeV and (ii) hard processes where the momentum transfer is high, $q \geqslant$ 1–2 GeV. Hard processes, in theory, are calculable from first principles, but the soft processes are not. The large number of low $p_{\rm T}$ particles in RHIC Au+Au collisions are from soft QCD processes. One builds models for these particles. As discussed earlier, one such model is the wounded nucleon model. In the wounded nucleon model, the soft process contribution to the multiplicity scales with the wounded nucleon or participant numbers,

$$\frac{dN_{\rm ch}}{d\eta} \propto N_{\rm part}. \tag{3.69}$$

In figure 3.7, experimentally measured charged particle multiplicity per participant pair

$$\left.\frac{1}{0.5N_{\rm part}}\frac{dN_{\rm ch}}{d\eta}\right|_{\eta=0},$$

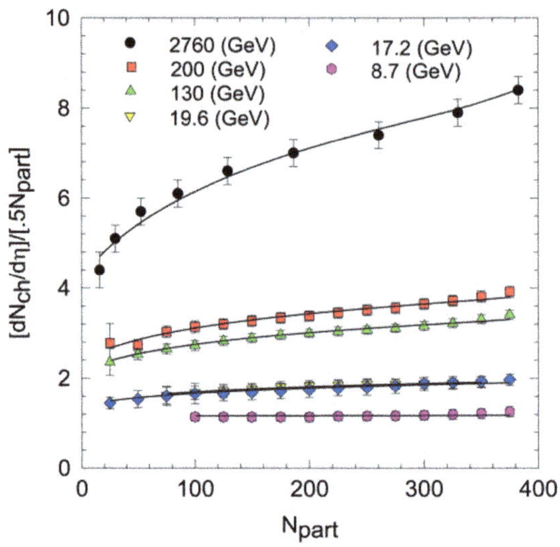

Figure 3.7. Centrality dependence of charged particle multiplicity $\frac{dN_{ch}}{d\eta}\Big|_{\eta=0}$, in the energy range $\sqrt{s}_{NN} = 8.7$–2760 GeV. $\sqrt{s}_{NN} = 2760$ GeV data are from [18]. $\sqrt{s}_{NN} = 8.7$–200 GeV data are compiled from PHENIX [19]. The solid lines in the figure are two-component model fits including soft and hard processes.

at various collision energies, is shown against the participant numbers. In the wounded nucleon model,

$$\frac{1}{0.5N_{part}}\frac{dN_{ch}}{d\eta}\Bigg|_{\eta=0}$$

is required to be constant. Deviation from the wounded nucleon prediction is small in $\sqrt{s}_{NN} = 8.7$ GeV collisions, but in higher-energy collisions the ratio increases slowly in more central collisions, indicating that multiparticle production in ultra-relativistic heavy ion collisions is not solely from soft processes. It has non-negligible contributions from hard processes. In the two-component model, one assumes that the hard process contribution scales with binary collision numbers. One can quantify the hard process contribution by parameterizing the rapidity distribution,

$$\frac{dN_{ch}^{AA}}{d\eta}\Bigg|_{\eta=0} = \frac{dN_{ch}^{pp}}{d\eta}\Bigg|_{\eta=0}\left[(1-f)\frac{1}{2}N_{part} + fN_{coll}\right]. \tag{3.70}$$

Experimental data at RHIC and LHC can be fitted with approximately constant values for the hard scattering fraction $f \approx 0.13$.

Bibliography

[1] Westfall G D, Gosset J, Johansen P J, Poskanzer A M, Meyer W G, Gutbrod H H, Sandoval A and Stock R 1976 *Phys. Rev. Lett.* **37** 1202

[2] Gosset J, Gutbrod H H, Meyer W G, Poskanzer A M, Sandoval A, Stock R, Westfall G D and Gutbrod H H 1977 *Phys. Rev.* C **16** 629

[3] Hufner J and Knoll J 1977 *Nucl. Phys.* A **290** 460
[4] Kapusta J I 1977 *Phys. Rev.* C **16** 1493
[5] Glauber R J 1959 *Lectures in Theoretical Physics* vol 1, ed W E Brittin and L G Dunham (New York: Interscience) p 315
[6] Franco V and Glauber R J 1966 *Phys. Rev.* **142** 1195
[7] Glauber R J and Matthiae G 1970 *Nucl. Phys.* B **21** 135
[8] Glauber R J 1955 *Phys. Rev.* **100** 242
[9] Glauber R J 2006 *Nucl. Phys.* A **774** 3
[10] Schiff L I 1968 *Quantum Mechanics* 3rd edn (New York: McGraw-Hill)
[11] Landau L D and Lifshitz E M 1991 *Quantum Mechanics* 3rd edn (Oxford: Pergamon)
[12] Shukla P 2003 *Phys. Rev.* C **67** 054607
[13] Miller M L, Reygers K, Sanders S J and Steinberg P 2007 *Annu. Rev. Nucl. Part. Sci.* **57** 205
[14] Alver B, Baker M, Loizides C and Steinberg P 2008 (arXiv: 0805.4411 [nucl-ex])
[15] Bearden I G *et al* (BRAHMS Collaboration) 2001 *Phys. Rev. Lett.* B **523** 227
Bearden I G *et al* 2002 *Phys. Rev. Lett.* **88** 202301
[16] Gottfried K 1974 *Phys. Rev. Lett.* **32** 957
[17] Bialas A, Bleszynski M and Czyz W 1976 *Nucl. Phys.* B **111** 461
[18] Aamodt K *et al* (ALICE Collaboration) 2011 *Phys. Rev. Lett.* **106** 032301
[19] Adler S S *et al* (PHENIX Collaboration) 2005 *Phys. Rev.* C **71** 034908
Adler S S *et al* (PHENIX Collaboration) 2005 *Phys. Rev.* C **71** 049901 (erratum)

Chapter 4

Classical theories for macroscopic systems

4.1 Introduction

The idea of colliding heavy ions with higher and higher energy is not just to proliferate the number of particles compared to those that are produced in lower-energy nucleus–nucleus (AA) collisions, but to produce a qualitatively different system. The qualitative difference is that unlike in lower-energy collisions, a high-energy AA collision is not just an incoherent superposition of nucleon–nucleon (NN) collisions. Rather, we expect to produce a *macroscopically* large *collective* system.

The basic difference between quarks and gluons inside a hadron and quarks and gluons in quark–gluon plasma—as existed in the early Universe, and as is found in the cores of neutron stars or produced in high-energy nuclear collisions—is that as opposed to the former, the latter can be treated as a macroscopic system. Generally, one calls a system microscopic if, as the name suggests, it is roughly of atomic dimensions or smaller, and a system macroscopic when it is large enough to be visible in the ordinary sense. However, this definition is rather inexact. A more appropriate definition depends on the number of particles in the system. A system is macroscopic if the number of particles (N) satisfies

$$\frac{1}{\sqrt{N}} \ll 1, \tag{4.1}$$

which means that statistical arguments can be applied with reasonable accuracy. Let us check the criterion *vis á vis* a $\sqrt{s}_{\mathrm{NN}} = 200\,\mathrm{GeV}$ Au+Au collision. In a $\sqrt{s}_{\mathrm{NN}} = 200\,\mathrm{GeV}$ Au+Au collision, approximately $N = 10^3$–10^4 particles are created for which $1/\sqrt{N} = 0.03$–0.01. It appears that the criterion in equation (4.1) is largely satisfied.

doi:10.1088/978-0-750-31060-4ch4

Now a macroscopic or many-body system is complex and it is impossible to gain an exact knowledge of it. For example, in classical mechanics, to completely specify an N-body system, $6N$ real numbers are required, as each particle will require three real numbers to specify its position and three for its momentum. Given the initial conditions for each of the particles, in principle, Newton's equations of motion can be solved for its future development. Now if N is very large (for example, one gram mole of gas contains $N \sim 10^{23}$ molecules), simultaneous solution of large numbers of differential equations is an impossible task.

It is also not needed for practical purposes. A macroscopic system can be characterized by a few macroscopic variables, e.g. energy, volume, pressure, temperature, etc, and the dynamics of the system is then obtained in terms of these macroscopic variables. The macroscopic variables depend too little on the exact 'microstate'[1] of the system. One of the fundamental assumptions of many-body physics is that an isolated many-body system, if left undisturbed for a sufficiently long time, will attain equilibrium. It is to be remembered that even if the system is in equilibrium, its constituents remain dynamic, for example, a glass of water left alone will soon reach equilibrium but the H_2O molecules will be in continual motion. By equilibrium, we understand that the macroscopic variables characterizing the system will be independent of time. More specifically, a system is in equilibrium if the probability of finding the system in a particular macrostate does not change with time. Thermodynamics and statistical physics are concerned with the equilibrium properties of the many-body system. They do not comment on the process of equilibration; that part is addressed in kinetic theory. Thermodynamics is a phenomenological theory of macroscopic systems at equilibrium. Its applicability range is wide. It can be applied to classical gases and liquids, quantum systems such as superconductors, black holes, the early Universe, etc. Thermodynamics tries to understand the macroscopic behavior of the many-body system, irrespective of the microscopic properties of the constituents and their interactions. Statistical physics on the other hand tries to understand macroscopic behavior in terms of the microscopic properties of the constituents and their interactions. Kinetic theory tries to understand the approach to equilibrium.

Alongside the following few important thermodynamics results, statistical physics and kinetic theory will be discussed. More insight can be gained from [1–5].

4.2 Brief review of thermodynamics

4.2.1 Thermodynamic laws

Thermodynamics is based on three laws, which are expressions of experimental findings.

(i) *The first law.* The first law is essentially an embodiment of James Prescott Joule's experimental demonstration of the equivalence of heat and energy. It is a statement of energy conservation, that energy can neither be created nor destroyed,

[1] Quantum mechanically, a microstate is a stationary state wave function of the many-body system.

and can only be changed from one form to another. The law can be formally stated as follows:

If in a thermodynamic transformation, ΔQ is the heat absorbed by the system and ΔW is the work done by the system, the quantity

$$\Delta E = \Delta Q - \Delta W \tag{4.2}$$

is the same for all transformations from one given initial state to a given final state.

The law immediately defines a state function called internal energy E

$$dE = dQ - PdV, \tag{4.3}$$

where we have replaced dW by PdV, the work done by a hydrostatic system with pressure P and volume V. By the first law, $\int dE$ depends only on the limits of the integration, not on the specific path of the integration and it is an exact differential.

(ii) *The second law.* The second law was derived from Carnot's theorem that 'No engine operating between two given temperatures is more efficient than a Carnot engine'. The efficiency of Carnot's engine, operating between temperature T_1 and T_2, $T_2 > T_1$, is given as

$$\eta = 1 - \frac{T_1}{T_2}. \tag{4.4}$$

There are several statements for the second law of thermodynamics. The Clausius statement is as follows.

There exists no thermodynamic transformation whose 'sole' effect is to extract a quantity of heat from a colder reservoir and to deliver it to a hotter reservoir.

A consequence of the second law is the Clausius theorem that in any cyclic process throughout which temperature is defined, the following inequality holds,

$$\oint \frac{dQ}{T} \leqslant 0, \tag{4.5}$$

a corollary of which is the statement that in a reversible process, the integral $\int \frac{dQ}{T}$ is independent of the path. The corollary defines the state variable 'entropy' which is an exact differential,

$$dS = \frac{dQ}{T}. \tag{4.6}$$

An alternative statement of the second law is that the entropy of an isolated system can never decrease.

In statistical physics, there is a more profound definition of entropy related to microstates (the state where all the parameters of the constituents are specified) of the system. Entropy is greater when more microstates are available. The law of increase of entropy states that the system evolves such that the number of microstates increases.

(iii) *The third law.* In 1905 Walther Nernst enunciated the third law of thermodynamics. It can be stated as follows:

Entropy approaches a constant value, which can be taken as zero, as the temperature approaches absolute zero.

The second law defines the difference between the entropy of two states connected by a reversible transformation. Such reversible transformations may not exist for all systems. The general statement of the third law, that entropy approaches a constant value as the temperature approaches absolute zero, irrespective of the system, uniquely determines the entropy of any system. An alternative statement of the third law is that absolute zero cannot be reached by any finite amount of processes.

Sometimes the zeroth law of thermodynamics is mentioned. This can be stated as follows.

If system A is in thermal equilibrium with system C, and system B is in thermal equilibrium with system C, then system A is in thermal equilibrium with system B.

An alternative statement is that an isolated system, left undisturbed, will reach equilibrium.

4.2.2 Thermodynamic potentials

A thermodynamic system is characterized by a number of state variables, either extensive or intensive[2]. Thermodynamic potentials are extensive state variables of the dimension of energy. Their purpose is to allow for simple treatment of equilibrium for systems interacting with the environment. As the name suggests, they also have exact differentials.

A differential form,

$$A(x_1, x_2, x_3, \ldots)\mathrm{d}x_1 + B(x_1, x_2, x_3, \ldots)\mathrm{d}x_2 + C(x_1, x_2, x_3, \ldots)\mathrm{d}x_3 \ldots, \tag{4.7}$$

is said to be exact differential if there exists a scalar function $\phi = \phi(x_1, x_2, x_3, \ldots)$ such that

$$\mathrm{d}\phi = \left(\frac{\partial\phi}{\partial x_1}\right)_{x_2, x_3, \ldots} \mathrm{d}x_1 + \left(\frac{\partial\phi}{\partial x_2}\right)_{x_1, x_3, \ldots} \mathrm{d}x_2 + \left(\frac{\partial\phi}{\partial x_3}\right)_{x_1, x_2, x_4 \ldots} \mathrm{d}x_3 + \cdots. \tag{4.8}$$

In thermodynamics, several exact differentials can be constructed and can serve as the thermodynamic potentials. In table 4.1, we have listed the thermodynamic potentials that are commonly used.

Let us consider the first law of thermodynamics, the mathematical statement for which can be written as

$$\mathrm{d}E = T\mathrm{d}S - P\mathrm{d}V. \tag{4.9}$$

[2] Extensive variables are additive, while intensive variables are not. For example, bring two bodies, each of volume (V) and pressure (P), together. The total volume will add up to $2V$ but the pressure will remain the same. Volume is an extensive variable, pressure is an intensive variable.

Table 4.1. List of commonly used thermodynamic potentials.

Thermodynamic potential	Symbol	Differential form	Natural variables
Internal energy	E	$Td S - Pd V + \mu d N$	S, V, N
Helmholtz free energy	G	$-Pd V - Sd T + \mu d N$	T, V, N
Gibbs free energy	F	$-Sd T + Vd P + \mu d N$	T, P, N
Enthalpy	H	$Td S + Vd P + \mu d N$	S, P, N
Grand potential	Ω	$-Pd V - Sd T - Nd\mu$	V, T, μ

It is an exact differential in variables (S,V), and consequently internal energy $E=E(S,V)$ can be considered as a thermodynamic potential. One can write

$$\mathrm{d}E = \left(\frac{\partial E}{\partial S}\right)_V \mathrm{d}S + \left(\frac{\partial E}{\partial V}\right)_S \mathrm{d}V, \tag{4.10}$$

and identify

$$\left(\frac{\partial E}{\partial S}\right)_V = T \tag{4.11}$$

$$\text{and} \quad \left(\frac{\partial E}{\partial V}\right)_S = -P. \tag{4.12}$$

Equations (4.11) and (4.12) are called Maxwell's relations. There are six more Maxwell's relations connecting various thermodynamic potentials and macroscopic variables.

When the energy of the system varies, rather than (S,V) it is more convenient to use the variables (T,V). The change can be effected through Legendre transformation (see appendix A.7),

$$F(T,\, V) = E - TS, \tag{4.13}$$

which defines the free energy (or more specifically the Helmholtz free energy) F. Free energy is also a thermodynamic potential as the differential of F is exact,

$$\begin{aligned} \mathrm{d}F &= \mathrm{d}E - T\mathrm{d}S - S\mathrm{d}T \\ &= -P\mathrm{d}V - S\mathrm{d}T. \end{aligned} \tag{4.14}$$

The differential of F can also be written as

$$\mathrm{d}F = \left(\frac{\partial F}{\partial V}\right)_T \mathrm{d}V + \left(\frac{\partial F}{\partial T}\right)_V \mathrm{d}T. \tag{4.15}$$

One can immediately identify that

$$\left(\frac{\partial F}{\partial V}\right)_T = -P \tag{4.16}$$

$$\text{and} \qquad \left(\frac{\partial F}{\partial T}\right)_V = -S. \tag{4.17}$$

The physical implication of free energy is also apparent from equation (4.14). For isothermal processes ($\mathrm{d}T = 0$), the change in free energy is the negative of the maximum possible work done by the system. To state it another way: at fixed temperature and volume, free energy is at a minimum at the equilibrium. At fixed temperature, free energy is the most important quantity. As will be discussed later, free energy is directly related to the partition function Z (which will be discussed shortly),

$$F = -T \log Z. \tag{4.18}$$

One can also define a Gibbs thermodynamic potential or Gibbs potential G,

$$G(T, P) = F + PV. \tag{4.19}$$

The differential G can be written as

$$\mathrm{d}G = \mathrm{d}F + P\mathrm{d}V + V\mathrm{d}P = -S\mathrm{d}T + V\mathrm{d}P \tag{4.20}$$

from which one can obtain the relations

$$\left(\frac{\partial G}{\partial T}\right)_P = -S \tag{4.21}$$

$$\text{and} \qquad \left(\frac{\partial G}{\partial P}\right)_T = V. \tag{4.22}$$

The physical meaning of the Gibbs potential is apparent from equation (4.20): if the system is kept at fixed temperature and pressure, the Gibbs potential remains unchanged. Or to put it another way: at constant temperature and pressure, the Gibbs potential is a minimum at equilibrium.

One can also define a quantity called enthalpy, $H(P, S)$,

$$\begin{aligned} H &= E + PV \\ \mathrm{d}H &= T\mathrm{d}S + V\mathrm{d}P \end{aligned} \tag{4.23}$$

and identify

$$\left(\frac{\partial H}{\partial S}\right)_P = T \tag{4.24}$$

$$\text{and} \qquad \left(\frac{\partial H}{\partial P}\right)_S = V. \tag{4.25}$$

From the definition of enthalpy, its physical significance is clear. It is a measure of total energy (internal as well as the work done by the system).

The relations equations (4.11), (4.12), (4.16), (4.17), (4.21), (4.22), (4.24) and (4.25) are called Maxwell's thermodynamic relations.

The statement of the first law in equation (4.9) assumes that the number of constituents of the system (molecules, atoms or whatever is appropriate) remains unchanged. We name the constituents particles. Indeed, the first law is from a pre-relativistic era when the equivalence of mass and energy was not known. In a relativistic system, allowing for a change in the particle number N, the mathematical statement for the first law is

$$\mathrm{d}E = T\mathrm{d}S - p\mathrm{d}V + \mu\mathrm{d}N \tag{4.26}$$

where we have introduced a quantity μ called chemical potential. The physical meaning of chemical potential is also clear. It is the energy cost of adding one more particle to the system while keeping both S and V fixed. Strictly speaking, in equation (4.26), $\mu\mathrm{d}N$ should be written as $\sum_i \mu_i \mathrm{d}N_i$, making an allowance for the production of different types of particles, but for now we will continue with $\mu\mathrm{d}N$. The results can be generalized as required.

In certain applications it is preferable to define energy density as $\varepsilon = E/V$, entropy density as $s = S/V$ and number density as $n = N/V$. One then obtains the well-used thermodynamic identity

$$\varepsilon = Ts - P + \mu n. \tag{4.27}$$

Equation (4.26) implies that E is a function of (S, V, N). When energy and number of particles both vary, it is more convenient to use the set of variables (T, V, μ). The change is again effected through a Legendre transformation,

$$\Omega(T, V, \mu) = E - TS - \mu N \tag{4.28}$$

where we have introduced the Landau potential or grand potential Ω. The differential $\mathrm{d}\Omega$ can be calculated as

$$\begin{aligned}\mathrm{d}\Omega &= \mathrm{d}E - T\mathrm{d}S - S\mathrm{d}T - \mu\mathrm{d}N - N\mathrm{d}\mu \\ &= -p\mathrm{d}V - S\mathrm{d}T - N\mathrm{d}\mu \\ &= \left(\frac{\partial\Omega}{\partial V}\right)_{T,\mu}\mathrm{d}V + \left(\frac{\partial\Omega}{\partial T}\right)_{V,\mu}\mathrm{d}T + \left(\frac{\partial\Omega}{\partial \mu}\right)_{T,V}\mathrm{d}\mu.\end{aligned} \tag{4.29}$$

One can identify the partial derivatives as

$$N = -\left(\frac{\partial\Omega}{\partial\mu}\right)_{T,V} \qquad S = -\left(\frac{\partial\Omega}{\partial T}\right)_{\mu,V} \qquad P = -\left(\frac{\partial\Omega}{\partial V}\right)_{T,\mu}. \tag{4.30}$$

It may be noted that when allowance is made for a change in the particle number, all the thermodynamic potentials discussed earlier should be supplemented by a term $\mu\mathrm{d}N$ without affecting their significance.

4.3 Brief review of statistical physics

Like thermodynamics, statistical physics also tries to understand the equilibrium properties of macroscopic systems. However, while thermodynamics makes no reference to constituent properties, statistical physics tries to understand equilibrium properties with reference to a system's constituents and their interactions.

4.3.1 Classical statistical physics

We mentioned earlier that classically, to completely specify a many-body system of N particles, $6N$ real numbers are needed. If H is the Hamiltonian of the system, the dynamics of the system will be completely specified by the equations of motion,

$$\dot{q}_i = \frac{\partial H(q, p)}{\partial p_i} \qquad i = 1, 2, \ldots 3N \tag{4.31a}$$

$$-\dot{p}_i = \frac{\partial H(q, p)}{\partial q_i} \qquad i = 1, 2, \ldots 3N, \tag{4.31b}$$

where $(q, p) \equiv (q_1, q_2 \ldots q_{3N}, p_1, p_2 \ldots p_{3N})$. In principle, given an initial condition for the system, its future is completely dictated by equation (4.31).

In a $6N$-dimensional phase space (called Γ-space), an N-body system can be represented by a point. As the system evolves with time following the equations of motion, the representative point will trace out a complex trajectory in Γ-space, and each point of the trajectory will indicate a definite state of the system. The trajectory will not cross itself as a crossing would indicate a double value for $\dot{p}_i$ and $\dot{q}_i$, but for a well behaved Hamiltonian $\frac{\partial H(q,p)}{\partial p_i}$ and $\frac{\partial H(q,p)}{\partial q_i}$ are single valued. For an isolated system, the energy is constrained $E = H(p, q)$, and the locus of the representative point will lie on a constant energy (E) surface in the Γ-space. Gibbs, Maxwell and Boltzmann realized very early that if left for a sufficiently long time, the trajectories will eventually fill out the entire energy surface. If the system is observed for some macroscopic observable, its average value can be found to be

$$\langle O \rangle = \frac{1}{t_2 - t_1} \int_{t_1}^{t_2} O \mathrm{d}t. \tag{4.32}$$

It is an experimental observation that for a sufficiently long interval $t_2 - t_1$, the average value $\langle O \rangle$ to a large extent does not depend on the initial and final time. The observation can be interpreted as the insensitivity of the macroscopic observable O on the details of the microscopic constituents. For example, if we require that the system of N particles in volume V has energy E, a large number of representative points will satisfy the criterion. Thus even after knowing the values of those quantities, one still does not know which among those representative points actually corresponds to the system under consideration. One then invokes the most important statistical assumption.

All possible states are equally likely.

Since all possible states are equally likely, instead of following a single macroscopic system with time, one can consider the macroscopic system as a collection of systems, all identical in composition and some 'macroscopic' condition, but existing in different microscopic states. For example, one can throw a dice N times to find out the probability of throwing a particular number, e.g. six, or alternatively, one can make N copies of the dice and throw them simultaneously. One understands that both processes will lead to the same answer.

Thus Gibbs introduced the key concept of ensemble in statistical physics. An ensemble is a large number of copies of a system, which do not interact with each other and each of which represents a possible state that the real system might be in. In Γ-space, an ensemble can be represented by a collection of points. One can introduce a density function $\rho(q, p, t)$, with (q, p) an abbreviation for $(\mathbf{q}_1, \mathbf{q}_2 \dots \mathbf{q}_N; \mathbf{p}_1, \mathbf{p}_2, \dots \mathbf{p}_N)$ such that $\rho(q, p, t)\mathrm{d}^{3N}p\mathrm{d}^{3N}q$ is the number of representative points in an infinitesimal volume element $\mathrm{d}^{3N}p\mathrm{d}^{3N}q$ at time t. An ensemble is completely specified by the density function $\rho(p, q, t)$, the subsequent values being determined by the microscopic dynamics. One can prove Liouville's theorem, $\frac{\mathrm{d}\rho}{\mathrm{d}t} = 0$, i.e. the distribution of representative points moves like a incompressible fluid. For a system in equilibrium, the density function will not depend explicitly on time (note that implicit dependence through the Hamiltonian will continue). Instead of following the system with time, the average value of the observable O can be obtained as an ensemble average,

$$\langle O \rangle = \frac{\int O\rho(p, q)\mathrm{d}p\mathrm{d}q}{\int \rho(p, q)\mathrm{d}p\mathrm{d}q}. \tag{4.33}$$

In kinetic theory, one comes across a one-body distribution function, $f(\mathbf{p}, \mathbf{q}, t)$. $f(\mathbf{p}, \mathbf{q}, t)\mathrm{d}^3p\mathrm{d}^3q$ is the number of particles with a momentum between $\mathbf{p}$ and $\mathbf{p} + \mathrm{d}\mathbf{p}$ and position vectors $\mathbf{q}$ and $\mathbf{q} + \mathrm{d}\mathbf{q}$. It is clear that the state of the many-body system is not uniquely determined from f. For example, in a system, if two particles are interchanged, it will not lead to any change in f, but to two distinct points in Γ-space. A given distribution function then corresponds to a volume in the Γ-space. An equilibrium distribution function will correspond to maximal volume in Γ-space.

In statistical physics, one generally constructs three types of ensemble.

(i) *Microcanonical ensemble*: for isolated systems, where total energy and total number, two macroscopic properties of the system, are constrained to be fixed.

(ii) *Canonical ensemble*: for a system in contact with the environment such that heat or energy can be exchanged. Only the total number of the system is constrained.

(iii) *Grand canonical ensemble*: without any constraint, the system can exchange energy as well as particles with the environment. For relativistic systems, grand canonical ensemble is the appropriate choice.

For a given ensemble, one can construct the density function $\rho(p, q)$ and calculate for the macroscopic observables. However, in nature, particles obey quantum mechanics and classical statistics misses some quantum aspects, most importantly, the indistinguishability of particles. It also misses the spin structure of the particles. Then classical statistics leads to some paradoxes, e.g. the Gibbs paradox, wherein contrary to experiment, on mixing two ideal gases of the same kind at the same temperature, entropy increases. The Gibbs paradox is resolved only in quantum statistics. Thus, in the following, we review certain aspects of quantum statistics.

4.3.2 Quantum statistics

In quantum mechanics, particle trajectories have no meaning (one cannot fix position and momentum simultaneously). The system will be specified by the wave function, which satisfies the Schrödinger equation,

$$H\Psi = E\Psi, \tag{4.34}$$

where H is the Hamiltonian of the many-body system and Ψ is the many-body wave function,

$$\Psi = \text{linear combination of } \psi_a(1)\psi_b(2)\psi_c(3)\ldots. \tag{4.35}$$

where a, b, c... are the quantum numbers. In a many-body system, a large (practically infinite) number of microstates (the states where all the parameters of the constituents are specified) is possible and the wave function can be a very complicated object. Here again, one can define an ensemble as an incoherent superposition of microstates.

Now, two important features distinguish a many-body quantum system from a classical one: (i) the density of energy levels is very high and (ii) the many-body system can exist only in a mixed state. The very high density of energy levels is understood. For example, a particle in a box of linear dimension L can have an infinite number of energy levels with the separation $\sim 1/L^2$ between them. The energy levels of N non-interacting particles are just the sum of the energy levels of N particles. Taking all possible combinations, one can have a very large number of energy levels in any appreciable finite part of the energy spectrum. Indeed, one can show that the separation between the levels for an N-body system is $\sim 10^{-N}$, which is a very small number.

The second feature arises from the fact that a truly isolated system does not exist in nature[3]. It must have some interaction with the external world. Even a weak interaction of the system with the external world will cause broadening of energy which will be large compared to the infinitesimal separation between the energy levels. There is an even more fundamental reason why a quantum many-body

[3] Possibly the Universe in its entirety can be considered a truly isolated system.

system can only be in a mixed state. A quantum mechanical wave function is a result of interaction between the system under consideration and an external system (e.g. the measurement apparatus) which obeys classical mechanics. One needs to distinguish between energy E before the interaction and energy E' after the interaction. The uncertainties in E and E' are related to the duration of the interaction Δt,

$$|\Delta E - \Delta E'| \sim \frac{\hbar}{\Delta t}. \tag{4.36}$$

ΔE and $\Delta E'$ are of the same order and one cannot make $\Delta E'$ arbitrarily small. Then $\Delta E' \sim \frac{\hbar}{\Delta t}$. For a state to be stationary, $\Delta E'$ must be smaller than the separation between the energy levels. For a many-body system, separation between the energy levels is very small and it will require a very long interaction time $\Delta t \sim \hbar/\Delta E'$ to bring the system into a stationary state.

Quantum mechanical description of mixed states is effected by means of the density matrix ρ (see appendix A.4). The expectation or mean value of any observable operator $\hat{O}$ at time t can be calculated as

$$\langle O \rangle = \mathrm{Tr}[\rho O]. \tag{4.37}$$

In quantum statistics, the density matrix plays a role analogous to the density function $\rho(p, q)$ in classical statistics. In classical statistics, $\rho(p, q)$ gives the probability distribution of the various values of coordinates and momenta of the particles of the body. In quantum statistics the diagonal matrix elements of density matrix $\rho_{nn} = \rho_n$ give the probability of finding the body in a particular state n. Determination of the density matrix is the main task in quantum statistics. Let us obtain the density matrix for the three types of ensemble considered here.

(i) *Microcanonical ensemble.* Consider a microcanonical ensemble with energy between E and $E + \Delta E$, $\Delta E \ll E$. It can exist in a large number of microstates. Since we are ignorant about the microstates, we assign equal probability to each possible microstate.

$$\begin{aligned} \rho_n &= \frac{1}{\sum_{E<E_n<E+\Delta E} 1} \qquad E < E_n < E + \Delta E \\ &= 0 \qquad \text{otherwise.} \end{aligned} \tag{4.38}$$

$\mathrm{Tr}\rho = \Gamma(E)$ then counts the number of mtextcrostates lying between E and $E + \Delta E$. The connection between the microcanonical ensemble and other thermodynamic variables is established by calculating entropy as

$$S(E, V) = k \log \Gamma(E). \tag{4.39}$$

Entropy can also be calculated in terms of probabilities of the microstates,

$$S(E, V) = -k \log\left[\frac{1}{\Gamma(E)}\right]$$
$$= -k \sum_n \frac{1}{\Gamma(E)} \log\left[\frac{1}{\Gamma(E)}\right] = -k \sum_n \rho_n \log \rho_n. \tag{4.40}$$

The last line is possible since all the terms in the summation are the same and there are exactly $\Gamma(E)$ terms.

(ii) *Canonical ensemble.* In a canonical ensemble, the system under consideration can exchange energy with the environment at temperature $T = \frac{1}{\beta}$. One assigns a weight factor $e^{-\beta E_n}$ to the probability to find the system at energy state E_n. The density matrix for the canonical ensemble is then written as

$$\rho_n = Ae^{-\beta E_n}. \tag{4.41}$$

The constant A satisfies the normalization condition, $\sum_n \rho_n = 1$, giving

$$A = \frac{1}{\sum_n e^{-\beta E_n}}. \tag{4.42}$$

One can calculate the entropy,

$$S = -\sum_n \rho_n \log \rho_n = -\langle \log \rho_n \rangle$$
$$= \frac{\langle E \rangle}{T} - \log A \tag{4.43}$$

where $\log A = \frac{\langle E \rangle - TS}{T}$. The mean energy $\langle E \rangle$ is the energy meant in thermodynamics. Noting that the Gibbs free energy $F = E - TS$, we obtain for the Gibbs free energy

$$F = -T \log \sum_n e^{-\beta E_n}. \tag{4.44}$$

This formula is of fundamental importance in statistical physics. It facilitates calculations of thermodynamic functions for systems whose energy spectrum is known. The sum in the logarithm is called the partition function and is generally denoted by the symbol Z. It is just the trace of the operator $e^{-\beta \hat{H}}$, $\hat{H}$ being the Hamiltonian of the system.

$$Z = \sum_n e^{-\beta E_n} = \mathrm{Tr} e^{-\beta \hat{H}}. \tag{4.45}$$

(iii) *Grand canonical ensemble*. In a grand canonical ensemble, both the particle number and energy can change. One assigns a weight factor $e^{-\beta(E_n-\mu N_n)}$ to the probability of finding the system of N_n particles at energy state E_n. The density matrix for the grand canonical ensemble is then written as

$$\rho_n = \frac{e^{-\beta(E_n-\mu N_n)}}{\sum_n e^{-\beta(E_n-\mu N_n)}}. \tag{4.46}$$

In equation (4.46), $\hat{H}$ and $\hat{N}$ are the energy and number operators. When they operate on the quantum state, $|\psi_n\rangle$ gives the energy of the state and the number of particles in the state,

$$\hat{H}|\psi_n\rangle = E_n|\psi_n\rangle \tag{4.47a}$$

$$\hat{N}|\psi_n\rangle = N_n|\psi_n\rangle, \tag{4.47b}$$

such that the grand canonical partition function

$$\begin{aligned} Z &= \mathrm{Tr} e^{-\beta(\hat{H}-\mu\hat{N})} = \sum_n \langle\psi_n|e^{-\beta(\hat{H}-\mu\hat{N})}|\psi_n\rangle \\ &= \sum_n e^{-\beta(E_n-\mu N_n)}. \end{aligned} \tag{4.48}$$

4.3.3 Thermodynamic parameters from the grand canonical ensemble

From the partition function, the observable or expectation value of any operator can be obtained as

$$\langle O\rangle = \frac{\mathrm{Tr} O e^{-\beta(\hat{H}-\mu\hat{N})}}{\mathrm{Tr} e^{-\beta(\hat{H}-\mu\hat{N})}}. \tag{4.49}$$

In the following, we derive the explicit relations for computing the particle number, energy, pressure and entropy of a system from the grand canonical partition function as given in equation (4.48).

If the logarithm of the partition function is differentiated by μ we obtain

$$\frac{\partial \ln Z}{\partial\mu} = \frac{1}{Z}\mathrm{Tr} e^{-\beta(\hat{H}-\mu\hat{N})}\left(\beta\hat{N}\right) = \beta\langle N\rangle. \tag{4.50}$$

Thus the average number of particles in the system is

$$\langle N\rangle = T\frac{\partial \ln Z}{\partial\mu}. \tag{4.51}$$

The average energy of the system can be obtained by differentiating ln Z by the inverse temperature β,

$$\frac{\partial \ln Z}{\partial \beta} = \frac{1}{Z}\mathrm{Tr}e^{-\beta(\hat{H}-\mu\hat{N})}\left[-\left(\hat{H} - \mu\hat{N}\right)\right] = -\langle E\rangle + \mu\langle N\rangle, \tag{4.52}$$

giving the average energy of the system as

$$\langle E\rangle = -\frac{\partial \ln Z}{\partial \beta} + \mu\langle N\rangle = T^2\frac{\partial \ln Z}{\partial T} + \mu\langle N\rangle. \tag{4.53}$$

The entropy of the system can be obtained from the relation

$$\begin{aligned} S &= -\sum_n \rho_n \ln \rho_n = -\left\langle \ln \rho_n \right\rangle \\ &= \beta(\langle E\rangle - \mu\langle N\rangle) + \ln Z = \frac{\partial T \ln Z}{\partial T}. \end{aligned} \tag{4.54}$$

Finally, using the thermodynamic relation, $E = TS - PV + \mu N$, the pressure P is obtained as

$$P = \frac{1}{V}T \ln Z. \tag{4.55}$$

In some applications, it is more convenient to use particle density, energy density, etc. In the following, we rewrite equations (4.51), (4.53), (4.54) and (4.55) in a different form.

$$\begin{aligned} \text{Particle density: } n &= \frac{N}{V} = \frac{1}{V}\frac{\mathrm{Tr}\hat{N}e^{-\beta(\hat{H}-\mu\hat{N})}}{\mathrm{Tr}e^{\beta(\hat{H}-\mu\hat{N})}} \\ &= \frac{1}{V}\frac{T\partial \ln Z}{\partial \mu} \end{aligned} \tag{4.56}$$

$$\begin{aligned} \text{Energy density: } \varepsilon &= \frac{E}{V} = \frac{1}{V}\frac{\mathrm{Tr}\hat{H}e^{-\beta(\hat{H}-\mu\hat{N})}}{\mathrm{Tr}e^{\beta(\hat{H}-\mu\hat{N})}} \\ &= \frac{1}{V}\frac{T^2\partial \ln Z}{\partial T} + \mu n \end{aligned} \tag{4.57}$$

$$\text{Pressure: } P = \frac{1}{V}T \ln Z \tag{4.58}$$

$$\text{Entropy density: } s = \frac{S}{V} = \frac{1}{V}\frac{\partial T \ln Z}{\partial T}. \tag{4.59}$$

Fluctuations in particle numbers or in energy can also be calculated from the partition function. For example, by doubly differentiating the partition function,

$$\begin{aligned} \frac{\partial^2 \ln Z}{\partial \mu^2} &= \frac{1}{Z}\mathrm{Tr}e^{-\beta(\hat{H}-\mu\hat{N})}\left(\beta\hat{N}\right)^2 - \frac{1}{Z^2}\left[\mathrm{Tr}e^{-\beta(\hat{H}-\mu\hat{N})}\left(\beta\hat{N}\right)\right]^2 \\ &= \beta^2\left[\langle N^2\rangle - \langle N\rangle^2\right] \end{aligned} \tag{4.60}$$

or

$$(\Delta N)^2 = \langle N^2 \rangle - \langle N \rangle^2 = T\frac{\partial \langle N \rangle}{\partial \mu}. \tag{4.61}$$

Since $\langle N \rangle$ and $\ln Z$ are extensive quantities, the relative inaccuracy in particle number is

$$\frac{\Delta N}{N} \sim O\left(\frac{1}{\sqrt{N}}\right). \tag{4.62}$$

The fluctuations in energy can be obtained as

$$\begin{aligned} \frac{\partial^2 \ln Z}{\partial \beta^2} &= \frac{\mathrm{Tr} e^{-\beta(\hat{H}-\mu\hat{N})}\left[\left(\hat{H}-\mu\hat{N}\right)^2\right]}{Z} - \left[\frac{\mathrm{Tr} e^{-\beta(\hat{H}-\mu\hat{N})}\left[-\left(\hat{H}-\mu\hat{N}\right)\right]}{Z}\right]^2 \\ &= [\langle E^2 \rangle - \langle E \rangle^2] + \mu^2\left[\langle N^2 \rangle - \langle N \rangle^2\right] = (\Delta E)^2 + \mu^2(\Delta N)^2. \end{aligned} \tag{4.63}$$

In a system with a fixed number of particles,

$$\frac{\partial \langle E \rangle}{\partial \beta} = -T^2\frac{\partial \langle E \rangle}{\partial T} = -\frac{\partial^2 \ln Z}{\partial \beta^2}, \tag{4.64}$$

and we obtain the relation for specific heat, $C_V = \frac{\partial \langle E \rangle}{\partial T}$,

$$T^2 C_V = \frac{\partial^2 \ln Z}{\partial \beta^2} = \langle E^2 \rangle - \langle E \rangle^2. \tag{4.65}$$

4.3.4 Partition function with quantum statistics

As opposed to integral spin bosons, half-integer spin fermions obey the Pauli exclusion principle, and two fermions cannot occupy the same state. The grand canonical partition function defined in equation (4.48) did not take into account the quantum nature of the particles. In the following, we derive the partition function explicitly accounting for the quantum nature of the particles.

Let us consider a system of N fermions. Let the single-particle energy spectrum be denoted by $\{\epsilon_n, n = 0, 1, 2, \ldots\}$. In a grand canonical ensemble, a given state can be populated by an arbitrary number of particles. Let us consider a state $|n\rangle$. It can be considered as a subsystem with particle number N_n and energy $N_n\epsilon_n$. The grand partition function for this state can be written as

$$\begin{aligned} Z_n &= \sum_{N_n} e^{-\beta(\epsilon_n N_n - \mu N_n)} = \sum_{N_n}\left[e^{-\beta(\epsilon_n - \mu)}\right]^{N_n} \\ &= 1 + e^{-\beta(\epsilon_n - \mu)}. \end{aligned} \tag{4.66}$$

The last line is obtained by using the fact that for fermions, N_n can be either 0 or 1.

The grand partition function for a system of fermions is then obtained as

$$Z = \prod_n Z_n = \prod_n \left[1 + e^{-\beta(\epsilon_n - \mu)}\right]. \tag{4.67}$$

To show that the partition function does correctly reproduce the observables, let us calculate the number of fermions,

$$\begin{aligned} \langle N \rangle &= \frac{T\partial \ln Z}{\partial \mu} = T\sum_n \frac{\partial}{\partial \mu} \ln\left[1 + e^{-\beta(\epsilon_n - \mu)}\right] \\ &= T\sum_n \frac{\beta e^{-\beta(\epsilon_n-\mu)}}{\left[1 + e^{-\beta(\epsilon_n-\mu)}\right]} = \sum_n \left[\frac{1}{e^{+\beta(\epsilon_n-\mu)} + 1}\right]. \end{aligned} \tag{4.68}$$

The bracketed quantity is just the occupation probability in Fermi–Dirac statistics. As expected, the average number is the sum of the occupation probability in each state.

The grand partition function for a system of bosons can also be obtained similarly. As for fermions, the partition function for the *n*th eigenstate in a bosonic system is

$$Z_n = \sum_{N_n} \left[e^{-\beta(\epsilon_n-\mu)}\right]^{N_n}. \tag{4.69}$$

However, unlike the fermions, any number of bosons can occupy a state—thus, $N_n = 0, 1, 2, \ldots. \infty$ and we have a sum of a geometric series. For $\epsilon_n - \mu > 0$, the series converges to

$$Z_n = \frac{1}{1 - e^{-\beta(\epsilon_n-\mu)}}. \tag{4.70}$$

The partition function for the total system can then be obtained as

$$Z = \prod_n Z_n = \prod_n \frac{1}{\left[1 - e^{-\beta(\epsilon_n-\mu)}\right]}. \tag{4.71}$$

Here again, we can calculate the average number of particles in the system,

$$\langle N \rangle = \frac{T\partial \ln Z}{\partial \mu} = \sum_n \left[\frac{1}{e^{+\beta(\epsilon_n-\mu)} - 1}\right] \tag{4.72}$$

where the bracketed quantity is just the occupation probability in Bose–Einstein statistics. It is particularly important to note that in summing up equation (4.69) we have used the condition $\epsilon_n - \mu > 0$. The *n*th eigenstate can be any state including the ground state. For the ground state, the energy is taken as zero. Then for bosons, the chemical potential can only be negative.

Partition functions facilitate the calculation of the thermodynamic properties of macroscopic systems. Sometimes it is useful to use the thermodynamic potential Ω which is related to the partition function as

$$\Omega = -T \log Z. \tag{4.73}$$

The thermodynamic potential for a system of fermions or a system of bosons can thus be written as

$$\Omega = -T\eta \sum_n \log\left[1 + \eta e^{-\beta(\epsilon_n - \mu)}\right] \tag{4.74}$$

with $\eta = +1$ for fermions and $\eta = -1$ for bosons.

In many-body systems, energy levels are approximately continuous and we can replace the sum by an integration,

$$\sum_n \to \int d^3n = \frac{V}{(2\pi)^3} \int_0^\infty d^3p = \frac{4\pi V}{(2\pi)^3} \int_0^\infty p^2 dp. \tag{4.75}$$

Replacing the sum by an integration, the thermodynamic potential in equation (4.74) can be written as

$$\Omega = -\eta T \frac{4\pi V}{(2\pi)^3} \int_0^\infty \log\left[1 + \eta e^{-\beta(\epsilon - \mu)}\right] p^2 dp. \tag{4.76}$$

Integrating by parts,

$$\begin{aligned} \Omega = -T \ln Z &= -\eta T \frac{4\pi V}{(2\pi)^3} \int_0^\infty \frac{\beta \eta e^{-\beta(\epsilon(p) - \mu)}}{\left[1 + \eta e^{-\beta(\epsilon - \mu)}\right]} \frac{d\epsilon}{dp} \frac{p^3}{3} dp \\ &= -\frac{V}{6\pi^2} \int_0^\infty \frac{p^4 dp}{\sqrt{p^2 + m^2}} \frac{1}{\left[e^{\beta\left(\sqrt{p^2+m^2} - \mu\right)} + \eta\right]}. \end{aligned} \tag{4.77}$$

From the thermodynamical potential all the relevant thermodynamic quantities can be calculated.

4.3.5 Physical quantities for fermionic/bosonic systems

In the following, some physical quantities, e.g. number density, energy density, etc, for a system of particles obeying Fermi–Dirac or Bose–Einstein distribution will be obtained. Evaluation of the physical quantities of macroscopic systems is facilitated in terms of the partition function (Z) or equivalently in terms of the thermodynamic potential (Ω). For a system of particles obeying Bose–Einstein/Fermi–Dirac statistics, the thermodynamic potential is written as

$$\Omega = -T \ln Z = -\frac{gV}{6\pi^2} \int_0^\infty \frac{p^4 dp}{\sqrt{p^2 + m^2}} \frac{1}{e^{\beta\left(\sqrt{p^2+m^2} - \mu\right)} + \eta} \tag{4.78}$$

with $\eta = -1$ for the Bose–Einstein distribution and $\eta = +1$ for the Fermi–Dirac distribution. The factor g in equation (4.78) is the degeneracy factor.

For antiparticles, the expression remains the same, but the sign of the chemical potential is reversed. The thermodynamic potential for a macroscopic system containing particles and antiparticles is

$$\Omega = -T\ln Z = -\frac{gV}{6\pi^2}\int_0^\infty \frac{p^4\mathrm{d}p}{\sqrt{p^2+m^2}}\left\{\frac{1}{\mathrm{e}^{\beta\left(\sqrt{p^2+m^2}-\mu\right)}+\eta}+\frac{1}{\mathrm{e}^{\beta\left(\sqrt{p^2+m^2}+\mu\right)}+\eta}\right\}. \tag{4.79}$$

Let us evaluate the integral,

$$I = \int_0^\infty \frac{p^4\mathrm{d}p}{\sqrt{p^2+m^2}}\frac{1}{\mathrm{e}^{\beta\left(\sqrt{p^2+m^2}-\mu\right)}+\eta}. \tag{4.80}$$

Now the Fermi–Dirac or Bose–Einstein distribution can always be written as an infinite sum of the Maxwell–Boltzmann distribution. For example, the Fermi–Dirac distribution can be written as

$$f(x) = \frac{1}{\mathrm{e}^x+1} = \mathrm{e}^{-x}[1+\mathrm{e}^{-x}]^{-1} = \mathrm{e}^{-x}\sum_{n=0}^{\infty}\mathrm{e}^{-nx} = \sum_{n=1}^{\infty}\mathrm{e}^{-nx}. \tag{4.81}$$

Similarly, the Bose–Einstein distribution can be written as

$$f(x) = \frac{1}{\mathrm{e}^x-1} = \mathrm{e}^{-x}\left[1-\mathrm{e}^{-x}\right]^{-1} = \mathrm{e}^{-x}\sum_{n=0}^{\infty}(-1)^n\mathrm{e}^{-nx} = \sum_{n=1}^{\infty}(-1)^{n+1}\mathrm{e}^{-nx}. \tag{4.82}$$

The results can be combined to write

$$\begin{aligned} f(E) &= \frac{g}{(2\pi)^3}\frac{1}{\mathrm{e}^{(E-\mu)/T}+\eta} \\ &= \frac{g}{(2\pi)^3}\sum_{n=1}^{\infty}(-\eta)^{n+1}\mathrm{e}^{-n(E-\mu)/T} \end{aligned} \tag{4.83}$$

with $\eta = -1$ for the Bose–Einstein distribution and $\eta = +1$ for the Fermi–Dirac distribution. Obviously, if the expansion is truncated at $n = 1$, it will correspond to the Maxwell–Boltzmann distribution function.

Using the expansion, the integral I in equation (4.80) can be written as

$$\begin{aligned} I &= \int_0^\infty \frac{p^4\mathrm{d}p}{\sqrt{p^2+m^2}}\sum_{n=1}^{\infty}(-\eta)^{n+1}\mathrm{e}^{n\mu/T}\mathrm{e}^{-n\left(\sqrt{p^2+m^2}\right)/T} \\ &= \int_0^\infty \frac{p^4\mathrm{d}p}{\sqrt{p^2+m^2}}\sum_{n=1}^{\infty}(-\eta)^{n+1}\mathrm{e}^{\mu/T'}\mathrm{e}^{-\left(\sqrt{p^2+m^2}\right)/T'} \end{aligned} \tag{4.84}$$

where in the last line we have defined a variable $T' = T/n$. Let us introduce the dimensionless variables, z and τ,

$$\tau = \frac{\sqrt{p^2 + m^2}}{T'} \tag{4.85a}$$

$$z = \frac{m}{T'} \tag{4.85b}$$

and write

$$p = T'\sqrt{\tau^2 - z^2} \tag{4.86a}$$

$$p\mathrm{d}p = T'^2 \tau \mathrm{d}\tau, \tag{4.86b}$$

$$p^4 \mathrm{d}p = T'^5 \tau(\tau^2 - z^2)^{3/2} \mathrm{d}\tau. \tag{4.86c}$$

In terms of the new variables the integral is

$$I = \sum_{n=1}^{\infty} (-\eta)^{n+1} \mathrm{e}^{\mu/T'} T'^4 \int_z^{\infty} \mathrm{d}\tau \left(\tau^2 - z^2\right)^{3/2} \mathrm{e}^{-\tau}. \tag{4.87}$$

A closed form expression can be obtained in terms of the modified Bessel function of the second kind [7],

$$K_n(z) = \frac{2^n n!}{(2n)!} \frac{1}{z^n} \int_z^{\infty} \mathrm{d}\tau \left(\tau^2 - z^2\right)^{n-1/2} \mathrm{e}^{-\tau} \tag{4.88}$$

$$\Rightarrow 3z^2 K_2(z) = \int_z^{\infty} \mathrm{d}\tau \left(\tau^2 - z^2\right)^{3/2} \mathrm{e}^{-\tau} \tag{4.89}$$

$$I = 3T^2 m^2 \sum_{n=1}^{\infty} (-\eta)^{n+1} \frac{1}{n^2} K_2\left(\frac{nm}{T}\right) \mathrm{e}^{n\mu/T}. \tag{4.90}$$

The thermodynamic potential in equation (4.78) is then obtained as

$$\Omega = -T \ln Z = -\frac{gV}{2\pi^2} T^2 m^2 \sum_{n=1}^{\infty} (-\eta)^{n+1} \frac{1}{n^2} K_2\left(\frac{nm}{T}\right) \mathrm{e}^{n\mu/T}. \tag{4.91}$$

We remind readers that the thermodynamic potential in equation (4.91) contains the contribution of the particles only. Antiparticles are not included.

Equation (4.91) can be used to compute various thermodynamic parameters.

(i) *Particle density*:

$$\begin{aligned} n &= \frac{T}{V} \frac{\partial \ln Z}{\partial \mu} \\ &= g \frac{T m^2}{2\pi^2} \sum_{k=1}^{\infty} \frac{(-\eta)^{k+1}}{k} K_2\left(\frac{km}{T}\right) \mathrm{e}^{k\mu/T}. \end{aligned} \tag{4.92}$$

(ii) *Pressure*:

$$P = \frac{T}{V}\ln Z$$
$$= g\frac{T^2m^2}{2\pi^2}\sum_{n=1}^{\infty}\frac{(-\eta)^{n+1}}{n^2}K_2\left(\frac{nm}{T}\right)e^{n\mu/T}. \tag{4.93}$$

Energy density (ε) needs a little algebraic work.

$$\varepsilon = \frac{T^2}{V}\frac{\partial \ln Z}{\partial T} + \mu n$$
$$= g\frac{T^2m^2}{2\pi^2}\left[\sum_{n=1}^{\infty}\frac{(-\eta)^{n+1}}{n^2}\frac{\partial}{\partial T}\left\{TK_2\left(\frac{nm}{T}\right)e^{n\mu/T}\right\}\right] + \mu n$$
$$= g\frac{T^2m^2}{2\pi^2}\sum_{n=1}^{\infty}\frac{(-\eta)^{n+1}}{n^2}\left[K_2\left(\frac{nm}{T}\right)e^{n\mu/T} + Te^{n\mu/T}\frac{\partial}{\partial T}K_2\left(\frac{nm}{T}\right)\right.$$
$$\left. + TK_2\left(\frac{nm}{T}\right)\frac{\partial}{\partial T}e^{n\mu/T}\right] + \mu n. \tag{4.94}$$

The derivative of the modified Bessel function can be written as

$$\frac{\partial K_n(z)}{\partial z} = -K_{n-1}(z) - \frac{n}{z}K_n(z), \tag{4.95}$$

which gives

$$\frac{\partial}{\partial T}K_2\left(\frac{nm}{T}\right) = \frac{nm}{T^2}K_1\left(\frac{nm}{T}\right) + \frac{2}{T}K_2\left(\frac{nm}{T}\right), \tag{4.96}$$

and equation (4.94) reduces to

$$\varepsilon = g\frac{T^2m^2}{2\pi^2}\sum_{n=1}^{\infty}(-\eta)^{n+1}\frac{1}{n^2}\left[3K_2\left(\frac{nm}{T}\right) + \left(\frac{nm}{T}\right)K_1\left(\frac{nm}{T}\right)\right.$$
$$\left. - K_2\left(\frac{nm}{T}\right)\frac{n\mu}{T}\right]e^{n\mu/T} + \mu n$$
$$= 3P + g\frac{Tm^3}{2\pi^2}\sum_{n=1}^{\infty}\frac{(\mp)^{n+1}}{n}K_1\left(\frac{nm}{T}\right)e^{n\mu/T}$$
$$\underline{- g\frac{Tm^2\mu}{2\pi^2}\sum_{n=1}^{\infty}\frac{(-\eta)^{n+1}}{n}K_2\left(\frac{nm}{T}\right)e^{n\mu/T}} + \mu n. \tag{4.97}$$

The underlined term is just μn. The final expression is as follows.

(iii) *Energy density*:

$$\varepsilon = 3P + g\frac{Tm^3}{2\pi^2}\sum_{n=1}^{\infty}g\frac{(-\eta)^{n+1}}{n}K_1\left(\frac{nm}{T}\right)e^{n\mu/T}. \tag{4.98}$$

For massless particles $m = 0$ the second term is identically zero and one obtains the well-known relation, $P = \frac{1}{3}\varepsilon$.

(iv) *Entropy density*:

$$\begin{aligned}
s &= \frac{1}{V}\frac{\partial T \ln Z}{\partial T} \\
&= g\frac{m^2}{2\pi^2}\frac{\partial}{\partial T}\left[\sum_{n=1}^{\infty}\frac{(-\eta)^{n+1}}{n^2}T^2K_2\left(\frac{nm}{T}\right)e^{n\mu/T}\right] \\
&= g\frac{m^2}{2\pi^2}\sum_{n=1}^{\infty}\frac{(-\eta)^{n+1}}{n^2}\left[2TK_2\left(\frac{nm}{T}\right)e^{n\mu/T}\right. \\
&\quad \left. + T^2\left\{\left[\frac{nm}{T^2}K_1\left(\frac{nm}{T}\right) + \frac{2}{T}K_2\left(\frac{nm}{T}\right)\right]e^{n\mu/T}\right\} + T^2K_2\left(\frac{nm}{T}\right)\frac{-n\mu}{T^2}e^{n\mu/T}\right] \\
&= g\frac{m^2}{2\pi^2}\sum_{n=1}^{\infty}\frac{(-\eta)^{n+1}}{n^2}\left[4TK_2\left(\frac{nm}{T}\right) + nmK_1\left(\frac{nm}{T}\right) - n\mu K_2\left(\frac{nm}{T}\right)\right]e^{n\mu/T}. \quad (4.99)
\end{aligned}$$

It is easy to check that equation (4.99) satisfies the thermodynamic relation,

$$s = \frac{P + \varepsilon - \mu n}{T}. \quad (4.100)$$

The expressions for thermodynamic variables are simplified in the massless ($m \to 0$) limit, when one can use the asymptotic relation for the modified Bessel function,

$$\lim_{z \to 0} z^n K_n(z) = 2^{n-1}(n-1)! \quad (4.101)$$

For massless particles, we rewrite the grand partition function as

$$T \ln Z = g\frac{VT^4}{2\pi^2}\sum_{n=1}^{\infty}\frac{(-\eta)^{n+1}}{n^4}\left(\frac{nm}{T}\right)^2 K_2\left(\frac{nm}{T}\right)e^{n\mu/T} \quad (4.102)$$

$$= g\frac{VT^4}{\pi^2}\sum_{n=1}^{\infty}\frac{(-\eta)^{n+1}}{n^4}e^{n\mu/T}. \quad (4.103)$$

We immediately obtain:

$$\text{Number density: } n = \frac{T}{V}\frac{\partial \ln Z}{\partial \mu} = g\frac{T^3}{\pi^2}\sum_{n=1}^{\infty}\frac{(-\eta)^{n+1}}{n^3}e^{n\mu/T} \quad (4.104a)$$

$$\text{Pressure: } P = \frac{T \ln Z}{V} = g\frac{T^4}{\pi^2}\sum_{n=1}^{\infty}\frac{(-\eta)^{n+1}}{n^4}e^{n\mu/T} \quad (4.104b)$$

$$\text{Energy density: } \varepsilon = \frac{T^2}{V}\frac{\partial \ln Z}{\partial T} = g\frac{3T^4}{\pi^2}\sum_{n=1}^{\infty}\frac{(-\eta)^{n+1}}{n^4}e^{n\mu/T} \tag{4.104c}$$

$$\text{Entropy density: } s = \frac{1}{V}\frac{\partial T \ln Z}{\partial T} = g\frac{1}{\pi^2}\sum_{n=1}^{\infty}(-\eta)^{n+1}\left[\frac{4T^3}{n^4} - \frac{\mu T^2}{n^3}\right]e^{n\mu/T}. \tag{4.104d}$$

4.3.6 Massless bosonic systems

In the massless limit, the chemical potential for bosons $\mu = 0$. Any number of bosons can occupy a state and for $m = 0$, it does not cost any energy. The thermodynamic variables are greatly simplified with the help of the Riemann zeta function $\zeta(s)$ (see appendix A.6 for more details):

$$\zeta(s) = \sum_{n=1}^{\infty}\frac{1}{n^s} \tag{4.105}$$

$$\text{Number density: } n = g\frac{T^3}{\pi^2}\sum_{n=1}^{\infty}\frac{1}{n^3} = g\frac{T^3}{\pi^2}\zeta(3) = g1.202\frac{T^3}{\pi^2} \tag{4.106a}$$

$$\text{Pressure: } P = g\frac{T^4}{\pi^2}\sum_{n=1}^{\infty}\frac{1}{n^4} = g\frac{T^4}{\pi^2}\zeta(4) = g\frac{\pi^2}{90}T^4 \tag{4.106b}$$

$$\text{Energy density: } \varepsilon = 3P = g\frac{\pi^2}{30}T^4 \tag{4.106c}$$

$$\text{Entropy density: } s = g\frac{2\cdot\pi^2}{45}T^3. \tag{4.106d}$$

4.3.7 Massless fermionic systems with $\mu = 0$

In the massless limit, $m \to 0$ chemical potential for a fermionic gas can be finite. Closed form equations for physical variables can be given in the limit $\mu = 0$, using Dirichlet functions $\eta(s)$,

$$\eta(s) = \sum_{n=1}^{\infty}\frac{(-1)^{n+1}}{n^s} = \left(1 - 2^{1-s}\right)\zeta(s) \tag{4.107}$$

$$\text{Number density: } n = g\frac{T^3}{\pi^2}\sum_{n=1}^{\infty}\frac{-1^{n+1}}{n^3} = g\frac{T^3}{\pi^2}\eta(3) = \frac{3}{4}\left(g1.202\frac{T^3}{\pi^2}\right) \tag{4.108a}$$

$$\text{Pressure: } P = g\frac{T^4}{\pi^2}\sum_{n=1}^{\infty}\frac{-1^{n+1}}{n^4} = g\frac{T^4}{\pi^2}\eta(4) = \frac{7}{8}\left(g\frac{\pi^2}{90}T^4\right) \tag{4.108b}$$

$$\text{Energy density: } \varepsilon = 3P = \frac{7}{8}\left(g\frac{\pi^2}{30}T^4\right) \tag{4.108c}$$

$$\text{Entropy density: } s = \frac{7}{8}\left(g\frac{2\cdot\pi^2}{45}T^3\right). \tag{4.108d}$$

If we compare equations (4.106) and (4.108), we note that in the limit, the $m \to 0$, $\mu \to 0$, pressure, energy density and entropy density of Fermi–Dirac and Bose–Einstein systems differ by a factor of only $\frac{7}{8}$.

For $m = 0$, $\mu \neq 0$ physical quantities for fermionic systems cannot be given any simpler form than in equation (4.104).

4.3.8 Massless fermionic systems of particles and antiparticles

If the system contains antiparticles as well, the partition function for the total system can be written as

$$\begin{aligned} T \ln Z &= g\frac{VT^4}{\pi^2}\sum_{n=1}^{\infty}\frac{(-1)^{n+1}}{n^4}\left[e^{n\mu/T} + e^{-n\mu/T}\right] \\ &= g\frac{VT^4}{\pi^2}\sum_{n=1}^{\infty}\frac{(-1)^{n+1}}{n^4}2\cosh\left(\frac{n\mu}{T}\right). \end{aligned} \tag{4.109}$$

Using the expansion,

$$\cosh x = 1 + \frac{x^2}{2!} + \frac{x^4}{4!} + \cdots \tag{4.110}$$

$$\begin{aligned} T \ln Z &= g\frac{2VT^4}{\pi^2}\sum_{n=1}^{\infty}(-)^{n+1}\left[\frac{1}{n^4} + \frac{1}{n^2}\frac{1}{2}\left(\frac{\mu}{T}\right)^2 + \frac{1}{n^0}\frac{1}{24}\left(\frac{\mu}{T}\right)^4\right] \\ &= g\frac{2VT^4}{\pi^2}\left[\eta(4) + \eta(2)\frac{1}{2}\left(\frac{\mu}{T}\right)^2 + \eta(0)\frac{1}{24}\left(\frac{\mu}{T}\right)^4\right] \\ &= g2V\left[\frac{7}{8}\frac{\pi^2}{90}T^4 + \frac{\mu^2T^2}{24} + \frac{\mu^4}{48\pi^2}\right]. \end{aligned} \tag{4.111}$$

For a system of particles and antiparticles obeying the Fermi–Dirac distribution we can easily compute:

$$\text{Net density: } n = n_{\text{particle}} - n_{\text{antiparticle}} = \frac{T}{V}\frac{\partial \ln Z}{\partial\mu} = 2g\left[\frac{\mu T^2}{12} + \frac{\mu^3}{12\pi^2}\right] \tag{4.112a}$$

$$\text{Pressure: } P = \frac{T\ln Z}{V} = 2g\left[\frac{7}{8}\frac{\pi^2}{90}T^4 + \frac{\mu^2T^2}{24} + \frac{\mu^4}{48\pi^2}\right] \tag{4.112b}$$

$$\text{Energy density:}\ \ \varepsilon = \frac{T^2}{V}\frac{\partial \ln Z}{\partial T} = 3P \tag{4.112c}$$

$$\text{Entropy density:}\ \ s = \frac{1}{V}\frac{\partial T \ln Z}{\partial T} = 2g\left[\frac{7}{8}\frac{2\pi^2}{45}T^3 + \frac{\mu^2 T}{12}\right]. \tag{4.112d}$$

In equations (4.112), we have kept factor 2 hanging out in order to indicate that it also contains contributions of antiparticles. The degeneracy factor g should be appropriately calculated, i.e. antiparticles should not be included in the degeneracy factor.

4.4 A brief review of relativistic kinetic theory

The properties of many-body systems depend on: (i) the interaction of the constituent particles and (ii) external constraints. As mentioned earlier, one characterizes the system in terms of macroscopic state variables, e.g. particle density, temperature, etc, and of the characteristic microscopic parameters of the system. One then tries to understand certain equilibrium/non-equilibrium properties of the macroscopic system. In kinetic theory this program is realized by means of a statistical description, in terms of the 'one-particle distribution function' and its transport equation. From the transport equation, on the basis of conservation laws, a hydrodynamic theory of perfect fluid can be constructed. Supplementing the conservation laws with entropy law, the hydrodynamics for a dissipative fluid can be constructed.

In the following, we briefly discuss relativistic Boltzmann or the kinetic equation. We then show that basic equations for hydrodynamics are obtained by coarse graining Boltzmann transport equations. Most of the discussions are from [1, 2, 5, 6].

4.4.1 Some basic definitions in kinetic theory

Let us first define μ-space. To completely specify a particle, we need six real coordinates, three for the positions and three for the momenta. The six-dimensional space is called μ-space. In μ-space, a particle can be represented as a point.

(1) *Distribution function* $f(x, p)$: in kinetic theory, a macroscopic system is generally studied in terms of the distribution function, $f(x, p)$. $f(x, p)\mathrm{d}^3x\mathrm{d}^3p$ is defined as the average number of particles in small volume d^3x, at time t, with momenta between $\mathbf{p}$, $\mathbf{p} + \mathbf{dp}$. It is implicitly understood that particle content in the volume element d^3x is large enough to apply concepts of statistical physics, yet d^3x is small in the macroscopic scale.

(2) *Particle four-flow* N^μ: is defined as the first moment of the distribution function.

$$N^\mu(x) = \int \frac{\mathrm{d}^3p}{p^0} p^\mu f(x, p). \tag{4.113}$$

Four-components of particle four-flow can be identified as follows:

$$\text{Particle density:}\ \ N^0(x) = \int \mathrm{d}^3p f(x, p) \tag{4.114}$$

$$\text{Particle flow: } N^i(x) = \int \mathrm{d}^3p\left(\frac{p^i}{p^0}\right)f(x, p)$$
$$= \int \mathrm{d}^3p u^i f(x, p) \qquad i = 1, 2, 3 \qquad (4.115)$$

where we have introduced the velocity $\mathbf{u} = \mathbf{p}/\mathrm{p}^0$.

(3) *Energy-momentum tensor* $T^{\mu\nu}$: is the second moment of the distribution function.

$$T^{\mu\nu}(x) = \int \frac{\mathrm{d}^3p}{p^0} p^\mu p^\nu f(x, p). \qquad (4.116)$$

The components can be identified as follows:

$$\text{Energy density: } T^{00}(x) = \int \mathrm{d}^3p p^0 f(x, p) \qquad (4.117a)$$

$$\text{Energy flow: } T^{0i}(x) = \int \mathrm{d}^3p p^0 u^i f(x, p) \qquad i = 1, 2, 3 \qquad (4.117b)$$

$$\text{Momentum density: } T^{i0}(x) = \int \mathrm{d}^3p p^i f(x, p) \qquad i = 1, 2, 3 \qquad (4.117c)$$

Momentum flow or

$$\text{pressure tensor: } T^{ij}(x) = \int \mathrm{d}^3p p^i u^j f(x, p) \qquad i = 1, 2, 3. \qquad (4.117d)$$

(4) *Entropy four-flow* S^μ:

$$S^\mu(x) = -\int \frac{\mathrm{d}^3p}{p^0} p^\mu f(x, p)\left[\log h^3 f(x, p) - 1\right]. \qquad (4.118)$$

$f(x, p)$ is a dimensionful quantity (dimension = $\mathrm{fm}^{-3}\,\mathrm{GeV}^{-3}$). Logarithmic function, however, is only defined for a number, i.e. for a dimensionless quantity. To make $f(x, p)$ dimensionless, one generally multiplies it by h^3 and subtracts unity. Note that the absolute value of entropy is not measurable, only the change in entropy is measurable. Then the observables remain unaffected.

(5) *Hydrodynamic four-velocity* u^μ: in each space–time point one defines a time-like vector, which is called hydrodynamic four-velocity u^μ. It is normalized as

$$u^2 = u^\mu(x)u_\mu(x) = 1. \qquad (4.119)$$

u^μ is time-like. At each space–time point x, one can define a proper Lorentz frame, which is called the local rest frame. In the local rest frame,

$$u^\mu_{\mathrm{LR}} = (1, 0, 0, 0). \qquad (4.120)$$

With the help of u^μ one also defines a tensor quantity,

$$\Delta^{\mu\nu}(x) = g^{\mu\nu} - u^\mu u^\nu. \qquad (4.121)$$

It is called the projector. One can easily check that

$$\Delta^{\mu\nu}u_\nu = 0. \tag{4.122}$$

When contracted with any arbitrary four-vector, the projector $\Delta^{\mu\nu}$ will annihilate the part of the four-vector parallel to four-velocity and project out the part perpendicular to it.

In the local rest frame,

$$\left(\Delta_{\mu\nu}\right)_{\rm LR} = \left(\Delta^{\mu\nu}\right)_{\rm LR} = {\rm diag}(0, -1, -1, -1) \tag{4.123}$$

$$\left(\Delta^{\mu}_{\nu}\right)_{\rm LR} = {\rm diag}(0, 1, 1, 1). \tag{4.124}$$

The definition of hydrodynamic velocity as in equation (4.119) is very general and one can construct various hydrodynamic velocities. In the literature, two particular choices are popular. They are described below:

(a) *Eckart's definition.* Hydrodynamic four-velocity is assumed to be proportional to the particle four-flow N^μ,

$$\begin{aligned} u^\mu &\propto N^\mu \\ &= \kappa N^\mu. \end{aligned} \tag{4.125}$$

The proportionality constant κ can be found from the normalization condition, $u^\mu u_\mu = 1$,

$$\kappa = \frac{1}{\sqrt{N_\nu N^\nu}}, \tag{4.126}$$

and the hydrodynamic four-velocity in Eckart's definition is

$$u^\mu = \frac{N^\mu}{\sqrt{N_\nu N^\nu}}. \tag{4.127}$$

Equation (4.127) can also be written as

$$u^\mu = \frac{N^\mu}{N^\nu u_\nu} \tag{4.128}$$

$$\text{or} \quad \Delta^{\mu\nu}N_\nu = 0. \tag{4.129}$$

With Eckart's definition of hydrodynamic velocity, the spatial components of the particle four-current vanish in the local rest frame.

$$\left(N^i\right)_{LR} = 0 \qquad i = 1, 2, 3. \tag{4.130}$$

(b) *Landau and Lifshitz's definition.* u^μ is proportional to the flow of energy,

$$\begin{aligned} u^\mu &\propto T^{\mu\nu}u_\nu \\ &= \frac{T^{\mu\nu}u_\nu}{\sqrt{u_\rho T^{\rho\sigma}T_{\sigma\tau}u^\tau}}. \end{aligned} \tag{4.131}$$

An alternative form can be

$$u^\mu = \frac{T^{\mu\nu}u_\nu}{u_\rho T^{\rho\sigma}u_\sigma}. \tag{4.132}$$

With the help of the projector, the Landau–Lifshitz definition can also be formulated as

$$\Delta^{\mu\nu}T_{\nu\sigma}u^\sigma = 0. \tag{4.133}$$

In the local rest frame, $u^\mu = (1, 0, 0, 0)$ and $\Delta^{\mu\nu} = \text{diag}(0, -1, -1, -1)$ and with Landau's definition of four-velocity, at the local rest frame, the spatial components of energy flow and momentum density vanish,

$$\left(T^{i0}\right)_{\text{LR}} = \left(T^{0i}\right)_{\text{LR}} = 0 \qquad i = 1, 2, 3. \tag{4.134}$$

It may be emphasized here that the choice of hydrodynamic velocity in no way affects the results of hydrodynamic calculation. One chooses either Eckart's velocity or Landau's velocity according to the given problem. For example, in the study of high-energy heavy ion collisions, the central rapidity region is essentially particle-free. It is difficult to define hydrodynamic four-velocity according to Eckart's definition. The Landau–Lifshitz choice of hydrodynamic velocity is preferred as it is related to energy flow. In other areas, Eckart's choice may be more appropriate.

4.4.2 Physical quantities of a simple system

We have mentioned earlier that a many-body or macroscopic system is characterized in terms of certain macroscopic variables, and that the dynamics of the system are studied in terms of those variables. With the help of hydrodynamic four-velocity $u^\mu(x)$, one can define the relevant macroscopic quantities, e.g. particle density, energy density, heat flow, pressure tensor and entropy density, in a covariant manner. We note that even though in the formulae below we have not shown x dependence explicitly, it is understood that the macroscopic variables defined below are a function of space and time.

(i) *Particle density.* We define the scalar quantity

$$n = N^\mu u_\mu \tag{4.135}$$

as the particle density. It is seen that n is the density of particles in the local rest frame $u^\mu = (1, 0, 0, 0)$: $n = N^0_{\text{LR}}$.

(ii) *Energy density.* The scalar energy density ε is defined as

$$\varepsilon = u_\mu T^{\mu\nu}u_\nu. \tag{4.136}$$

In the local rest frame, $\varepsilon = T^{00}_{\text{LR}}$.

(iii) *Heat flow*. Heat flow q^μ is defined as the difference of the energy flow and flow of enthalpy h carried by the particles,

$$q^\mu = \left(u_\nu T^{\nu\sigma} - hN^\sigma\right)\Delta^\mu_\sigma \tag{4.137}$$

where enthalpy per particle is defined as

$$h = \frac{\varepsilon + p}{n} \tag{4.138}$$

with p the local hydrostatic pressure, to be defined shortly. Heat flow has the property that it is transverse to the hydrodynamic four-velocity,

$$q^\mu u_\mu = 0. \tag{4.139}$$

(iv) *Pressure tensor*. The pressure tensor is defined as

$$P^{\mu\nu} = \Delta^\mu_\sigma T^{\sigma\tau}\Delta^\nu_\tau. \tag{4.140}$$

It is symmetric when $T^{\mu\nu}$ is symmetric. In the local rest frame, it is purely spatial,

$$P^{00}_{\text{LR}} = 0 \qquad P^{0i}_{\text{LR}} = P^{i0}_{\text{LR}} = 0 \qquad P^{ij}_{\text{LR}} = T^{ij} \qquad i, j = 1, 2, 3. \tag{4.141}$$

The pressure tensor has a 'reversible' and an 'irreversible' part,

$$P^{\mu\nu} = \Delta^\mu_\sigma T^{\sigma\tau}\Delta^\nu_\tau = -p\Delta^{\mu\nu} + \Pi^{\mu\nu}. \tag{4.142}$$

p is the hydrostatic pressure. The quantity $\Pi^{\mu\nu}$ is called the viscous pressure tensor. Writing equations (4.136), (4.137) and (4.142) in a slightly different manner, we find

$$\varepsilon = u_\mu T^{\mu\nu}u_\nu \tag{4.143a}$$

$$q^\mu + h\Delta^{\mu\nu}N_\nu = u_\nu T^{\nu\sigma}\Delta^\mu_\sigma \tag{4.143b}$$

$$-p\Delta^{\mu\nu} + \Pi^{\mu\nu} = \Delta^\mu_\sigma T^{\sigma\tau}\Delta^\nu_\tau. \tag{4.143c}$$

The terms can be collected to obtain the expression for an energy–momentum tensor,

$$T^{\mu\nu} = T^{\mu\nu}_{\text{rev}} + T^{\mu\nu}_{\text{irr}}, \tag{4.144}$$

with the 'reversible' and 'irreversible' parts,

$$T^{\mu\nu}_{\text{rev}} = \varepsilon u^\mu u^\nu - p\Delta^{\mu\nu} \tag{4.145a}$$

$$T^{\mu\nu}_{\text{irr}} = [(q^\mu + h\Delta^{\mu\sigma}N_\sigma)u^\nu + (q^\nu + h\Delta^{\nu\sigma}N_\sigma)u^\mu] + \Pi^{\mu\nu}. \tag{4.145b}$$

As mentioned earlier, two choices of hydrodynamic velocity are popular. In the Eckart frame, $u^\mu = N^\mu / \sqrt{N^\nu N_\nu}$ and $\Delta^{\mu\nu} N_\nu = 0$. Heat flow is

$$q^\mu = u_\nu T^{\nu\sigma} \Delta^\mu_\sigma \tag{4.146}$$

and the irreversible part of the energy–momentum tensor reduces to

$$T^{\mu\nu}_{\text{irr}} = \left[q^\mu u^\nu + q^\nu u^\mu\right] + \Pi^{\mu\nu}. \tag{4.147}$$

In the Landau–Lifshitz frame, $u^\mu = T^{\mu\nu} u^\nu / \sqrt{u_\rho T^{\rho\sigma} T_{\sigma\tau} u^\tau}$. Using the property $\Delta^{\mu\nu} u_\nu = 0$, heat flow is

$$q^\mu = -h \Delta^{\mu\nu} N_\nu. \tag{4.148}$$

The irreversible part of the energy–momentum tensor then has the simplified form

$$T^{\mu\nu}_{\text{irr}} = \Pi^{\mu\nu}. \tag{4.149}$$

In the Landau–Lifshitz frame, the energy–momentum tensor does not contain heat flow q^μ. However, as is manifest in equation (4.148), heat flow is not zero. It is manifested in particle flow. Using the projector, $\Delta^{\mu\nu}$, the particle four-flow can be split into two parts, in the direction of the four-velocity and in the direction perpendicular to it,

$$N^\mu = n u^\mu + V^\mu = n u^\mu + \Delta^{\mu\nu} N_\nu. \tag{4.150}$$

In the Landau–Lifshitz frame,

$$q^\mu = -h \Delta^{\mu\nu} N_\nu \Rightarrow V^\mu = \frac{q^\mu}{h}. \tag{4.151}$$

The particle four-flow and energy–momentum tensor in the Landau–Lifshitz frame and in Eckart's frame are then as follows.

Landau–Lifshitz frame:

$$N^\mu = n u^\mu + \left\{\frac{q^\mu}{h}\right\} \tag{4.152a}$$

$$T^{\mu\nu} = [\varepsilon u^\mu u^\nu - p \Delta^{\mu\nu}] + \left\{\Pi^{\mu\nu}\right\}. \tag{4.152b}$$

Eckart's frame:

$$N^\mu = n u^\mu \tag{4.153a}$$

$$T^{\mu\nu} = [\varepsilon u^\mu u^\nu - p \Delta^{\mu\nu}] + \left\{\Pi^{\mu\nu} + [q^\mu u^\nu + q^\nu u^\mu]\right\}. \tag{4.153b}$$

The terms inside the curly brackets refer to the irreversible parts of the particle-flow and energy–momentum tensor. Irreversible parts lead to dissipation[4]. In a realistic

[4] By dissipation we mean that energy is transferred from collective motion to random motion.

system, the irreversible parts can be small but never identically zero. However, if this is ignored, the fluid is called an ideal or inviscid fluid. Otherwise the fluid is called a dissipative fluid.

4.4.3 The relativistic kinetic (transport) equation

In kinetic theory, a macroscopic system is described in terms of the one-body distribution function $f(x, p, t)$. The kinetic or transport equation gives the space–time development of the distribution function. It was originally derived by Boltzmann and called the Boltzmann equation[5]. In the following, we derive the relativistic version of the Boltzmann equation. For more complete exposure to Boltzmann equations see [6].

The transport equation without collisions. We have defined particle four-flow,

$$N^\mu = \int \frac{d^3p}{p^0} p^\mu f(x, p). \tag{4.154}$$

The time component ($\mu = 0$) is the particle density and the space components ($\mu = 1, 2, 3$) are the particle flow, both measured with respect to the observer frame of reference. If $d\sigma_\mu$ is an oriented three-surface element of a plane space-like surface (i.e. a surface whose tangent vector is time) and $\Delta^3\sigma$ is a small segment situated at x, then we can contract N^μ with $d\sigma_\mu$ to obtain the scalar quantity.

$$\Delta N(x) = \int_{\Delta^3\sigma} d\sigma^3_\mu N^\mu(x) = \int_{\Delta^3\sigma} \int d^3\sigma_\mu \frac{d^3p}{p^0} p^\mu f(x, p). \tag{4.155}$$

In the Lorentz frame, $d^3\sigma_\mu = (d^3x, 0, 0, 0)$ is purely time-like and

$$\Delta N(x) = \int_{\Delta^3x} \int d^3x d^3p f(x, p). \tag{4.156}$$

From the definition of the distribution function $\Delta N(x)$ is just the number of particles in a volume element Δ^3x.

In Minkowski space, a particle is represented by a world line. $\Delta N(x)$ in equation (4.155) then has a simple geometrical interpretation, the average number of world lines crossing the three-segment $\Delta^3\sigma$. Restricting the momentum integration in equation (4.155) within a narrow range Δ^3p around p, we can define $\Delta N(x, p)$,

$$\Delta N(x, p) = \int_{\Delta^3\sigma} \int_{\Delta^3p} d^3\sigma_\mu \frac{d^3p}{p^0} p^\mu f(x, p) \tag{4.157}$$

[5] Ludwig Eduard Boltzmann (1844–1906) was an Austrian physicist. He made major contributions in statistical thermodynamics. He was an early proponent of atomic theory, before it had gained ground. During his lifetime, his works were not appreciated. Famous physicists like Ernest Mach and Wilhelm Ostwald opposed his atomic view. Rejection of his views made Boltzmann depressed, and in 1906 he committed suicide.

as the average number of world lines crossing a segment $\Delta^3\sigma$ with momenta in the range $\Delta^3 p$ around p. If the particles do not suffer collision, some time later the same particles will cross a surface element $\Delta^3\sigma'$. Then we have the identity

$$\int_{\Delta^3\sigma}\int_{\Delta^3 p} d^3\sigma_\mu \frac{d^3p}{p^0} p^\mu f(x, p) - \int_{\Delta^3\sigma'}\int_{\Delta^3 p} d^3\sigma_\mu \frac{d^3p}{p^0} p^\mu f(x, p) = 0. \tag{4.158}$$

Consider the four-volume $\Delta^4 x$ enclosed by surface $\Delta^3\sigma$ and $\Delta^3\sigma'$ and the surface of the tube of the world lines (see figure 4.1). Equation (4.158) can be interpreted such that no particle world lines cross the tube surface. Thus net flow of particles through the surface $\Delta^3 x$ of four-volume $\Delta^4 x$ vanishes.

$$\int_{\Delta^3\sigma}\int_{\Delta^3 p} d^3\sigma_\mu \frac{d^3p}{p^0} p^\mu f(x, p) = 0. \tag{4.159}$$

Apply the Gauss theorem,

$$\int_{\Delta^4 x}\int_{\Delta^3 p} d^4x \frac{d^3p}{p^0} p^\mu \partial_\mu f(x, p) = 0. \tag{4.160}$$

Since $\Delta^4 x$ and $\Delta^3\sigma$ are arbitrary

$$p^\mu \partial_\mu f(x, p) = 0. \tag{4.161}$$

This is the Boltzmann transport equation for a collisionless system. In three-vector notation, it can be written in the more familiar form

$$\left[\partial_t + \frac{\mathbf{p}}{p^0} \cdot \nabla\right] f(x, p) = [\partial_t + \mathbf{v} \cdot \nabla] f(x, p) = 0 \tag{4.162}$$

with $\mathbf{v} = \frac{\mathbf{p}}{p^0}$, the three-velocity.

Transport equation with collisions. In the previous section, we established that when the constituents of the macroscopic body do not suffer interaction, the net flow of particles through three-surface $\Delta^3 x$ of four-volume $\Delta^4 x$ is zero. Net flow will not be zero if the particles suffer collisions. If the amount of change in the number of

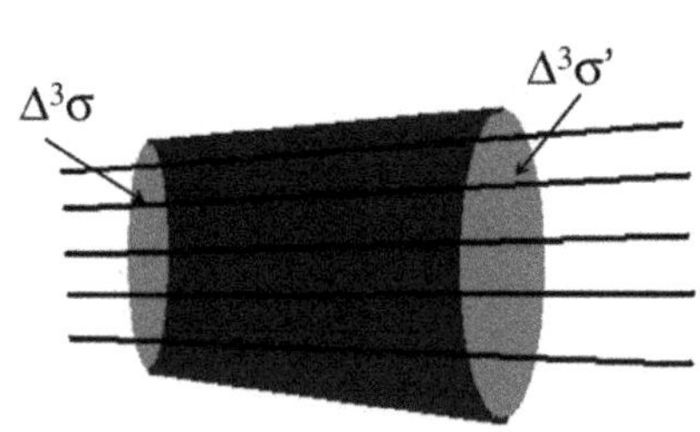

Figure 4.1. World lines crossing a three-surface $\Delta^3 x$ in Minkowski space. After some time the same world lines emerge through the surface $\Delta^3\sigma'$. No world lines cross the tube-like three-surface of the four-volume enclosed by surfaces $\Delta^3\sigma$ and $\Delta^3\sigma'$.

particles in the momentum range $(\mathbf{p}, \mathbf{p} + \Delta\mathbf{p})$ in a small volume element $\Delta^4 x$ is written as

$$\Delta^4 x \frac{\Delta^3 p}{p^0} C(x, p) \tag{4.163}$$

the equation (4.160) for a collisionless system will be changed to

$$\int_{\Delta^4 x} \int_{\Delta^3 p} \mathrm{d}^4 x \frac{\mathrm{d}^3 p}{p^0} p^\mu \partial_\mu f(x, p) = \Delta^4 x \frac{\Delta^3 p}{p^0} C(x, p) \tag{4.164}$$

where $C(x, p)$ is an invariant function whose form is to be found. Boltzmann makes the following assumptions to calculate $C(x, p)$.

(a) The system is dilute enough such that only two-particle collisions need to be considered.
(b) The momenta of the colliding particles are uncorrelated and independent of position. This is a key assumption of Boltzmann and is called 'Stosszahlansatz' (collision number hypothesis) or the 'molecular chaos' hypothesis.
(c) Distribution function $f(x, p)$ varies slowly in space–time.
(d) The collisions are instantaneous.

By assumption (a), only two-particle collisions are important. Consider a collision (figure 4.2) in which two particles with initial momenta p^μ and p_1^μ collide to a final state p'^μ and $p_1'^\mu$,

$$\underset{\text{initial state}}{p^\mu + p_1^\mu} \rightarrow \underset{\text{final state.}}{p'^\mu + p_1'^\mu} \tag{4.165}$$

This type of collision in four-volume $\Delta^4 x$ will reduce the number of particles in the momentum range $p^\mu + \Delta p^\mu$. According to the molecular chaos hypothesis, the average number ΔN_{loss} of such collisions is proportional to:

(i) the average number of particles per unit volume with three-momentum between $\mathbf{p}$ and $\mathbf{p} + \Delta\mathbf{p}$,

$$f(x, p)\Delta^3 p, \tag{4.166}$$

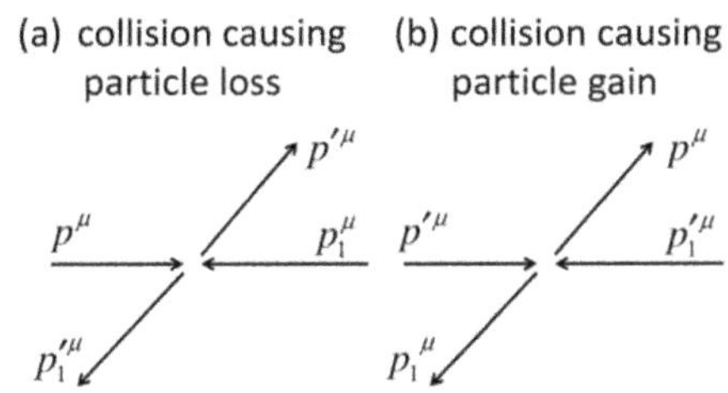

Figure 4.2. Schematic diagram of the two-particle collision considered in Boltzmann transport equation. (a) The particle with momentum p^μ is lost through the collision; (b) the particle with momentum p^μ is gained through the collision.

(ii) the average number of particles per unit volume with three-momentum between $\mathbf{p}_1$ and $\mathbf{p}_1 + \Delta\mathbf{p}_1$,

$$f(x, p_1)\Delta^3 p_1, \tag{4.167}$$

and

(iii) the intervals $\Delta^3 p'$, $\Delta^3 p_1'$ and $\Delta^4 x$,

$$\Delta^3 p' \Delta^3 p_1' \Delta^4 x. \tag{4.168}$$

$$\begin{aligned}\Delta N_{\text{loss}} &\propto f(x, p)\Delta^3 p f(x, p_1)\Delta^3 p_1 \Delta^3 p' \Delta^3 p_1' \Delta^4 x \\ &= \frac{W\left(pp_1 | p'p_1'\right)}{p^0 p_1^0 p'^0 p_1'^0} f(x, p)\Delta^3 p f(x, p_1)\Delta^3 p_1 \Delta^3 p' \Delta^3 p_1' \Delta^4 x\end{aligned} \tag{4.169}$$

where we have written the proportionality factor as

$$\frac{W\left(pp_1 | p'p_1'\right)}{p^0 p_1^0 p'^0 p_1'^0}. \tag{4.170}$$

$W(pp_1|p'p_1')$ is called the transition rate. Note that we have neglected difference in the space–time coordinate in $f(x, p_1)$ and $f(x, p_2)$. This is due to the assumption (c) that the distribution function varies slowly. For the same reason the transition probability also depends only on the four-momenta of the particles, not on the space–time coordinates.

The average number of particles in the range $\Delta^4 x$ of Minkowski space and with momentum in the range $(\mathbf{p}, \mathbf{p} + \mathrm{d}\mathbf{p})$ lost through collisions is then obtained by integrating the above defined number of collisions,

$$N_{\text{loss}} = \frac{1}{2}\Delta^4 x \frac{\Delta^3 p}{p^0} \int \frac{\mathrm{d}^3 p'}{p'^0} \frac{\mathrm{d}^3 p_1}{p_1^0} \frac{\mathrm{d}^3 p_1'}{p_1'^0} f(x, p) f(x, p_1) \times W\left(pp_1 | p'p_1'\right). \tag{4.171}$$

The factor $\frac{1}{2}$ accounts for the fact that the final state $(p_1'^\mu, p_2'^\mu)$ is indistinguishable from the state $(p_2'^\mu, p_1'^\mu)$. In a similar manner, the gain term due to restitution collisions $p'^\mu + p_1'^\mu \to p^\mu + p_1^\mu$ can be calculated,

$$N_{\text{gain}} = \frac{1}{2}\Delta^4 x \frac{\Delta^3 p}{p^0} \int \frac{\mathrm{d}^3 p'}{p'^0} \frac{\mathrm{d}^3 p_1}{p_1^0} \frac{\mathrm{d}^3 p_1'}{p_1'^0} f(x, p') f\left(x, p_1'\right) \times W\left(p'p_1' | pp_1\right). \tag{4.172}$$

The net change of the number of particles in the interval $\Delta^4 x$ and $\Delta^3 p$ is $N_{\text{gain}} - N_{\text{loss}}$ and is to be equated with equation (4.163). The invariant collision function in equation (4.163) then has the following form,

$$C(x, p) = \frac{1}{2} \int \frac{d^3 p'}{p'^0} \frac{d^3 p_1}{p_1^0} \frac{d^3 p_1'}{p_1'^0} \Big(f(x, p') f(x, p_1') \times W(p' p_1' | p p_1) \\ - f(x, p) f(x, p_1) \times W(p p_1 | p' p_1') \Big) \tag{4.173}$$

and the transport equation with collision can be written as

$$p^\mu \partial_\mu f = \frac{1}{2} \int \frac{d^3 p'}{p'^0} \frac{d^3 p_1}{p_1^0} \frac{d^3 p_1'}{p_1'^0} \left[f' f_1' \, W(p' p_1' | p p_1) - f f_1 \, W(p p_1 | p' p_1') \right] \tag{4.174}$$

where we have used the abbreviations $f = f(x, p)$, $f_1 = f(x, p_1)$, $f' = f(x, p')$ and $f_1' = f(x, p_1')$. The transition rate in equation (4.174) is a scalar and, in principle, can be expressed in terms of any two of the three Mandelstam variables defined earlier. For example, if one uses

$$s = (p + p_1)^2 \tag{4.175}$$

$$t = (p - p')^2 \tag{4.176}$$

it is easy to see that the following identity, which is called the detailed balance, holds,

$$W(p p_1 | p' p_1') = W(p' p_1' | p p_1). \tag{4.177}$$

The left hand side of equation (4.174), which is called the collision integral, can be written as

$$C(x, f) = \frac{1}{2} \int \frac{d^3 p'}{p'^0} \frac{d^3 p_1}{p_1^0} \frac{d^3 p_1'}{p_1'^0} \left[f' f_1' - f f_1 \right] W(p p_1 | p' p_1'). \tag{4.178}$$

For a two-body collision $p + p_1 \to p' + p_1'$, the transition rate $W(p p_1 | p' p_1')$ can be related to the differential cross section,

$$W(p p_1 | p' p_1') = s\sigma(s, \theta) \delta^4(p + p_1 - p' - p_1'), \tag{4.179}$$

and using the detailed balance property, the transport equation can be written as

$$p^\mu \partial_\mu f = \frac{1}{2} \int \frac{d^3 p_1}{p_1^0} (f' f_1' - f f_1) F \sigma \tag{4.180}$$

or in a more familiar form,

$$(\partial_t + \mathbf{u} \cdot \nabla) f = \frac{1}{2} \int d^3 p (f' f_1' - f f_1) v \sigma \tag{4.181}$$

where we have defined the Lorentz invariant flux,

$$F = \sqrt{(p \cdot p_1) - m_1^2 m_2^2}, \tag{4.182}$$

and v is the Møller velocity,

$$v = \frac{F}{p^0 p_1^0} = \sqrt{(\mathbf{v} - \mathbf{v}_1)^2 - (\mathbf{v} \times \mathbf{v}_1)^2}. \tag{4.183}$$

There are established procedures for solving the Boltzmann transport equation, e.g. the Chapman–Enskog method or Grad's 14-moment method. These methods, however, will not be discussed here.

4.4.4 Boltzmann equation with external force

Let us consider that an external force $F^\mu(x)$ acts on the system. If particle trajectories are not largely affected by the external force, such that two particles still approach each other on a rectilinear path, the effect of the force can be included in the Boltzmann equation.

Let

$$\Delta N(x, p) = \int_{\Delta^3\sigma} \int_{\Delta^3 p} d^3\sigma_\mu \frac{d^3p}{p^0} p^\mu f(x, p) \tag{4.184}$$

be the average number of world lines crossing a three-surface $\Delta^3\sigma$ with momenta in the range $[p^\mu, p^\mu + \Delta p^\mu]$. In a collisionless plasma, some time later, the same particles will cross a surface element $\Delta^3\sigma'$. However, due to the external force F^μ, during the passage from surface $\Delta^3\sigma$ to $\Delta^3\sigma'$ the momentum p^μ will change to $p^\mu + F^\mu\Delta\tau$ giving

$$\int_{\Delta^3\sigma'} \int_{\Delta^3 p} d^3\sigma_\mu \frac{d^3p}{p^0} p^\mu f(x, p + F\Delta\tau) - \int_{\Delta^3\sigma} \int_{\Delta^3 p} d^3\sigma_\mu \frac{d^3p}{p^0} p^\mu f(x, p) = 0. \tag{4.185}$$

Taylor expanding $f(x, p + F\Delta\tau)$ and arranging terms,

$$\left[\int_{\Delta^3\sigma'} \int_{\Delta^3 p} d^3\sigma_\mu \frac{d^3p}{p^0} p^\mu f(x, p) - \int_{\Delta^3\sigma} \int_{\Delta^3 p} d^3\sigma_\mu \frac{d^3p}{p^0} p^\mu f(x, p)\right]$$
$$+ \int_{\Delta^3\sigma'} \int_{\Delta^3 p} d^3\sigma_\mu \frac{d^3p}{p^0} p^\mu F^\nu \Delta\tau \frac{\partial f(x, p)}{\partial p^\nu} = 0. \tag{4.186}$$

The bracketed term is the same as that obtained in the collisionless plasma. Using the Gauss theorem the term gives

$$\int_{\Delta^4 x} \int_{\Delta^3 p} d^4x \frac{d^3p}{p^0} p^\mu \partial_\mu f = \Delta^4 x \int_{\Delta^3 p} \frac{d^3p}{p^0} p^\mu \partial_\mu f. \tag{4.187}$$

The proper time $\Delta\tau$ can be obtained from the identity,

$$\Delta^4 x = \Delta\tau \frac{p^\mu}{m} \int_{\Delta^3\sigma} d^3\sigma_\mu \tag{4.188}$$

and the unbracketed term yields

$$\int_{\Delta^3 p} \frac{d^3p}{p^0}\left(\Delta\tau \frac{p^\mu}{m}\int_{\Delta^3\sigma'} d^3\sigma_\mu\right) mF^\nu \frac{\partial f(x, p)}{\partial p^\nu} = \Delta^4 x \int \frac{d^3p}{p^0} mF^\nu \frac{\partial f(x, p)}{\partial p^\nu}. \quad (4.189)$$

Collecting all the terms, equation (4.185) can be written as

$$\Delta^4 x \frac{\Delta^3 p}{p^0}\left[p^\mu \partial_\mu + mF^\mu \frac{\partial}{\partial p^\mu}\right] f(x, p) = 0. \quad (4.190)$$

Since $\Delta^4 x$ as well as $\Delta^3 p$ are arbitrary, the Boltzmann equation with external force for collisionless plasma can be written as

$$\left[p^\mu \partial_\mu + mF^\mu \frac{\partial}{\partial p^\mu}\right] f(x, p) = 0. \quad (4.191)$$

If collisions are included, the zero on the right hand side is to be replaced by the collision integral,

$$\left[p^\mu \partial_\mu + mF^\mu \frac{\partial}{\partial p^\mu}\right] f(x, p) = C(x, p). \quad (4.192)$$

4.4.5 Boltzmann equation with quantum effects

In the Boltzmann equation derived here, the quantum nature of the particles is not accounted for. For example, if the particles are fermions and if there is a particle in the momentum state p'^μ, the collision

$$p^\mu + p_1^\mu \to p'^\mu + p_1'^\mu, \quad (4.193)$$

will be Pauli blocked. The Pauli blocking effect for fermions can be accounted for by re-writing the collision term as

$$\begin{aligned} C_{\text{fermions}}(x, p) = \frac{1}{2}\int \frac{d^3p_1}{p_1^0}\frac{d^3p'}{p'^0}\frac{d^3p_1'}{p_1'^0}\Big[& f'f_1'\left(1 - h^3 f\right)\left(1 - h^3 f_1\right) W\left(p'p_1' | pp_1\right) \\ & - ff_1\left(1 - h^3 f'\right)\left(1 - h^3 f_1'\right) W\left(p, p_1 | p'p_1'\right)\Big] \end{aligned} \quad (4.194)$$

where the term $(1 - h^3 f)(1 - h^3 f_1)$ in the first term takes account of the reduced probability of scattering $p'^\mu + p_1'^\mu \to p^\mu + p_1^\mu$. Similarly, the term $(1 - h^3 f')$ $(1 - h^3 f_1')$ accounts for the reduced probability of scattering $p^\mu + p_1^\mu \to p'^\mu + p_1'^\mu$.

In an analogous way, the enhanced probability of scattering for the bosonic nature of the particles is accounted for as

$$\begin{aligned} C_{\text{bosons}}(x, p) = \frac{1}{2}\int \frac{d^3p_1}{p_1^0}\frac{d^3p'}{p'^0}\frac{d^3p_1'}{p_1'^0}\Big[& f'f_1'\left(1 + h^3 f\right)\left(1 + h^3 f_1\right) W\left(p'p_1' | pp_1\right) \\ & - ff_1\left(1 + h^3 f'\right)\left(1 + h^3 f_1'\right) W\left(p, p_1 | p'p_1'\right)\Big]. \end{aligned} \quad (4.195)$$

Writing

$$\bar{f} = \left(1 + \epsilon h^3 f\right) \tag{4.196}$$

where $\epsilon = -1$ for fermions and $\epsilon = +1$ for bosons, the Boltzmann equation with quantum statistics can be written as

$$p^\mu \partial_\mu f(x, p) = \frac{1}{2} \int \frac{d^3 p_1}{p_1^0} \frac{d^3 p'}{p'^0} \frac{d^3 p_1'}{p_1'^0} \Big[f' f_1' \bar{f} \bar{f}_1 W\left(p' p_1' | p p_1\right)$$
$$- f f_1 \bar{f}' \bar{f}_1' W\left(p, p_1 | p' p_1'\right)\Big]. \tag{4.197}$$

Equation (4.197) is also called the Boltzmann–Uehling–Uhlenbeck collision.

4.4.6 Boltzmann equation for mixtures

The Boltzmann transport equation for single species gas can be immediately generalized for mixtures. Each type of particle can be endowed with the distribution function $f_i(x, p_i)$, $i = 1,2,\ldots N$ and the transport equation for the kth species can be written as

$$p_k^\mu \partial_\mu f_k(x, p_k) = \sum_{l=1}^{N} C_{kl}(x, p_k) \tag{4.198}$$

$$C_{kl}(x, p_k) = \left(1 - \frac{1}{2}\delta_{kl}\right) \int \frac{d^3 p_l}{p_l^0} \frac{d^3 p_k'}{p_k'^0} \frac{d^3 p_l'}{p_l'^0} \Big(f\left(x, p_k'\right) f\left(x, p_l'\right) \times W\left(p_k' p_l' | p_k p_l\right)$$
$$- f(x, p_k) f(x, p_l) \times W\left(p_k p_l | p_k' p_l'\right)\Big). \tag{4.199}$$

The equations take into account elastic collisions between the constituents. In relativistic energy inelastic collisions are possible,

$$k + l \to i + j. \tag{4.200}$$

If inelastic collisions are included, the collision term is modified as

$$C_{kl}(x, p_k) = \frac{1}{2} \int \frac{d^3 p_l}{p_l^0} \frac{d^3 p_i'}{p_i'^0} \frac{d^3 p_j'}{p_j'^0} \Big(f(x, p_i) f\left(x, p_j\right) \times W\left(p_k p_l | p_i p_j\right)$$
$$- f(x, p_k) f(x, p_l) \times W\left(p_k p_l | p_i p_j\right)\Big). \tag{4.201}$$

The H-theorem. The Boltzmann transport equation is manifestly time irreversible. However, the microscopic interactions are reversible. How does this qualitative change come about? One of the major assumptions in the transport equation is the molecular chaos hypothesis: that colliding particles are uncorrelated. This hypothesis makes a distinction between past and future. The relativistic transport equation thus describes an irreversible process. This property is manifested most clearly in the Boltzmann H-theorem. In simple terms, the H-theorem (or second law of thermodynamics) states that entropy production at any space–time point is never negative, $\partial_\mu S^\mu \geqslant 0$.

H-theorem also defines the equilibrium state of a macroscopic system. In the equilibrium state,

$$\partial_\mu S^\mu = 0. \tag{4.202}$$

This condition with the transport equation determines the equilibrium distribution function: it is called the Jüttner distribution.

4.4.7 Equilibrium distribution function

The Boltzmann equation describes the space–time evolution of the distribution function for a macroscopic system. One of the fundamental assumptions for a macroscopic system is that left undisturbed, the system will reach equilibrium. In equilibrium,

$$\frac{\mathrm{d}f(x, p, t)}{\mathrm{d}t} = 0. \tag{4.203}$$

It is useful to distinguish between global and local equilibrium. The total system can be divided into several small yet macroscopic subsystems (see figure 4.3). In global equilibrium, all the subsystems can be characterized by the same thermodynamic variables, i.e. thermodynamic variables are independent of spatial positions. In contrast, in local equilibrium, thermodynamic variables are a function of the spatial position of the subsystem. If the macroscopic system is in local thermal equilibrium, at each space–time point x we can specify, in addition to hydrodynamic velocity $v(x)$, a temperature $T(x)$ and for each particle species a chemical potential $\mu_i(x)$, which controls the particle density at x. The equilibrium distribution function can be obtained from kinetic theory under the condition that

$$\partial_\mu S^\mu = 0, \tag{4.204}$$

i.e. it is the distribution which extremizes the entropy four-flow. We have defined entropy current as

$$S^\mu(x) = -\int \frac{\mathrm{d}^3p}{p^0} p^\mu f(x, p)\left[\log h^3 f(x, p) - 1\right], \tag{4.205}$$

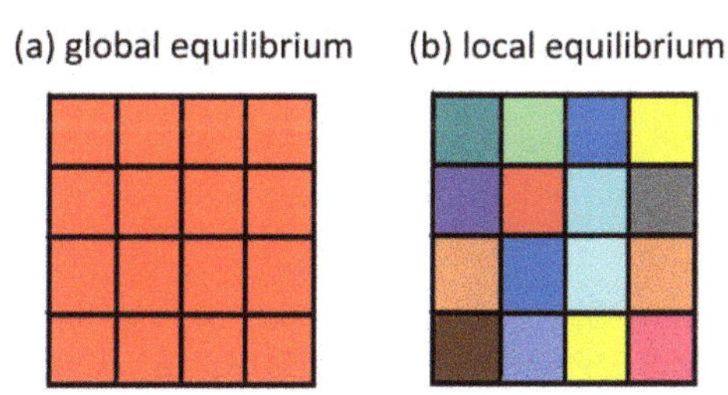

Figure 4.3. The boxes represent small, yet macroscopic, subsystems of the total macroscopic system. (a) In global equilibrium the macroscopic variables characterizing the subsystems are independent of space; (b) in local equilibrium, the macroscopic variables are a function of space. The different colors of the boxes are representative of different macroscopic variables.

for which the entropy four-flow is

$$\partial_\mu S^\mu(x) = -\int \frac{d^3p}{p^0}\left[\log h^3 f(x, p)\right] p^\mu \partial_\mu f(x, p)$$
$$= -\int \frac{d^3p}{p^0}\left[\log h^3 f(x, p)\right] C(x, f) \tag{4.206}$$

and in the last line $p^\mu \partial_\mu f(x, p)$ is replaced by the collision integral $C(x, p)$. Previously, we have determined the form of the collision integral (see equation (4.178)). The collision integral is identically zero when

$$f(x, p) f(x, p_1) - f(x, p') f(x, p_1') = 0. \tag{4.207}$$

Equation (4.207) is then the condition for maximizing entropy four-flow. Denoting the local equilibrium distribution function by $f^{(0)}(x, p)$, the relation equation (4.207) can be written as

$$\log h^3 f^{(0)}(x, p) + h^3 \log f^{(0)}(x, p_1) = \log h^3 f^{(0)}(x, p') + \log h^3 f^{(0)}(x, p_1') \tag{4.208}$$

where we have multiplied $f^{(0)}$ by Planck's constant to render it a dimensionless quantity such that one can take a logarithm.

In a binary collision,

$$p^\mu + p_1^\mu = p'^\mu + p_1'^\mu, \tag{4.209}$$

a function $F(x, p)$ satisfying

$$F(x, p) + F(x, p_1) = F(x, p') + F(x, p_1') \tag{4.210}$$

is called the summational invariant. The most general form for a summational invariant is a linear combination of a constant and the four-momentum p^μ,

$$F(x, p) = a(x) + b_\mu(x) p^\mu. \tag{4.211}$$

Comparing equation (4.208) and equation (4.210) one concludes that $\log h^3 f^{(0)}(x, p)$ is a summational invariant and the local equilibrium distribution function can be written as

$$f^{(0)}(x, p) = \frac{1}{h^3} e^{a(x) + b_\mu(x) p^\mu}. \tag{4.212}$$

Further insight can be obtained by inserting $f^{(0)}(x, p)$ into the Boltzmann equation, $p^\mu \partial_\mu f^{(0)}(x, p) = 0$. Simple calculations indicate that for any arbitrary p^μ, the parameters $a(x)$ and $b^\mu(x)$ must satisfy

$$p^\mu \partial_\mu a(x) + p^\mu p^\nu \partial_\nu b_\mu(x) = 0. \tag{4.213}$$

The condition can be re-written as

$$\partial_\mu a(x) = 0 \tag{4.214a}$$

$$\partial_\nu b_\mu(x) + \partial_\mu b_\nu(x) = 0. \tag{4.214b}$$

The second equation is called the Killing's equation, the general solution for which can be written as a superposition of translation and rigid rotation,

$$b^\mu(x) = b^\mu + \omega^{\mu\nu} x_\nu. \tag{4.215}$$

If rigid rotation is neglected, both the parameters a and b_μ are constant and the equilibrium distribution function can be written as

$$f^{(0)}(p) = \frac{1}{h^3} \mathrm{e}^{a+b_\mu p^\mu}. \tag{4.216}$$

Writing

$$a = \frac{\mu}{T} \qquad b_\mu = -\frac{u_\mu}{T} \tag{4.217}$$

the Lorentz-covariant global equilibrium function can be written as

$$f^{(0)}(p) = \frac{1}{(2\pi\hbar)^3} \mathrm{e}^{\frac{\mu - u_\mu p^\mu}{T}}. \tag{4.218}$$

It is possible to show explicitly that μ, T and u^μ are indeed global chemical potential, temperature and hydrodynamic four-velocity. The form (4.218) is an example of the Jüttner distribution function.

A Lorentz-covariant local equilibrium distribution function is obtained when the parameters T, μ and u^μ depend on the spatial position,

$$f^{(0)}(x, p) = \frac{1}{(2\pi)^3} \mathrm{e}^{\frac{\mu(x) - u_\mu(x) p^\mu}{T(x)}}. \tag{4.219}$$

In the rest frame $u_\mu p^\mu = E$ and we can identify the Jüttner distribution equation (4.218) with the familiar Maxwell–Boltzmann distribution function,

$$f_{MB}(p) = \frac{1}{(2\pi\hbar)^3} \mathrm{e}^{\frac{\mu - E}{T}}. \tag{4.220}$$

The Boltzmann transport equation without quantum statistics gives the Maxwell–Boltzmann distribution as the equilibrium distribution function. If quantum effects are accounted for, the collision integral becomes identically zero (see equation (4.197)) when,

$$ff_1\overline{f}'\overline{f}_1' = f'f_1'\overline{f}\,\overline{f}_1$$

or

$$\frac{\overline{f}}{f}\frac{\overline{f}_1}{f_1} = \frac{\overline{f}'}{f'}\frac{\overline{f}_1'}{f_1'} \tag{4.221}$$

with $\overline{f} = 1 + \epsilon h^3 f$. $\epsilon = +1(-1)$ for the boson (fermion).

Taking logarithm,

$$\log\frac{\overline{f}}{h^3 f} + \log\frac{\overline{f}_1}{h^3 f_1} = \log\frac{\overline{f}'}{h^3 f'} + \log\frac{\overline{f}_1'}{h^3 f_1'} \tag{4.222}$$

one can identify $\log\frac{\overline{f}}{h^3 f}$ as the summational invariant whose general form, as before, can be written as

$$\log\frac{\overline{f}}{h^3 f} = \log\left(\frac{1}{h^3 f} + \epsilon\right) = a(x) + b_\mu(x)p^\mu \tag{4.223}$$

and the local equilibrium distribution function with quantum statistics is obtained as

$$f(x, p) = \frac{1}{h^3}\frac{1}{\mathrm{e}^{a(x)+b_\mu(x)p^\mu} - \epsilon}. \tag{4.224}$$

Identifying $a(x) = -\frac{\mu(x)}{T}$ and $b_\mu(x) = \frac{u_\mu(x)}{T(x)}$, the local equilibrium distribution for a system boson or fermion is obtained as

$$f(x, p) = \frac{1}{(2\pi\hbar)^3}\frac{1}{\mathrm{e}^{\frac{u_\mu(x)p^\mu - \mu(x)}{T(x)}} - \epsilon} \tag{4.225}$$

with $\epsilon = +1$ for bosons and $\epsilon = -1$ for fermions.

In the rest frame, $p \cdot u = E$ and one obtains a global equilibrium distribution function for fermions as

$$\text{Maxwell–Boltzmann: } f(x, p) = \frac{1}{(2\pi\hbar)^3}\frac{1}{\mathrm{e}^{\frac{E-\mu}{T}}} \tag{4.226a}$$

$$\text{Bose–Einstein: } f(x, p) = \frac{1}{(2\pi\hbar)^3}\frac{1}{\mathrm{e}^{\frac{E-\mu}{T}} - 1} \tag{4.226b}$$

$$\text{Fermi: } f(x, p) = \frac{1}{(2\pi\hbar)^3}\frac{1}{\mathrm{e}^{\frac{E-\mu}{T}} + 1}. \tag{4.226c}$$

It is instructive to compare the Bose–Einstein and Fermi–Dirac distributions. In figure 4.4 we have shown the functions as a function of $\frac{E-\mu}{T}$. Bose–Einstein distribution tends to ∞ as $\frac{E-\mu}{T} \to 0$. Asymptotically, for large $\frac{E-\mu}{T}$ both Bose–Einstein and Fermi–Dirac distributions coincide with the Maxwell–Boltzmann distribution function.

4.4.8 Conservation equations

An important property of the collision term in the Boltzmann equation is that if multiplied by a summational invariant $\psi(x) = a(x) + b_\mu(x)p^\mu$ and integrated over all possible momentum values, it identically reduces to zero,

$$\int \frac{\mathrm{d}^3p}{p^0}\psi(x, p)C(x, p) = 0. \tag{4.227}$$

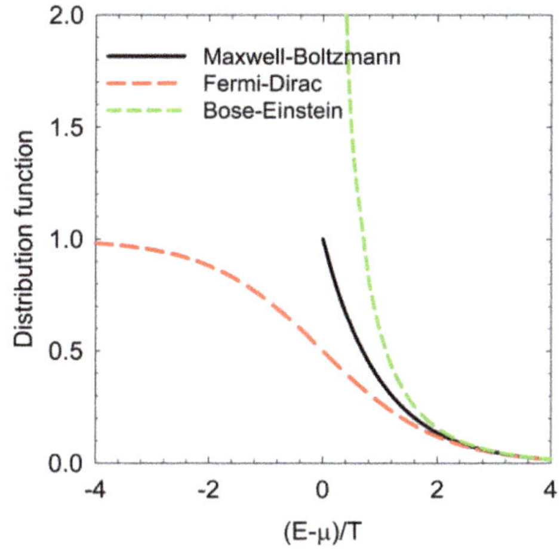

Figure 4.4. Bose–Einstein and Fermi–Dirac distribution functions as a function of $(E - \mu)/T$. For comparison, the Maxwell–Boltzmann distribution function is also shown.

Equation (4.227) can be used to derive conservation equations. For example, consider the summational invariant $\psi(x, p) = a(x)$. One obtains

$$0 = \int \frac{d^3p}{p^0} a(x)C(x, p) = \int \frac{d^3p}{p^0} a(x)p^\mu \partial_\mu f(x, p). \tag{4.228}$$

Since we have defined particle four-current as $N^\mu = \int \frac{d^3p}{p^0} p^\mu f(x, p)$, the above equation can be written as the macroscopic conservation law of the total particle number,

$$\partial_\mu N^\mu = 0. \tag{4.229}$$

In a system where the number of particles of each component is conserved separately, one can write

$$\partial_\mu N_k^\mu = 0 \qquad K = 1, 2, \ldots N. \tag{4.230}$$

For the summational invariant $\psi(x) = b_\mu(x)p^\mu$, one obtains

$$0 = \int \frac{d^3p}{p^0} b_\mu(x)p^\mu C(x, p) = \int \frac{d^3p}{p^0} b_\mu(x)p^\mu p^\nu \partial_\nu f(x, p). \tag{4.231}$$

The energy–momentum tensor is defined as $T^{\mu\nu} = \int \frac{d^3p}{p^0} p^\mu p^\nu f(x)$ and the above equation then gives the energy–momentum conservation law,

$$\partial_\mu T^{\mu\nu} = 0. \tag{4.232}$$

For a system with a singly conserved charge, the five equations,

$$\partial_\mu N^\mu = 0, \tag{4.233}$$

$$\partial_\mu T^{\mu\nu} = 0 \tag{4.234}$$

govern the motion of the fluid. They are called hydrodynamic equations. Explicit decomposition of the energy–momentum tensor and particle four-flow were given earlier, see equations (4.152) and (4.153). They must be supplemented by the H-theorem or the second law of thermodynamics,

$$\partial_\mu S^\mu \geqslant 0. \tag{4.235}$$

Bibliography

[1] Huang K 1997 *Statistical Mechanics* (New Delhi: Wiley Eastern Private)

[2] Landau L D and Lifshitz E M 1969 *Statistical Physics* (Oxford: Pergamon)

[3] Reif F 2008 *Fundamentals of Statistical and Thermal Physics* (New York: McGraw-Hill)

[4] Mandl F 1988 *Statistical Physics* (New York: Wiley)

[5] Chandrasekhar S 1967 *An Introduction to the Study of Stellar Structure* (New York: Dover) chapter 10

[6] de Groot S R, van Leeuwen W A and Van Weert C G 1957 *Relativistic Kinetic Theory* (Amsterdam: North Holland)

[7] Abramowitz M and Stegun I A (ed) 1969 *Handbook of Mathematical Functions* (Washington, DC: US Department of Commerce)

IOP Publishing

A K Chaudhuri

Chapter 5

Quantum chromodynamics

5.1 Introduction

The modern theory of strong interaction is quantum chromodynamics (QCD). QCD is a rather old theory. In 1964, Murray Gell-Mann [1], and independently George Zweig [2], postulated the existence of quarks (Zweig called them 'aces'). Gell-Mann was awarded the Nobel prize in 1969 for his discovery of quarks. Unfortunately, Zweig's contribution was not recognized by the Nobel committee. Color as a three-valued charge degree of freedom was introduced by Oscar Wallace Greenberg [3] in 1964 and in 1965 as a gauge symmetry by Moo Young Han and Yoichiro Nambu [4]. A basic framework for a field theory based on color already existed in 1954, mainly due to Chen Ning Yang and Robert Mills [5]. Gerard 't Hooft and Martinus Veltman [6] showed in 1971 that such 'non-abelian' gauge theories are renormalizable. In 1973, David Jonathan Gross, Frank Anthony Wilczek [7] and Hugh David Politzer [8] discovered asymptotic freedom (interaction between particles gets weaker with increasing energy or decreasing distance). The discovery led them to receive the Nobel prize in 2004.

Formally, QCD can be defined as a field theoretical scheme for describing strong interaction. QCD is built on three major concepts: (i) colored quarks, (ii) interaction between colored quarks resulting from exchange of spin 1 colored gluon fields and (iii) local gauge symmetry.

(i) *Quarks*. Quarks are fundamental constituents of matter. They are spin half, structureless point particles. In table 5.1, the properties of the six members of the quark family are listed. Quarks have various intrinsic properties, e.g. electric charge, color charge, spin and mass. They are grouped into three generations, the lightest pair making the first generation and so on. Properties, except for the mass, remain unchanged between the generations. Similarly, three generations are also seen in the lepton family. The experimental fact that quarks (fractional charges) are not observed in isolation was accommodated in the theory by postulating 'color confinement'. Quarks combine to form physically observable, 'color-neutral',

doi:10.1088/978-0-750-31060-4ch5

Table 5.1. Summary of properties of quarks. Current quark masses are from [9].

Generation	Quark flavor	Symbol	Charge Q/e	Color charge	Constituent mass (MeV)	Current mass (GeV)
I	up	u	2/3	3	~350	0.015–0.003
	down	d	−1/3	3	~350	0.035–0.006
II	charm	c	2/3	3	~1800	$1.27^{+0.07}_{-0.11}$
	strange	s	−1/3	3	~550	$0.105^{+0.025}_{-0.035}$
III	top	t	2/3	3	$\sim 170 \times 10^3$	$171.2 \pm 1.1 \pm 1.2$
	bottom	b	−1/3	3	$\sim 4 \times 10^3$	$4.2^{+0.17}_{-0.07}$

particles; mesons (pion, kaon, etc) and hadrons (protons, neutrons, etc). From table 5.1, one can identify protons as composites of (uud) and neutrons as composite of (ddu). It may be mentioned here that the mechanism of color confinement is not properly understood as yet. The QCD Lagrangian is highly singular at small momentum (large distance limit). Numerical simulation of QCD on a lattice does indicate confinement. Some clarification is needed as to quark mass. We have listed two types of mass, constituent quark mass and current quark mass. Since quarks are not observed, their mass cannot be measured directly. The mass of a quark is a parameter of the QCD Lagrangian which describes the interaction of the quarks. These masses are called current quark mass. It is on the same footing as other parameters of the Lagrangian, e.g. fields, coupling constant, etc, and depends on the momentum scale and renormalization scheme. In the constituent quark model, hadrons are bound states of quasi-particles, which are valence quarks, dressed up with gluons and quark–antiquark pairs. Constituent quarks have the same spin, color and flavor as the current quarks, but their masses are phenomenological parameters to be fitted to experimental hadron mass spectra. Naturally, the constituent quark masses are also model dependent.

(ii) *Gluons*. Gluons are the mediators of strong interaction. They are massless bosons (spin 1), with two polarizations (left handed and right handed). Indeed, the role of photons in quantum electrodynamics (QED) is played by gluons in QCD. But unlike photons, which are not self-interacting, gluons are. There are eight types of gluons. This can be understood if we note that quarks (anti-quarks) can carry three color charges. They can be combined in nine different ways: one (singlet) colorless state and eight (octet) colored states ($3 \times \bar{3} = 1 + 8$). Gluons cannot occur in a singlet state (color singlet states cannot interact with colored states). Hence there can only be eight types of gluons.

(iii) *Gauge theory*. QCD is a gauge theory, i.e. the Lagrangian is invariant under a continuous group of local transformations. The gauge group corresponding to QCD is SU(3).

There are several textbooks on QCD and related physics [10–19]. Additionally, there are a large number of review articles, e.g. [20–28]. The reader can also see [29] for a comprehensive list of references. In the following, some selected aspects of QCD will be discussed. To be self-contained, we first discuss briefly the gauge theory. Some elements of the SU(3) group are discussed in appendix A.5.

5.2 Brief introduction to gauge theory

QCD is based on the principle of local gauge symmetry of color interaction. Consider a complex scalar field $\phi(x)$, with Lagrangian density,

$$\mathcal{L}_0\big(\phi(x), \partial^\mu\phi(x)\big) = \partial_\mu\phi^*\partial^\mu\phi - V\big(\phi\phi^*\big). \tag{5.1}$$

The Lagrangian is invariant under a constant phase change,

$$\phi(x) \to \phi'(x) = U\phi(x) \qquad U = \mathrm{e}^{-\mathrm{i}\alpha} \tag{5.2}$$

where α is an arbitrary real constant. This transformation is called 'global' gauge transformation. Note $U = \mathrm{e}^{-\mathrm{i}\alpha}$ is a unitary matrix in one dimension, $UU^\dagger = 1$. The theory is said to be invariant under global gauge transformation under the unitary group[1].

If the complex field is written as

$$\phi = \frac{1}{\sqrt{2}}\big(\phi_1 + \mathrm{i}\phi_2\big) \tag{5.3a}$$

$$\phi^* = \frac{1}{\sqrt{2}}\big(\phi_1 - \mathrm{i}\phi_2\big) \tag{5.3b}$$

the transformation: $\phi \to \mathrm{e}^{-\mathrm{i}\alpha}\phi$, $\phi^* \to \mathrm{e}^{\mathrm{i}\alpha}\phi^*$ gives

$$\phi' = \frac{1}{\sqrt{2}}\big(\phi_1' + \mathrm{i}\phi_2'\big) = \mathrm{e}^{-\mathrm{i}\alpha}\big(\phi_1 + \mathrm{i}\phi_2\big) \tag{5.4a}$$

$$\phi^{*\prime} = \frac{1}{\sqrt{2}}\big(\phi_1' - \mathrm{i}\phi_2'\big) = \mathrm{e}^{\mathrm{i}\alpha}\big(\phi_1 - \mathrm{i}\phi_2\big)\big), \tag{5.4b}$$

which is equivalent to

$$\begin{pmatrix} \phi' \\ \phi^{*\prime} \end{pmatrix} = \begin{pmatrix} \cos\alpha & \sin\alpha \\ -\sin\alpha & \cos\alpha \end{pmatrix}\begin{pmatrix} \phi_1 \\ \phi_2 \end{pmatrix}. \tag{5.5}$$

[1] A complex square $(n \times n)$ matrix A is called unitary when it satisfies the condition $AA^\dagger = A^\dagger A = I$, I being the unit matrix of dimension n. The group of $n \times n$ unitary matrices with group operation of multiplication is called a unitary group of dimension n and is denoted by $U(n)$. The simplest case, $U(1)$, corresponds to all the complex numbers of absolute value 1, i.e. unit circle on the complex plane.

The transformation $\phi(x) \to e^{-i\alpha}\phi(x)$ can be thought of as a rotation in some internal space by an angle α. Thus the U(1) group is isomorphic[2] to O(2), the group of rotation in two dimensions.

In a global gauge transformation, $\phi(x)$ must be rotated by the same angle α in all the space–time points. This is contrary to the spirit of relativity, according to which signal speed is limited by the velocity of light. Then without violating causality, ϕ in all the spatial positions cannot be rotated by the same angle at the same time. This inconsistency is corrected in 'local' gauge transformation, where freedom is given to choose the phase locally, and the phase angle α becomes space–time dependent,

$$\phi(x) \to \phi'(x) = U(x)\phi(x) = e^{-i\alpha(x)}\phi(x). \tag{5.6}$$

Under such a transformation,

$$\partial^\mu \phi(x) \to U(x)\partial^\mu \phi(x) + \underline{\phi(x)\partial^\mu U(x)} \tag{5.7}$$

and the Lagrangian is not invariant under the gauge transformation. The underlined term must be compensated. This can be done by introducing a gauge field $A^\mu(x)$, which under the local gauge transformation transforms as

$$A^\mu(x) \to A^\mu(x) + \frac{1}{e}\partial^\mu \alpha(x), \tag{5.8}$$

and replacing the partial derivative $\left(\partial_\mu\right)$ with covariant derivative $\left(D_\mu\right)$ defined as

$$D^\mu \phi(x) = \left[\partial^\mu + ieA^\mu(x)\right]\phi(x). \tag{5.9}$$

While the Lagrangian is now invariant under local gauge transformation, it is not the same Lagrangian as before. A gauge field $A^\mu(x)$ is now present as an external field. To obtain a closed system, we need to add a kinetic energy term, to be constructed from A_μ and its derivatives. The only term which is invariant under the gauge transformation is

$$F^{\mu\nu} = \partial^\mu A^\nu - \partial^\nu A^\mu. \tag{5.10}$$

Thus we arrive at a Lagrangian density for a closed dynamical system, invariant under local $U(1)$ gauge transformation,

$$\mathcal{L} = -\frac{1}{4}F^{\mu\nu}F_{\mu\nu} + D_\mu \phi^* D^\mu \phi - V\left(\phi\phi^*\right). \tag{5.11}$$

The Lagrangian in equation (5.11) is essentially for QED, which is a local gauge theory with U(1) group symmetry. The symmetry group for QCD on the other hand

[2] In group theory, two groups are called isomorphic when there is one-to-one correspondence between the group elements. Isomorphic groups have the same properties and need not be distinguished.

is SU(3). In contrast to U(1), which is an abelian group[3] (group elements commute), SU(3) is non-abelian (group elements do not commute). The non-abelian nature of the SU(3) group introduces additional complications.

5.3 QCD Lagrangian

Quarks are fermions and obey the Dirac equation. Classical Lagrangian density for free quarks of N_f flavors can be written as

$$\mathcal{L}_{\text{classical}} = \sum_{f=1}^{N_f} \overline{\psi}_f \left[i\gamma^\mu \partial_\mu - m_f \right] \psi_f \tag{5.12a}$$

$$= \sum_{f=1}^{N_f} \overline{\psi}_f [i\partial\!\!\!/ - m_f] \psi_f \tag{5.12b}$$

where in the second line, we have used Dirac's[4] slashed notation, $\partial\!\!\!/\psi = \gamma^\mu \partial_\mu \psi$. In equation (5.12), ψ_f is the wave function for the quark of flavor f. In the SU(3) color space, it can be written as a column matrix,

$$\psi = \begin{pmatrix} \psi_r \\ \psi_g \\ \psi_b \end{pmatrix}, \tag{5.13}$$

where each ψ_i, $i = r, g, b$ is a four-column Dirac spinor.

Analogously to the scalar fields, if we introduce eight-gluon potential

$$A_a^\mu(x) \qquad a = 1, 2, \ldots 8 \tag{5.14}$$

and associated covariant derivatives,

$$D_\mu = \partial_\mu - igA_\mu^a \frac{\lambda_a}{2}, \tag{5.15}$$

the SU(3) gauge invariant Lagrangian can be written as

$$\mathcal{L}_{\text{classical}} = \sum_{f=1}^{N_f} \overline{\psi}_f \left[i\gamma^\mu D_\mu - m_f \right] \psi_f - \frac{1}{4} F_a^{\mu\nu} F_{\mu\nu}^a \tag{5.16}$$

[3] The abelian group is named after the Norwegian mathematician Niels Henrik Abel. He invented group theory independently of the French mathematician, Evariste Galois. Apart from group theory, Abel is famous for his 'impossibility theorem', that general algebraic equations of order five or more cannot be solved analytically. Abel was largely unrecognized during his lifetime. In 1829, he died at the early age of 26 years.

[4] Richard Feynman invented the notation and it is also known as Feynman's slashed notation.

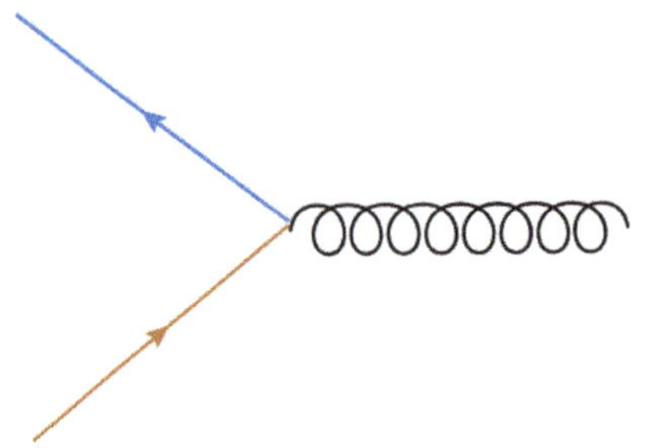

Figure 5.1. A quark interacts with a gluon and changes its color.

where we have included the kinetic energy term $\frac{1}{4}F_a^{\mu\nu}F_{\mu\nu}^a$ for the gluon fields. The gauge invariance restricts the field tensor to the form

$$F_{\mu\nu}^a = \partial_\mu A_\nu^a - \partial_\nu A_\mu^a + g f_{bc}^a A_\mu^b A_\nu^c. \tag{5.17}$$

One can identify several important aspects of equation (5.16). It contains terms like, (i) $\overline{\psi}_f[i\gamma^\mu\partial_\mu - m_f]\psi_f$ describing free quark fields, (ii) $\partial_\mu A_\nu^a \partial^\mu A_a^\nu$ describing free gluon fields and (iii) several terms describing interactions between quark fields and gluon fields and between gluon–gluon fields. For example, it contains terms like

$$\begin{aligned} \mathcal{L}_{qg} &= -ig\overline{\psi}_f \gamma_\mu A_a^\mu \frac{\lambda^a}{2}\psi_f, \\ \text{or more precisely,} \quad &= -ig\overline{\psi}_{f_j}\gamma_\mu A_a^\mu \frac{\lambda_{ji}^a}{2}\psi_{f_i}, \end{aligned} \tag{5.18}$$

describing interactions between quark and gluon fields. It can be understood as follows: a quark changes its color from i to j by interacting with a gluon of color a through the SU(3) generator $\frac{\lambda_{ji}^a}{2}$. The process is depicted in figure 5.1. The factor g is the coupling strength between the quark and gluon fields. Equation (5.16) also contains terms like

$$\mathcal{L}_{ggg} = g f_{abc} A_{\mu b} A_{\nu c}(\partial^\mu A_a^\nu) \tag{5.19a}$$

$$\mathcal{L}_{gggg} = g^2 f_{abc} f_{ade} A_{\mu b} A_{\nu c} A_d^\mu A_e^\nu. \tag{5.19b}$$

These terms represent gluon self-interaction. Equations (5.19*a*) and (5.19*b*), respectively, correspond to three- and four-gluon interaction, a schematic representation of which is given in figure 5.2. These terms constitute the major differences between QCD and QED. Unlike the photons in QED, in QCD, gluons, the mediator particles, can self-interact. Gluon self-interaction leads to two important properties of QCD: (i) asymptotic freedom and (ii) infrared slavery.

Equation (5.12) is the classical Lagrangian. It needs to be quantized. We will not discuss the quantization procedure in detail. Briefly, the quark part of the

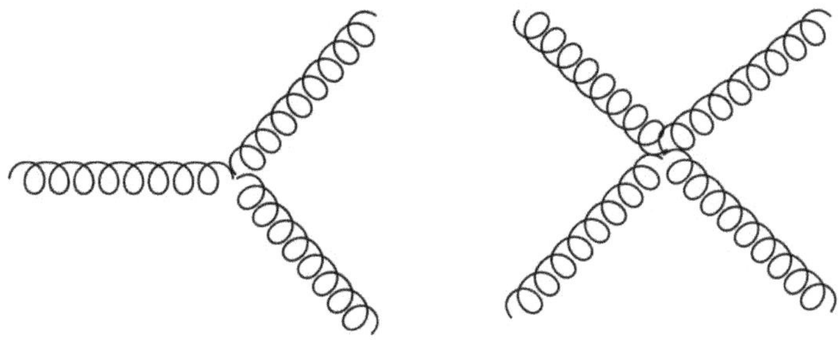

Figure 5.2. Gluon self-interaction. Three-gluon and flour-gluon vertices are shown.

Lagrangian can be quantized in the usual procedure, quantizing spinor fields in terms of creation and annihilation operators. As in QED, the gauge invariance of the Lagrangian poses some problems in quantization of the gauge fields. While the gauge field A^μ has four degrees of freedom, physical, massless gluon fields have only two degrees of freedom. Following QED, two redundant degrees of freedom are fixed by introducing the gauge fixing term, but unlike in QED, the extra fields introduced by the gauge fixing terms do not cancel each other, requiring further introduction of massless, unphysical Faddeev–Popov ghost fields.

5.4 Perturbative QCD

QCD has two distinguishing features: (i) asymptotic freedom and (ii) color confinement. The interaction strength $\alpha_s = \frac{g^2}{4\pi}$ between quarks becomes smaller as the distance between them gets shorter and in the limit $r \to 0$, $\alpha_s \to 0$. This is the asymptotic freedom; in the short distance or large momentum scale quarks are asymptotically free. At large distances however the interaction strength between two quarks increases such that they cannot be separated, a phenomenon known as color confinement. Asymptotic freedom allows one to use perturbative techniques to solve for certain problems in QCD, involving large momentum transfer (or short distance processes). The cross section for those processes can be expanded in terms of the coupling constant,

$$\sigma(p) = \sum_n A_n(p)\alpha_s^n \tag{5.20}$$

and smallness of coupling ensures that only a few terms contribute to the cross section.

5.4.1 Asymptotic freedom and infrared slavery

The QCD coupling constant runs. A running coupling constant is not particular to QCD, but is in all the quantum field theories, the underlying reason being the quantum fluctuations. To understand it better, consider the interaction between two electrons in a vacuum,

$$F = \frac{1}{4\pi}\frac{e^2}{r^2} = \frac{\alpha_{em}}{r^2}, \tag{5.21}$$

where we have introduced the QED coupling constant $\alpha_{em} = \frac{e^2}{4\pi}$. In a medium with dielectric constant ϵ, the interaction changes to

$$F = \frac{1}{4\pi\epsilon}\frac{e^2}{r^2}. \tag{5.22}$$

Interaction in the medium can also be written in the vacuum form (equation (5.21)) by redefining electron charges as $\tilde{e} = \frac{e}{\sqrt{\epsilon}}$, or by renormalizing the coupling constant, $\alpha_{em} \to \frac{\alpha_{em}}{\epsilon}$. In a quantum theory, a vacuum can be considered as a polarizable medium and due to fluctuations, even in a vacuum, one can create virtual pairs of e^+e^-. The virtual charges can effectively screen a test charge. A probe will reveal more and more of the charge as it approaches nearer to the test charge, making the effective coupling constant dependent on the separation, and the interaction can be written as

$$F = \frac{\alpha_{em}(r)}{r^2}. \tag{5.23}$$

More formally, in quantum field theory, the vacuum is just the lowest energy state of a field system and not necessarily empty. A QED vacuum can be considered as the state where the Dirac negative energy sea is filled with electrons. A photon passing through the vacuum can induce the transition of an electron from a negative to a positive energy state, virtually creating a pair of electrons and positrons (see I of figure 5.3). The phenomenon is known as vacuum fluctuation or vacuum polarization. In any electromagnetic process (or more precisely, in any quantum mechanical process), such loop diagrams contribute. However, they lead to divergences, called 'ultraviolet divergency'. The divergences in loop diagrams can be understood as follows: the electrons in the loop can have arbitrary momentum and upon integrating over the momentum give infinity.

In order to obtain a finite result, these divergences need to be eliminated. In renormalization theory, the divergences are absorbed by redefining various parameters of the theory, e.g. mass, coupling constant, etc. In other words, we abandon the idea of using or observing the parameters of the initial Lagrangian, the so-called 'bare' quantities and re-express everything in terms of the finite, 'renormalized' and observable parameters. We shall not discuss the renormalization procedure in detail.

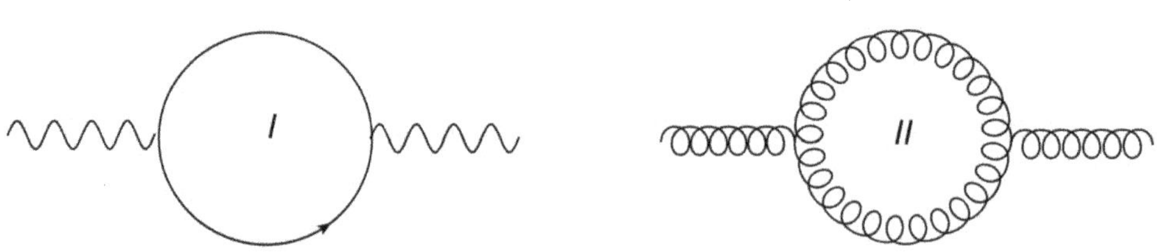

Figure 5.3. Diagram showing vacuum fluctuations. In I, a photon (gluon) excites the vacuum by producing a virtual electron–position (quark–antiquark) pair. Diagram II is possible only in QCD, where the vacuum is excited by gluon self-interaction.

It proceeds in two steps. The first step is the regularization: the divergent integrals are regularized by separating finite results and infinite results. There are several schemes for regularization, e.g. dimensional regularization or the Pauli–Villar regularization scheme. In the second step, fields and parameters in the model, e.g. mass, coupling constant, etc, are redefined to absorb these infinities resulting in a renormalized Lagrangian.

Dependence of the coupling constant α on distance r or equivalently on the momentum scale $\mu \sim \frac{1}{r}$ is given by the renormalization group equation,

$$\mu \frac{\partial \alpha}{\partial \mu} = \frac{\partial \alpha}{\partial \ln \mu} = \beta(\alpha) \tag{5.24}$$

where the function driving the momentum dependence is called the beta function and can be written as

$$\beta(\alpha) = -\alpha^2 \left[b_0 + b_1 \alpha + b_2 \alpha^2 + \cdots \right]. \tag{5.25}$$

In leading order, equation (5.24) can be integrated to obtain

$$\alpha(\mu) = \frac{\alpha(\mu_0)}{1 + \frac{b_0}{2} \frac{\alpha(\mu_0)}{3\pi} \ln \frac{\mu^2}{\mu_0^2}}. \tag{5.26}$$

In equation (5.26), μ_0 is a reference momentum scale. In QED, the leading order contribution to the beta function can be calculated as $b_0 = -\frac{2}{3\pi}$ and the QED running coupling constant can be written as

$$\alpha_{em}(\mu) = \frac{\alpha_{em}(\mu_0)}{1 - \frac{\alpha_{em}(\mu_0)}{3\pi} \ln \frac{\mu^2}{\mu_0^2}}. \tag{5.27}$$

QED coupling increases as the momentum scale is increased. It is called the screening effect. At finite distance, vacuum polarization screens a test charge. The vacuum effect is lessened at smaller distances and effective charge becomes larger. The reader may note that the coupling constant diverges at the momentum scale,

$$\Lambda_{\text{Landau}} = \mu_0 \exp\left(\frac{3\pi}{2\alpha(\mu_0)} \right). \tag{5.28}$$

It is called the Landau pole. It occurs at very large energy (beyond the Planck energy) and is of no consequence in practical applications.

The leading order contribution to the beta function in QCD can be calculated as

$$b_0 = \frac{1}{4\pi} \left(11 - \frac{2}{3} N_f \right). \tag{5.29}$$

The second term $-\frac{2N_f}{12\pi}$ in equation (5.29) comes from the quark–antiquark pair effect in the first diagram of figure 5.3 and produces a screening effect as in QED. The first term $\frac{11}{4\pi}$ comes from the second diagram of figure 5.3, vacuum excitation by gluon self-interaction. The opposite sign indicates that gluon self-coupling has an opposite or anti-screening effect. For $N_F \leqslant \frac{33}{2}$, the effect of vacuum polarization is to anti-screen the QCD interaction. The running coupling in QCD now can be written as

$$\alpha_s(\mu) = \frac{g^2}{4\pi} = \frac{\alpha_s(\mu_0)}{1 + \frac{\alpha_s(\mu_0)}{4\pi}\left(11 - \frac{2N_f}{3}\right)\ln\frac{\mu^2}{\mu_0^2}}. \tag{5.30}$$

Equation (5.30) demonstrates the property of asymptotic freedom. If $N_f < 17$, $\alpha_s(\mu)$ will asymptotically decrease to zero as $\mu \to \infty$, QCD is asymptotically free. On a short distance scale, quarks behave as free particles. Equation (5.30) also demonstrates the property of infrared slavery: in a low momentum scale or at large distance, the coupling constant diverges to infinity. For example, taking $\alpha_s(M_z) = 0.12$, and for typical values of $N_f = 2,3\ldots5$, $\alpha_s(\mu)$ exceeds unity for $\mu \leqslant O$ (100 MeV…1 GeV). Clearly, this is the region where perturbative expansions in α_s are not meaningful. Therefore energy scales below the order of 1 GeV are regarded as the non-perturbative region where confinement sets in.

Introducing a dimensionful parameter,

$$\Lambda_{\mathrm{QCD}}^2 = \frac{\mu_0^2}{\mathrm{e}^{1/b_0\alpha_s(\mu_0)}} \qquad b_0 = \frac{1}{4\pi}\left(11 - \frac{2}{3}N_f\right). \tag{5.31}$$

Equation (5.30) can be written as

$$\alpha_s(\mu) = \frac{1}{\frac{1}{4\pi}\left(11 - \frac{2}{3}N_f\right)\ln\frac{\mu^2}{\Lambda_{\mathrm{QCD}}^2}}. \tag{5.32}$$

Λ_{QCD} is called the QCD scale parameter. At this momentum scale the QCD coupling constant diverges to infinity. Technically, this is the minimum momentum scale, below which QCD is inapplicable. Typically, $\Lambda_{\mathrm{QCD}} \approx 200$ MeV, however, its value also depends on the renormalization scheme.

5.4.2 Factorization theorem

Ultraviolet divergences, associated with large momentum or short distance processes, are regularized in the renormalization scheme. In QCD there is another type of divergence called 'infrared divergence', associated with low momentum or large distance processes. The renormalization procedure introduces a renormalization scale μ and any physical quantity will be a function of three variables of dimension mass, (i) the kinematical energy scale Q of scattering, (ii) the renormalization scale μ and (iii) the mass. They appear as ratios, $\frac{Q}{\mu}$ and $\frac{\mu}{m}$. If μ is chosen to be large such that asymptotic freedom is applicable, the ratio $\frac{\mu}{m}$ becomes large. $\frac{\mu}{m} \to \infty$ is called infrared divergence. In a loop diagram, the coupling constant generally appears in combination $\alpha^2(Q)\ln^a\frac{\mu}{m}$, $a = 1$ or 2 and perturbative expansion is no longer an expansion in terms of a small parameter. In summary, a general cross section is a combination of short and long distance behavior, and is hence not computable directly in perturbation theory for QCD. We must mention here that there are some processes, e.g. e^+e^- annihilation to hadrons or into jets, which are infrared safe. Factorization theorem allows us to write the cross section for a QCD process in terms of a long distance (small momentum) part, which is not directly calculable, times a short distance (large momentum) part, which is calculable in perturbative QCD (pQCD). Factorization theorem is best explained in the parton model.

5.4.3 Parton model

In the pre-QCD era, Feynman introduced the parton model to explain the scaling behavior[5]. Basic assumptions of the parton model are as follows.

(i) A fast hadron is made up of point-like constituents called partons (later identified with quarks and gluons).

(ii) One can define parton distribution function $f_{i/A}(x)$ which gives the probability of finding a parton i with momentum $xP_A(x \leqslant 1)$ inside the hadron A with momentum P_A.

(iii) The inclusive cross sections are given by the convolution of 'tree level' parton cross sections with the distribution functions.

Consider the collision between two hadrons A and B, producing, say, a large invariant mass lepton pair. For simplicity, consider only one type of parton i. In the factorization theorem, the differential cross section can be written as

$$d\sigma^{AB\to X}(p_A, p_B) = \left(dx_1 f_{i/A}(x_1)dx_2 f_{i/B}(x_2)\right) \times \hat{\sigma}_{ii}(x_1 p_A, x_2 p_A). \tag{5.33}$$

The bracketed term gives the probability of finding one parton in hadron A and one parton in hadron B, respectively with momenta $x_1 p_A$ and $x_2 p_B$. All the large distance or small momentum effects are encoded in the parton distribution function, which cannot be calculated theoretically. The short distance or large momentum part is encoded in the 'tree level' cross section $\hat{\sigma}_{ii}(p, q)$, which is perturbatively calculable.

[5] In deep inelastic lepton–proton scattering processes $e + p \to e + X$, it is expected that the cross section will depend on both the energy loss ν of the lepton and four-momentum transfer $q^2 = -Q^2$. Experimental data however show dependence only on the scaled variable (called Bjorken's variable), $x = \frac{Q^2}{2M\nu}$.

In the parton model, partons are interaction-free, which is only approximately true in QCD. Also the parton model does not include gluons, which are an integral part of QCD. The parton model equations are then modified. The most important modification is due to the infrared singularities mentioned above. These infinities are regularized in the same fashion as for the case of renormalizing QCD. The procedure introduces a new momentum scale, μ_F, called the factorization scale. It represents the scale where the perturbative physics factorizes from the non-perturbative part. The parton distribution function thus modifies to $f_i(x) \to f_i(x, \mu_F^2)$. The cross section for the process $A + B \to X$ can be obtained as

$$\sigma^{AB\to X}(p_1, p_2) = \sum_{i,j=\text{quarks,gluons}} \int_0^1 dx_1 dx_2 f_{i/A}\left(x_1, \mu_F^2\right) f_{j/B}\left(x_2, \mu_F^2\right) \hat{\sigma}_{ij}(x_1 p_A, x_2 p_A). \quad (5.34)$$

A remarkable consequence of factorization is that measuring parton distributions for one scale μ_F allows their prediction for any other scale μ_F', as long as both μ_F and μ_F' are large enough that both $\alpha_s(\mu_F)$ and $\alpha_s(\mu_F')$ are small. The equation describing parton evolution as a function of momentum transfer is called the Dokshitzer–Gribov–Lipatov–Altarelli–Parisi (DGLAP) equation [30]. The equation increases the power of perturbative QCD enormously. The parton distribution function $f_{i/h}(x, \mu^2)$ cannot be calculated theoretically, but has to be obtained from experiments. The best source is deep inelastic electron scattering. One generally parameterizes the parton distribution function at some reference factorization scale μ_0 with a flexible functional form, e.g.

$$f_{i/h}\left(x, \mu_F^2\right) = Ax^B(1 - x)^C(1 + Ex)^D \quad (5.35)$$

and evolves them by DGLAP equations to the experimental scale [31]. The parameters A, B, C and D are then obtained by fitting the experimental data. In practice, a large number of parameters (20 or more) are used in the fitting function. In the literature, several parton distribution functions are available, e.g. CTEQ [32], GRV/GJR [33], MRST [34], etc. For more details about parton distribution functions see [35].

5.4.4 Fragmentation function

The final state of any QCD process is hadrons. QCD deals with partons (quarks and gluons). The process of the conversion of partons into hadrons is unknown, but basically, it is a low momentum or non-perturbative process. Physically, an energetic parton fragments (or showers) into many partons, which on a later time scale undergo hadronization. In the parton model, the hadronization process may be parameterized by introducing a fragmentation function $D_{h/i}(z)$ that gives the probability that parton i fragments into a hadron h with a fraction z of the parent parton momentum. The fragmentation function essentially encodes the non-perturbative process of hadronization of quarks. They cannot be calculated theoretically, but are obtained only from experiments. The best experimental source is electron–positron annihilation to a hadron, $e^+ + e^- \to (\gamma, Z) \to h + X$. The cross section for the process can be obtained as

$$\frac{1}{\sigma_{\text{tot}}}\frac{\mathrm{d}\sigma^h}{\mathrm{d}z} \approx \sum_i \sigma_0 D_{h/i}(z) \tag{5.36}$$

where z is the energy fraction, $z = E_h/E_{\text{beam}}$ and σ_0 is the cross section for producing a parton i.

In a simple parton model, the fragmentation function is a function of the momentum fraction $z = p_h/p_i$, but QCD correction introduces a scale dependence, $D_{h/i}(z) \to D_{h/i}(z, \mu_F^2)$. Fragmentation functions also obey the DGLAP type of evolution equation. General procedure [36] is to parameterize the fragmentation function at some momentum scale μ_{F}, with a flexible function, e.g.

$$D_{h/i}(z, \mu_{\text{F}}) = N_i z^{\alpha_i}(1 - z)^{\beta_i}, \tag{5.37}$$

and use the DGLAP equation to evolve it to the experimental scale and obtain the parameters by fitting the experimental data.

5.5 Non-perturbative QCD

In the high momentum regime, the QCD coupling constant is small, and one can apply perturbative methods to calculate the physical observables from the QCD Lagrangian. However, the QCD coupling constant runs and in the low momentum regime its value exceeds unity and perturbative approaches fail. Running coupling constants reflect the change in the underlying force law, as the energy/momentum scale at which physical processes occur varies.

In lattice QCD, the space–time is discretized to reduce the infinite degrees of freedom of 'field variables' to a finite and (numerically) tractable number. One immediately notices that due to the finite dimensions of the lattice, Lorentz invariance is broken. Gauge invariance, however, is kept explicitly by parallel transportation of the gauge fields between adjacent lattice sites. In the continuum limit, lattice spacing $a \to 0$, Lorentz invariance can be restored.

In the following, some selected aspects of lattice QCD will be discussed. For more informative exposure to lattice QCD, see [37–40]. Also, there is a large number of reviews on lattice QCD, e.g. [41–47].

Lattice QCD is intimately related to Feynman's path integral formulation of quantum mechanics. Below, we briefly sketch the ideas behind the path integral method. A brief introduction to parallel transport will also be given.

5.5.1 The path integral method

Richard Feynman is one of the most celebrated physicists of the twentieth century. Apart from the path integral formulation of quantum mechanics, he made pioneering contributions in quantum electrodynamics, superfluidity and particle

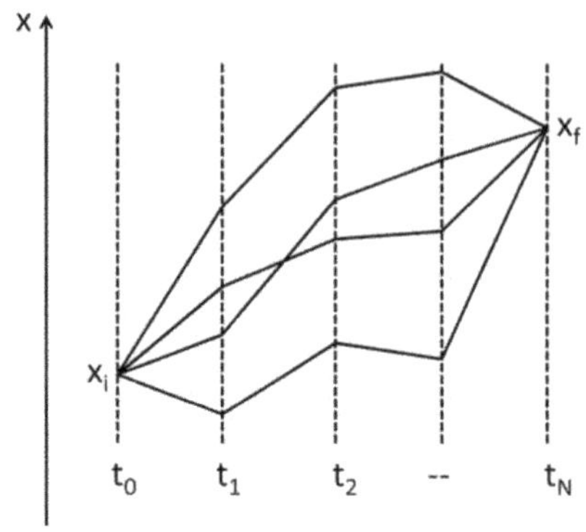

Figure 5.4. Particle trajectories or paths between (x_i, t_0) and (x_f, t_N) for discrete time steps.

physics. He invented the diagrammatic approach to QED (the Feynman diagrams). In 1965, Feynman, Julian Schwinger and Sin-Itiro Tomonaga were awarded the Nobel prize for their contributions to QED.

Consider the propagation of a particle from position x_i at time t_0 to position x_f at time t_N. For a given trajectory, $(x, \dot{x})$, the action is

$$S = \int_{t_0}^{t_N} \mathrm{d}t \mathcal{L}(x, \dot{x}). \tag{5.38}$$

$\mathcal{L}(x, \dot{x})$ is the Lagrangian. The path integral method states that the transition probability from (x_i, t_0) to (x_f, t_N) can be expressed as the weighted sum of all the possible paths or trajectories,

$$\langle x_f(t_N)|x_i(t_0)\rangle \sim \sum_{\text{paths(P)}} \exp\left[\mathrm{i}S(x(t), \dot{x}(t))\right]. \tag{5.39}$$

Let us discretize the time interval into N steps, $t_N - t_0 = N\Delta t$. In figure 5.4, the discretized paths are shown. It is clear that for small enough time steps, any continuous path can be adequately traced. Now one can sum the trajectories at a particular time step, say t_n:

$$\sum_{P(t=t_n)} \mathrm{e}^{\mathrm{i}S(x(t),\dot{x}(t))} \sim \int_{-\infty}^{\infty} \mathrm{d}x(t_n)\mathrm{e}^{\mathrm{i}S(x(t_n),\dot{x}(t_n))}. \tag{5.40}$$

The procedure can be repeated for each time step. In the limit $N \to \infty$,

$$\langle x_f(t_N)|x_i(t_0)\rangle \sim \mathcal{N} \int \prod_{n=1}^{N-1} \mathrm{d}x(t_n)\mathrm{e}^{\mathrm{i}S(x(t_n),\dot{x}(t_n))} \tag{5.41}$$

where $\mathcal{N}$ is some normalization factor.

Table 5.2. Relations in Minkowski and Euclidean space are compared.

	Minkowski space	Euclidean space	Relation
Coordinates $\left(x^0, x^1, x^2, x^3\right)$	(t, x, y, z)	(τ, x, y, z)	$t = -\mathrm{i}\tau$
Volume element d^4x	$\mathrm{d}t\mathrm{d}x\mathrm{d}y\mathrm{d}z$	$\mathrm{d}\tau\mathrm{d}x\mathrm{d}y\mathrm{d}z$	$\mathrm{d}t = -\mathrm{i}\mathrm{d}\tau$
Metric tensor	$g_{\mu\nu} = \mathrm{diag}(1, -1, -1, -1)$	$\delta_{\mu\nu} = \mathrm{diag}(1, 1, 1, 1)$	–
Line element $\mathrm{d}s^2$	$g_{\mu\nu}\mathrm{d}x^\mu \mathrm{d}x^\nu$	$\delta_{\mu\nu}\mathrm{d}x^\mu \mathrm{d}x^\nu$	–
Derivative ∂_μ	$\left(\partial_t, \partial_x, \partial_y, \partial_z\right)$	$\left(\partial_\tau, \partial_x, \partial_y, \partial_z\right)$	$\partial_t = \mathrm{i}\partial_\tau$
Gauge field	$\left(A_t, A_x, A_y, A_z\right)$	$\left(A_\tau, A_x, A_y, A_z\right)$	$A_t = \mathrm{i}A_\tau$
Gamma matrices γ^μ	$\gamma^0, \gamma^1, \gamma^2, \gamma^3$ $\left\{\gamma^\mu, \gamma^\nu\right\} = 2g^{\mu\nu}$	$\gamma_E^0, \gamma_E^1, \gamma_E^2, \gamma_E^3$ $\left\{\gamma_E^\mu, \gamma_E^\nu\right\} = 2\delta^{\mu\nu}$	$\gamma^0 = \gamma_E^0$ $\gamma^j = \mathrm{i}\gamma_{\mathrm{E}}^j$
Action	$S_{\mathrm{M}} = \int \mathrm{d}t\mathrm{d}^3x\mathcal{L}_{\mathrm{M}}$	$S_E = \int \mathrm{d}\tau\mathrm{d}^3x\mathcal{L}_{\mathrm{E}}$	$\mathrm{i}S_{\mathrm{M}} = -S_{\mathrm{E}}$, $\mathcal{L}_{\mathrm{M}} = -\mathcal{L}_{\mathrm{E}}$

It is easy to extend the formalism to fields. Consider a one-dimensional field $\phi(x, t)$. Again, consider the transition amplitude for the field $\phi_i(x, t_i)$ to $\phi_f(x, t_f)$,

$$\left\langle \phi_i(x, t_i) \middle| \phi_f\left(x, t_f\right)\right\rangle \sim \sum_{\phi_P} \mathrm{e}^{\mathrm{i}S\left(\{\phi_P\}, \{\partial_\mu \phi_P\}\right)}. \tag{5.42}$$

As before, we discretize the time intervals in N steps. Additionally, we discretize the space coordinates into N_x steps. Note that space is infinite in dimension. Thus, discretization can only be an approximation of the infinite space.

$$\left\langle \phi_i(x, t_i) \middle| \phi_f\left(x, t_f\right)\right\rangle \sim \mathcal{N} \lim_{N_x, N \to \infty} \int \prod_{m=1}^{N_x} \prod_{n=1}^{N-1} \mathrm{d}\phi(x_m, t_n) \mathrm{e}^{\mathrm{i}S\left(\{\phi(x_m, t_n)\}, \{\partial_\mu \phi(x_m, t_n)\}\right)}. \tag{5.43}$$

It is convenient to make a Wick's rotation, $t = -\mathrm{i}\tau$, so that the space is Euclidean. Relations between Euclidean and Minkowski space are noted in table 5.2. In Euclidean space,

$$\mathrm{i}S = \mathrm{i}\int \mathrm{d}t\mathrm{d}^3x\mathcal{L} = -\int \mathrm{d}\tau\mathrm{d}^3x\mathcal{L}_{\mathrm{E}} = -S_{\mathrm{E}}. \tag{5.44}$$

This can be explicitly seen by considering the action for a simple scalar field,

$$\begin{aligned} S &= \int \mathrm{d}t\mathrm{d}^3x\left[\frac{1}{2}\partial_\mu\phi\partial^\mu\phi - V(\phi)\right] = \int \mathrm{d}t\mathrm{d}^3x\left[\frac{1}{2}\left(\partial_t\phi\partial_t\phi - \nabla\phi\nabla\phi\right) - V(\phi)\right] \\ &= -\mathrm{i}\int \mathrm{d}\tau\mathrm{d}^3x\left[\frac{1}{2}\left(-\partial_\tau\phi\partial_\tau\phi - \nabla\phi\nabla\phi\right) - V(\phi)\right] \\ &= \mathrm{i}\int \mathrm{d}\tau\mathrm{d}^3x\left[\frac{1}{2}\left(\partial_\tau\phi\partial_\tau\phi + \nabla\phi\nabla\phi\right) + V(\phi)\right] = \mathrm{i}S_{\mathrm{E}}. \end{aligned} \tag{5.45}$$

In terms of the Euclidean action (S_E), the transition probability can be written as

$$\left\langle \phi_i(x, t_i) \middle| \phi_f(x, t_f) \right\rangle \sim \mathcal{N} \lim_{N_x, N \to \infty} \int \prod_{m=1}^{N_x} \prod_{n=1}^{N-1} d\phi(x_m, t_n) e^{-S_E(\{\phi(x_m, t_n)\}, \{\partial_\mu \phi(x_m, t_n)\})}$$
$$\sim \mathcal{N} \int \mathcal{D}[\phi] e^{-S_E}. \tag{5.46}$$

$\mathcal{D}[\phi]$ is the shorthand notation of the integration measure,

$$\mathcal{D}[\phi] = \prod_{m=1}^{N_x} \prod_{n=1}^{N-1} d\phi(x_m, t_n). \tag{5.47}$$

Now in statistical mechanics, the central problem is to compute the partition function, defined as

$$Z = Tr e^{-\beta H} = \sum_\phi \langle \phi(x) | e^{-\beta H} | \phi(x) \rangle. \tag{5.48}$$

The summation is over all the possible states $|\phi(x)\rangle$. $\beta = 1/T$ is the inverse temperature. It can be rewritten as

$$Z = \sum_\phi \langle e^{-\beta H} \phi(x) | \phi(x) \rangle$$
$$= \sum_\phi \langle \phi(x, t = -i\beta) | \phi(x, t = 0) \rangle. \tag{5.49}$$

The partition function in statistical mechanics then corresponds to the path integral formulation for the transition probability.

$$Z = \mathcal{N} \int \mathcal{D}[\phi] e^{-S_E}. \tag{5.50}$$

This is an important realization. All the tools of statistical mechanics can be applied to field theory problems. The expectation value of any observable can be obtained as

$$\langle O \rangle = \frac{\int \mathcal{D}[\phi] e^{-S_E} O}{\int \mathcal{D}[\phi] e^{-S_E}}. \tag{5.51}$$

5.5.2 Parallel transport

One of the problems in general relativity is the derivative of a vector (or more generally, a tensor) quantity. In flat space–time, the derivative of a vector can be computed easily,

$$V'^\mu(x) = \lim_{h \to 0} \frac{V^\mu(x + h) - V^\mu(x)}{h}. \tag{5.52}$$

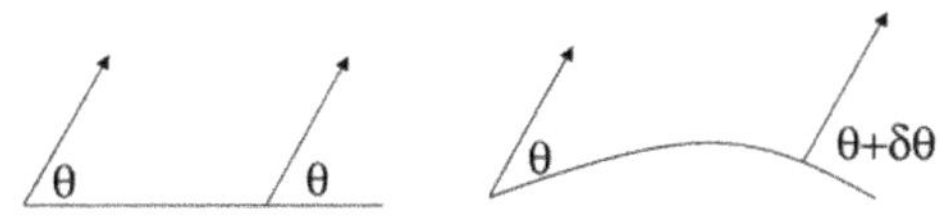

Figure 5.5. In a parallel transport, the tangent angle of a vector is changed in curved space–time, but not in flat space–time.

However, in curved space–time, since the metric tensor $g^{\mu\nu}$ depends on space, additional terms arise. This can be understood from figure 5.5. In flat space, with a vector at x, when transported to $x + h$ the tangent angle remains the same. But in a curved space, the tangent angle is changed. In general relativity, this is accommodated by defining the covariant (or semicolon) derivative,

$$\partial_{;\mu} V^{\nu} = \partial_{\mu} V^{\nu} + \Gamma^{\mu}_{\nu\alpha} V^{\alpha} \tag{5.53}$$

where $\Gamma^{\mu}_{\nu\alpha}$ is the Christoffel symbol and is defined as

$$\Gamma^{\mu}_{\nu\alpha} = \frac{1}{2} g^{\mu m} \left(\frac{\partial g_{m\nu}}{\partial x^{\alpha}} + \frac{\partial g_{m\alpha}}{\partial x^{\nu}} - \frac{\partial g_{\nu\alpha}}{\partial x^{m}} \right). \tag{5.54}$$

$\Gamma^{\mu}_{\nu\alpha} V^{\alpha}$ in equation (5.53) accounts for the change in the vector's coordinate representation during the transport ($\Gamma^{\mu}_{\nu\alpha} = 0$ is flat space–time).

Covariant derivative

$$D^{\mu}\phi(x) = \left[\partial^{\mu} + \mathrm{i}eA^{\mu}(x)\right]\phi(x), \tag{5.55}$$

defined in equation (5.9), is analogous to parallel transport; $\mathrm{i}eA^{\mu}(x)$ is the change in the field's representation during transport from x^{μ} to $x^{\mu} + \mathrm{d}x^{\mu}$. Then,

$$\begin{aligned} \phi(x^{\mu} + \mathrm{d}x^{\mu}) &= \phi(x^{\mu}) + \mathrm{d}x^{\mu} D_{\mu}\phi(x^{\mu}) \\ &= \phi(x^{\mu}) + \mathrm{d}x^{\mu}(\partial_{\mu} + \mathrm{i}eA_{\mu})\phi(x^{\mu}) \\ &= \mathrm{d}x^{\mu}\partial_{\mu}\phi + \left[1 + \mathrm{i}eA_{\mu}\mathrm{d}x^{\mu})\right]\phi(x^{\mu}). \end{aligned} \tag{5.56}$$

The first term in equation (5.56) is essentially a translational term. The second term containing the A^{μ} describes the transport of the gauge field between two close points x and $x + \mathrm{d}x$. For infinitesimal distance, the second term in equation (5.56) can be written as, $\mathrm{e}^{\mathrm{i}eA_{\mu}\mathrm{d}x^{\mu}}\phi(x^{\mu})$. By repeated application of infinitesimal transport, the current (gauged) value of the phase of the wave function ϕ, at the four-dimensional space–time point y, is related to its value at some reference point x by the parallel transport,

$$\phi(y) = \mathrm{e}^{\mathrm{i}e\int_x^y \mathrm{d}\zeta^{\mu} A_{\mu}(\zeta)}\phi(x). \tag{5.57}$$

The integration in the exponent goes along some path C_{xy} that connects x and y. For the non-abelian gauge group SU(3), a quark can alter its color under parallel transport.

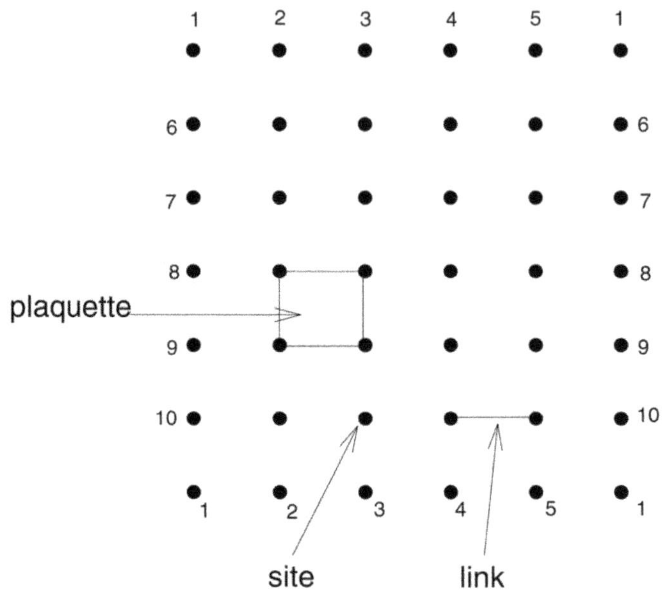

Figure 5.6. Schematic representation of a lattice, in two dimensions.

Then for SU(3) gauge fields, the exponential or the phase factor is a 3×3 unitary matrix. An extension of the above equation can be written as

$$\phi(y) = P\mathrm{e}^{\mathrm{i}g\int_x^y \mathrm{d}\zeta^\mu A_{\mu a}(\zeta)\frac{\lambda_a}{2}}\phi(x) = U\phi(x). \tag{5.58}$$

The symbol P means path ordering. U is the path dependent representation of an element of the gauge group SU(3). To construct the matrix of parallel transport at finite distance, one has to subdivide the path C_{xy} into small parts and form ordered product of parallel transport along these small parts:

$$U = P\mathrm{e}^{\mathrm{i}g\int_x^y \mathrm{d}\zeta^\mu A_{\mu a}(\zeta)\frac{\lambda_a}{2}} = \prod_\zeta \left(1 + \mathrm{i}g\mathrm{d}\zeta^\mu A_{\mu a}(\zeta)\frac{\lambda_a}{2}\right). \tag{5.59}$$

5.5.3 QCD Lagrangian on a lattice

A lattice is a regular set of space–time points. A schematic representation of a lattice in two dimensions is given in figure 5.6. For our purposes, we define:

(i) site (node): the lattice points, characterized by the coordinate x, generally in the unit of the lattice spacing,

(ii) link: the shortest distance connecting two sites, characterized by coordinates and direction,

(iii) plaquette: an elementary square bounded by four links, characterized by coordinates and two directions.

In general, one also imposes periodic boundary conditions for bosons, $x_{n+1} = x_1$ and anti-periodic boundary condition for fermions, $x_{n+1} = -x_1$.

Lattice QCD simulations are computer intensive. The total number of degrees of freedom is very large on a lattice. The fermions are defined on the nodes (site), $\psi_\alpha^a(i)$, where $a(=1, 2, 3)$ is the color index and $\alpha(=1, 2, 3, 4)$ is the Dirac index. They are complex, requiring 24 real variables per node. One associates the gauge fields with the links, $U_{i\to j}^{ab}$, where i and j are the neighboring points and a,b are color indices. U is a unitary 3×3 matrix, a total of nine complex variables times four possible directions, i.e. 72 real variables per node for the link variables. In total in each node we have $(24N_f + 72)$ variables. For two-flavor QCD, even a small lattice 16^4 will deal with 7864320 real variables. Effectively, one has to compute a 7864320-fold integration.

The relation between the matrices U and the gauge field A_μ^α is the following,

$$U_\mu(x) = \exp\left(ig \int \hat{A}_\mu dx^\mu\right) \qquad \hat{A}_\mu = \sum_\alpha \frac{\lambda_\alpha}{2} A_\mu^\alpha. \tag{5.60}$$

$U_\mu(x)$ is the SU(3) matrix attached to the lattice link connecting the sites at x and $x + dx$, in the direction μ. The inverse of the matrix connects the sites in the opposite direction,

$$U_{-\mu}(x + dx) = U_\mu^{-1}(x) = U_\mu^\dagger(x). \tag{5.61}$$

In lattice QCD, one evaluates the partition function,

$$Z = \int d[U]d[\psi]d[\psi']e^{-S_E(U,\psi,\psi')} \tag{5.62}$$

where the action $S_E(U, \psi, \psi') = \int d^4x\mathcal{L}(\psi, \psi', A)$ and $d[\psi] = \prod_n d\psi_n$ represents all the possible paths.

Gauge invariance is explicitly maintained in lattice QCD. As mentioned earlier, quark fields are placed on the nodes and gauge fields are associated with the links. One then parallel transports the gauge fields from lattice site n to $n + 1$, maintaining gauge invariance. Gauge invariant objects are made from gauge links between quarks and anti-quarks or products of gauge fields in a closed loop.

5.5.4 Gauge fields on a lattice

In figure 5.7, the simplest closed loop of the gauge field is shown. It is called a plaquette, the product of four links connecting four adjacent nodes. It can be shown that continuum pure gauge action can be effectively discretized by the sum over all the plaquettes.

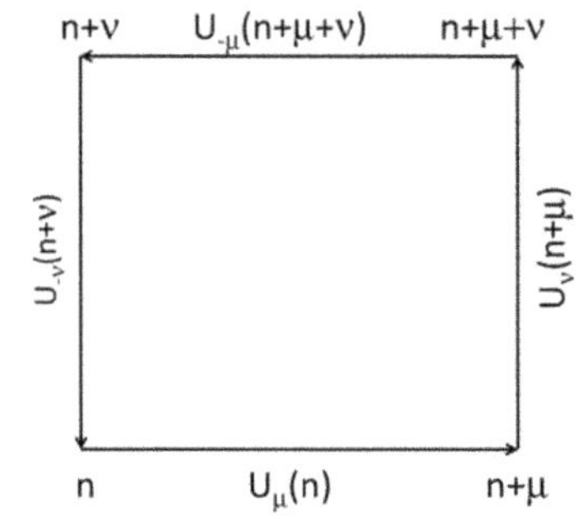

Figure 5.7. A plaquette on the lattice. The unit vectors μ, ν are two generic directions.

Let us write down the product of the link variables associated with the plaquette in figure 5.7,

$$P_{\square} = U_{-\nu}(n+\nu)U_{-\mu}(n+\mu+\nu)U_{\nu}(n+\mu)U_{\mu}(n), \tag{5.63}$$

and consider the terms separately. $U_\mu(n)$ connects the site n with its nearest neighbor (one lattice unit away) in the μ-direction. For small lattice spacing a, the trapezoidal rule, $\int_A^B dx f(x) = \frac{B-A}{2}[f(A)+f(B)]$ can be used to compute it (suppressing the 'hat' for clarity, $\hat{A} = \frac{\lambda_a}{2}A^a \equiv A$),

$$\begin{aligned} U_\mu(n) &= \exp\left[ig\int_n^{n+\mu} A_\mu(x)dx^\mu\right] = \exp\left[ig\frac{a}{2}\left(A_\mu(n) + A_\mu(n+\mu)\right)\right] \\ &= \exp\left[ig\frac{a}{2}\left(A_\mu(n) + \left\{A_\mu(n) + a\partial_\mu A_\mu(n)\right\}\right)\right] \\ &= \exp\left[ig\frac{a}{2}\left(2A_\mu(n) + a\partial_\mu A_\mu(n)\right)\right]. \end{aligned} \tag{5.64}$$

$U_\nu(n+\mu)$ links the sites $n+\mu$ and $n+\mu+\nu$ in the ν-direction. We compute it as

$$\begin{aligned} U_\nu(n+\mu) &= \exp\left[ig\int_{n+\mu}^{n+\mu+\nu} A_\nu(x)dx^\nu\right] = \exp\left[ig\frac{a}{2}\left(A_\nu(n+\mu) + A_\nu(n+\mu+\nu)\right)\right] \\ &= \exp\left[ig\frac{a}{2}\left(\left\{A_\nu(n) + a\partial_\mu A_\nu\right\} + \left\{A_\nu(n) + a\partial_\mu A_\nu(n) + a\partial_\nu A_\nu(n)\right\}\right)\right] \\ &= \exp\left[ig\frac{a}{2}\left(2A_\nu(n) + 2a\partial_\mu A_\nu(n) + a\partial_\nu A_\nu(n)\right)\right]. \end{aligned} \tag{5.65}$$

Link variables $U_{-\mu}(n+\mu+\nu)$ and $U_{-\nu}(n+\nu)$ can be computed similarly,

$$\begin{aligned} U_{-\mu}(n+\mu+\nu) &= \exp\left[-ig\int_{n+\mu+\nu}^{n+\nu} A_\mu(x)dx^\mu\right] \\ &= \exp\left[-ig\frac{a}{2}\left(2A_\mu(n) + a\partial_\mu A_\mu(n) + 2a\partial_\nu A_\mu(n)\right)\right] \end{aligned} \tag{5.66}$$

and

$$U_{-\nu}(n+\nu) = \exp\left[-\mathrm{i}g\int_{n+\nu}^{n} A_\nu(x)\mathrm{d}x^\nu\right]$$
$$= \exp\left[-\mathrm{i}g\frac{a}{2}\left(2A_\nu(n) + a\partial_\nu A_\nu(n)\right)\right]. \tag{5.67}$$

Using the Baker–Campbell–Hausdorff formula,

$$\mathrm{e}^A\mathrm{e}^B = \mathrm{e}^{A+B+\frac{1}{2}[A,B]+\dots}, \tag{5.68}$$

we evaluate the product of the operators $U_\nu(n+\mu)U_\mu(n)$ and $U_{-\nu}(n+\nu)$ $U_{-\mu}(n+\mu+\nu)$. Keeping the terms of the order of a^2 only,

$$U_\nu(n+\mu)U_\mu(n) = \exp\left[\mathrm{i}g\frac{a}{2}\{2A_\mu(n) + 2A_\nu(n) + a\partial_\mu A_\mu(n) + a\partial_\nu A_\nu(n)\right.$$
$$\left. + 2a\partial_\mu A_\nu(n)\} - \frac{1}{2}g^2a^2\left[A_\nu(n), A_\mu(n)\right]\right] \tag{5.69}$$

$$U_{-\nu}(n+\nu)U_{-\mu}(n+\mu+\nu) = \exp\left[-\mathrm{i}g\frac{a}{2}\{2A_\nu(n) + 2A_\mu(n) + a\partial_\mu A_\mu(n) + a\partial_\nu A_\nu(n)\right.$$
$$\left. + 2a\partial_\nu A_\mu(n)\} - \frac{1}{2}g^2a^2\left[A_\nu(n), A_\mu(n)\right]\right]. \tag{5.70}$$

The plaquette in equation (5.63) can be calculated as

$$P_\square = \exp\left[\mathrm{i}ga^2\left(\partial_\mu A_\nu(n) - \partial_\nu A_\mu(n)\right) - a^2g^2\left[A_\nu(n), A_\mu(n)\right]\right]$$
$$= \exp\left[\mathrm{i}ga^2\left(\partial_\mu A_\nu(n) - \partial_\nu A_\mu(n) - \mathrm{i}g\left[A_\mu(n), A_\nu(n)\right]\right)\right]$$
$$= \exp\left[\mathrm{i}ga^2F_{\mu\nu}\right] \approx 1 + \mathrm{i}ga^2F_{\mu\nu} - \frac{g^2a^4}{2}F_{\mu\nu}F^{\mu\nu}. \tag{5.71}$$

When we take trace of $P_\square$, the linear term $F_{\mu\nu} = \frac{\lambda_i}{2}F^i_{\mu\nu}$ vanishes as the trace of Gell-Mann matrices is zero. The trace of unity is a unit matrix and is irrelevant for the action. The trace of the quadratic term can be calculated as

$$Tr\left[F_{\mu\nu}F^{\mu\nu}\right] = Tr\left[\frac{\lambda_a}{2}\frac{\lambda_b}{2}F^a_{\mu\nu}F^{\mu\nu b}\right] = Tr\left[\frac{1}{2}\delta_{ab}F^a_{\mu\nu}F^{\mu\nu b}\right] = \frac{1}{2}F^a_{\mu\nu}F^{\mu\nu a}. \tag{5.72}$$

Comparing with pure gauge action in the continuum,

$$S = \int \mathrm{d}^4x\left[\frac{1}{4}F^a_{\mu\nu}F^{\mu\nu a}\right], \tag{5.73}$$

the simplest gauge action (called Wilson action) on the lattice can be written as

$$S_G(U) = \frac{6}{g^2}\sum_{\Box}\left[1 - \frac{1}{6}Tr\left(P_{\Box} + P_{\Box}^{\dagger}\right)\right] = \beta\sum_{\Box}\left[1 - \frac{1}{3}\mathrm{ReTr}P_{\Box}\right] \tag{5.74}$$

with $\beta = \frac{6}{g^2}$. The sum over $\Box$ in equation (5.74) indicates that the sum extends over all possible elementary plaquettes in a four-dimensional lattice. For a lattice of size $N_s^3 N_t$, there will be $6N_s^3 N_t$ elementary plaquettes.

5.5.5 Fermions on a lattice

Adding quarks to lattice action needs additional effort. Quark fields are defined on the nodes. Quarks are fermions and obey the Pauli exclusion principle. Thus they have to be included as anti-commuting Grassmann numbers. See appendix A.9 for some elementary information on Grassmann numbers.

In continuous Euclidean space–time, a fermion field $\psi(x)$ has the action,

$$S_F = \int \mathrm{d}^4x\overline{\psi}(x)\left(\gamma^\mu D^\mu + m\right)\psi(x) \tag{5.75}$$

with covariant derivative, $D^\mu = \partial^\mu - \mathrm{i}gA^\mu$. On lattice ψ and $\overline{\psi}$ are now Grassmann numbers and naively, the action can be discretized as

$$S_F = a^4\sum_n\left[\overline{\psi}(n)\gamma^\mu D^\mu\psi(n) - m\overline{\psi}(n)\psi(n)\right]. \tag{5.76}$$

Writing,

$$\partial^\mu\psi(n) = \frac{1}{2a}[\psi(n+\mu) - \psi(n-\mu)] \tag{5.77a}$$

$$\psi(n) = \frac{1}{2}[\psi(n+\mu) + \psi(n-\mu)], \tag{5.77b}$$

the first term of equation (5.76) can be written as

$$\begin{aligned}
\overline{\psi}(n)\gamma^\mu D^\mu\psi(n) &= \frac{1}{2}\overline{\psi}(n)\gamma^\mu\left[2\partial^\mu\psi(n) - 2\mathrm{i}gA^\mu\psi(n)\right] \\
&= \frac{1}{2}\overline{\psi}(n)\gamma^\mu\left[\frac{\psi(n+\mu) - \psi(n-\mu)}{a} - \mathrm{i}gaA^\mu\frac{\psi(n+\mu) + \psi(n-\mu)}{a}\right] \\
&= \frac{1}{2a}\overline{\psi}(n)\gamma^\mu\left[\left(1 - \mathrm{i}gaA^\mu\right)\psi(n+\mu) - \left(1 + \mathrm{i}gaA^\mu\right)\psi(n-\mu)\right] \\
&= \frac{1}{2a}\overline{\psi}(n)\gamma^\mu\left[U_{-\mu}(n+\mu)\psi(n+\mu) - U_\mu(n-\mu)\psi(n-\mu)\right].
\end{aligned} \tag{5.78}$$

In the last line, we have used

$$U_{-\mu}(n+\mu) \approx e^{-igaA_\mu(n)} \approx 1 - igaA_\mu(n) \tag{5.79a}$$

$$U_\mu(n-\mu) \approx e^{igaA_\mu(n)} \approx 1 + igaA_\mu(n). \tag{5.79b}$$

Rescaling, $m \to m/a$, $\psi \to \psi/a^{3/2}$, the fermionic action can be written as

$$\begin{aligned} S_F(\overline{\psi}, \psi, U) &= \frac{1}{2}\sum_n \overline{\psi}(n)\gamma^\mu\left[U_{-\mu}(n+\mu)\psi(n+\mu) - U_\mu(n-\mu)\psi(n-\mu)\right] \\ &\quad + m\sum_n \overline{\psi}(n)\psi(n) \\ &= \sum_{n,m} \overline{\psi}(n)M_{nm}(U)\psi(m). \end{aligned} \tag{5.80}$$

The Dirac operator M is a matrix with,

$$n_{\text{sites}} \times n_{\text{color}} \times n_{\text{Dirac}} = 12N_s^3 N_t \tag{5.81}$$

rows and columns. For free fermion ($U = 1$),

$$M_{nm} = \frac{1}{2}\gamma^\mu\left(\delta_{m,n+\mu} - \delta_{m,n-\mu}\right) + m\delta_{mn}. \tag{5.82}$$

The Sommer scale. In lattice gauge theory, action is written in terms of dimensionless quantities. One needs to fix one dimensionful quantity in order to set the overall scale. In the pure gauge theory, this is usually done through the string tension, defined as $K = \lim_{r\to\infty} F(r)$, where $F(r)$ is the force between two static quarks at distance r. The limiting procedure is not easy to do because the statistical and systematic errors on the force increase with the distance. R Sommer [48] introduced a hadronic scale r_0 through the force $F(r)$ between static quarks at intermediate distances r, $F(r_0)r_0^2 = 1.65$, which corresponds to $r_0 = 0.49$ fm in phenomenologically successful potentials, e.g. the Cornell potential and the Richardson potential. The scale $r_0 = 0.49$ fm is called the Sommer scale.

Lines of constant physics. While lattice simulations are done with finite lattice spacing a, physical results are obtained only in the continuum limit $a \to 0$. The parameters of the action are the fermion masses m_q and the coupling β. To take the continuum limit, these parameters have to be tuned such that on the one hand the lattice spacing a goes to zero while on the other hand a certain set of user defined dimensionful physical observables remain constant. These trajectories in the parameter space of β and m_q along which a set of physical observables remains constant as the limit $a \to 0$ are called lines of constant physics.

5.5.6 The fermion problem

Lattice QCD with dynamical quarks has the well known 'fermion doubling' problem: for each physical quark on the lattice one obtains $2^{d=4} = 16$ quarks. One can easily compute from equation (5.80) the inverse fermion propagator,

$$D_F^{-1}(p) = \mathrm{i}\sum \frac{\gamma_\mu \sin(p^\mu a)}{a} + m. \tag{5.83}$$

The momentum space of the lattice theory is a four-dimensional Brillouin zone $B = [-\pi/a, \pi/a]^4$ with anti-periodic boundary conditions. At each corner of the $2^4 = 16$ Brillouin zone, $\sin(p_\mu a) = 0$ and the propagator has 16 poles instead of the physical one. The origin of the doubling problem is deeply connected with chiral symmetry and we should just mention the Nielsen and Ninomiya [49] theorem that a local, real, free fermion lattice action, having chiral and translational invariance, necessarily has fermion doubling.

There are many ways to formulate fermion action on a lattice, e.g. Wilson fermions [50], staggered fermions [51–53], domain wall fermions [54, 55], overlap fermions [56, 57], etc. However, the current treatment of fermion action on a lattice is inadequate. Below, without much elaboration, we will briefly discuss two popular approaches, Wilson's fermion method and the staggered fermion method, to circumvent the fermion problem.

Wilson's fermion method. Wilson [50] introduced an additional term $\frac{r}{2}\partial_\mu\partial_\mu$ to the fermion action,

$$S^W = \int \mathrm{d}^4x \bar{\psi}\left(\gamma^\mu D_\mu + m + \frac{r}{2}\partial_\mu\partial^\mu\right)\psi. \tag{5.84}$$

r is called the Wilson parameter. The additional term can be interpreted as a momentum dependent mass term. In the continuum limit, the effect of the additional term is to makes the doubler modes infinitely heavy, effectively removing them from the spectrum. Evidently, Wilson fermions destroy the chiral symmetry.

Staggered fermion method. In the staggered fermion approach [51–53], chiral symmetry is preserved at the expense of partially reducing the fermion problem. One uses the naive discretization and diagonalizes the action with respect to the spinor degrees of freedom. The spinor degrees of freedom are then decoupled and three of the four degrees can be ignored. Staggered fermions then have a single one-component pair of Grassmann variables $\psi(n)$ and $\bar{\psi}(n)$ per lattice point n. This procedure effectively decreases the number of doublers from 16 to 4. The remaining four doublers are regarded as flavors and are called 'tastes'.

5.5.7 Partition function on a lattice

The functional integral for the partition function in lattice QCD can be written as

$$Z = \int [\mathrm{d}U][\mathrm{d}\overline{\psi}][\mathrm{d}\psi]\mathrm{e}^{-S_{\mathrm{G}}(U)-S_{\mathrm{F}}(U,\psi,\overline{\psi})} \tag{5.85}$$

with,

$$S_{\mathrm{G}}(U) = \beta\sum_{\square}\left[1 - \frac{1}{3}\mathrm{ReTr}P_{\square}\right] \tag{5.86}$$

$$S_{\mathrm{F}}(U, \overline{\psi}, \psi) = \sum_{n,m}\overline{\psi}(n)M_{nm}(U)\psi(m). \tag{5.87}$$

Computing numerically with Grassmann variables is non-trivial. Using integration rules for Grassmann numbers, one can integrate out the fermion fields,

$$\int [\mathrm{d}\overline{\psi}][\mathrm{d}\psi]\mathrm{e}^{-S_{\mathrm{F}}} = \int [\mathrm{d}\overline{\psi}][\mathrm{d}\psi]\mathrm{e}^{-\sum_{n,m}\overline{\psi}(n)M_{nm}\psi(m)} = \det[M(U)], \tag{5.88}$$

leaving only the gauge fields, weighted by the determinant of the Dirac matrix M,

$$Z = \int [\mathrm{d}U]\mathrm{e}^{-S_{\mathrm{G}}(U)}\det[M(U)]. \tag{5.89}$$

The partition function in equation (5.89) is a many-fold integration. One generally uses Monte-Carlo (MC) sampling to evaluate the partition function. One generates gauge field configurations with Boltzmann weight $\det M(U)\mathrm{e}^{-\beta S_{\mathrm{G}}(U)}$ and evaluates the expectation value of an operator as

$$\langle O\rangle = \frac{1}{N_{\mathrm{sample}}}\sum_{i=1}^{N_{\mathrm{sample}}} O(U_i). \tag{5.90}$$

A simple MC technique however is inadequate to integrate the partition function efficiently and one requires importance sampling. There are several algorithms for importance sampling. One such algorithm is by Metropolis.

The Metropolis algorithm. The Metropolis algorithm [58] is based on the principle of detail balance.

The algorithm proceeds as follows:

(i) start from arbitrary configuration (e.g. randomly distributed),

(ii) look at the value of the field (say ϕ) at any given point and change it: $\phi \to \phi'$,

(iii) calculate the variation in action: $\delta S = S(\phi') - S(\phi)$. If δS is negative, it is a lower energy state and desirable. One replaces the old value ϕ with the new value ϕ'. If δS is positive, one accepts the new value with the probability $\exp(-\delta S)$.

The procedure, after many iterations, will produce an equilibrium distribution. Any physically relevant observable can be computed from the equilibrium partition function,

$$\langle f(U)\rangle = \frac{\int d[U]e^{-S_{eq}}f(U)}{\int d[U]e^{-S_{eq}}}. \tag{5.91}$$

5.5.8 The Wilson loop

Consider a $q\bar{q}$ pair at a distance r. A schematic representation of the evolution of the pair is shown in figure 5.8(a). In quantum mechanics, the time evolution of the pair is governed by

$$\psi(t) = e^{-iE_{q\bar{q}}t}\psi(t = 0). \tag{5.92}$$

For confining quark potential ($V(r) \approx kr$), as kinetic energy goes as $1/m$, for infinitely heavy quarks, $E_{q\bar{q}} \approx kr$. In Euclidean space–time ($t \to -i\tau$), the time evolution of the pair is then governed by

$$e^{-iE_{q\bar{q}}t} \to e^{-kr\tau} = e^{-kA} \tag{5.93}$$

where $A = r \times \tau$ is the area spanned by the $q\bar{q}$ system during its evolution.

The Wilson loop is defined as the trace of the gauge fields along the world line. A typical Wilson loop is shown in figure 5.8(b). It is just the product of link variables along the contour

$$w(r, T) = trU_1U_2\ldots U_N. \tag{5.94}$$

In the continuum, the expectation value of the Wilson loop for large T and r is

$$\langle w(r, T)\rangle \sim \left\langle e^{-i\int_c dx^\mu A_\mu}\right\rangle \sim e^{-k\cdot(\text{Area})} \tag{5.95}$$

The area law is a manifestation of confinement.

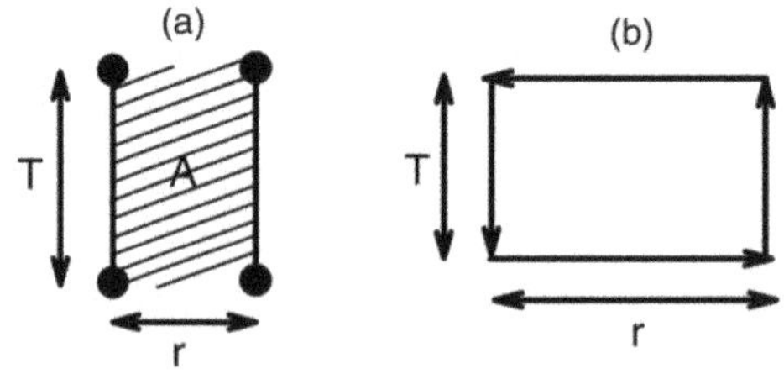

Figure 5.8. (a) Area spanned by the $q\bar{q}$ pair at a relative distance r, as a function of time; (b) a Wilson loop.

5.5.9 Lattice QCD at finite temperatures

QCD at finite temperatures can also be simulated on a lattice. One reformulates the partition function as

$$Z = Tr e^{-\beta H} = \int d\phi \exp\left[-\frac{1}{T}\int_0^{1/T} dt \int d^3x \mathcal{L}(\phi, \dot{\phi})\right]. \tag{5.96}$$

From the above equation, one realizes that finite temperature effects can be accounted for in a lattice which extends to infinity in the three space-like directions, but only to a finite extent in the temporal direction. Periodic boundary conditions in time must be imposed on the dynamical variables and the temperature is related to temporal size,

$$\text{Temperature} = \frac{1}{\text{Time}} = \frac{1}{N_t a}. \tag{5.97}$$

On a lattice, even the spatial extents are finite. Finite temperature QCD is then studied on an anisotropic lattice with

$$N_t \leqslant N_x = N_y = N_z. \tag{5.98}$$

5.5.10 The Polyakov loop

The central role in QCD at finite temperature is played by the trace of the product $U_{x,\mu}$ along a line parallel to the time axis (see figure 5.9). The trace is called the Polyakov loop.

Consider two Polyakov loops separated by the distance d. Gauge invariance is ensured by periodicity of boundary condition which allows us to 'close' the loops. The points denoted by $A(B)$ are physically the same points due to boundary conditions. The correlation of the two loops as a function of their separation d decreases as

$$C(d) \sim e^{-E_{q\bar{q}}(d)\cdot t} \sim e^{-E_{q\bar{q}}(d)/T} \tag{5.99}$$

where $E_{q\bar{q}}(d)$ is the potential energy of the quark pair. Now imagine that one separates the two loops more and more such that one of the loops goes out of the

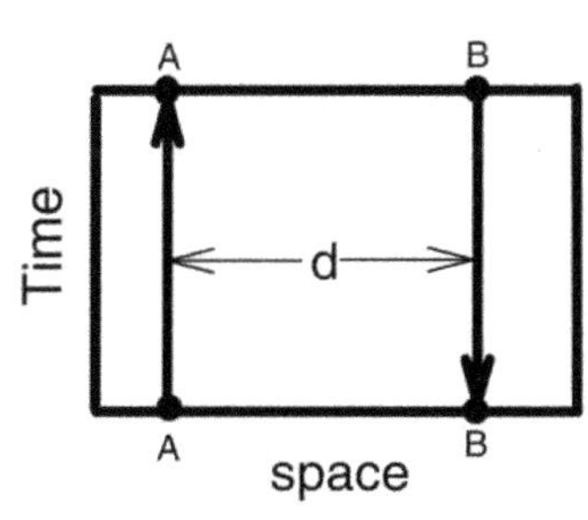

Figure 5.9. Schematic representation of two Polyakov loops separated by a distance d.

lattice volume. Then one measures $E_{q\bar{q}}(d \to \infty)$, i.e. the energy of a free quark. Therefore, the expectation value of 'one' Polyakov loop behaves as

$$\langle L \rangle \sim \mathrm{e}^{-E_{q\bar{q}}(d=\infty)/T}. \tag{5.100}$$

In a pure gauge (infinitely heavy quarks) theory the Polyakov loop can be identified as the order parameter of a confinement–deconfinement phase transition:

$$\begin{aligned} \text{confinement: } & E_{q\bar{q}}(d = \infty) = \infty \Rightarrow \langle L \rangle = 0 \\ \text{deconfinement: } & E_{q\bar{q}}(d = \infty) = \text{finite} \Rightarrow \langle L \rangle \neq 0. \end{aligned} \tag{5.101}$$

5.5.11 $Z(3)$ symmetry

Now whenever there is a phase transition, some internal symmetry is broken. What is the symmetry that is broken in confinement–deconfinement phase transition? Pure gauge QCD has a hidden, discrete symmetry called $Z(3)$ symmetry. To understand the symmetry, let us define:

$Z(G)$: The center of a group G is the set of elements that commutes with every element of G,

$$Z(G) = [Z \in G | zg = gz,\ g \in G]. \tag{5.102}$$

For SU(3), the center of group $Z(3)$ has elements, $(1, \mathrm{e}^{i2\pi/3}, \mathrm{e}^{i3\pi/3})$. One understands $Z(3)$ symmetry as the group of discrete rotation around the unit circle in the complex plane. The Euclidean action is invariant under these groups of rotation, but the Polyakov loop is not. The issue of confinement–deconfinement is then related to the breaking of $Z(3)$ symmetry. In the confined phase $\langle trL \rangle = 0$ and $Z(3)$ symmetry is preserved. In the deconfined phase, $\langle trL \rangle \neq 0$ and $Z(3)$ symmetry is broken.

With dynamical quarks, the interpretation of the Polyakov loop as an order parameter is more complex. Not the Polyakov loop, rather the correlator $\langle L(x) L^{\dagger}(x) \rangle$ is the order parameter and indicates confinement or deconfinement as $x \to \infty$. However, its absolute value can be related to the quark–antiquark free energy at infinite separation:

$$|\langle L \rangle|^2 = \exp\left[-\frac{F_{q\bar{q}}(r \to \infty)}{T}\right]. \tag{5.103}$$

The free energy has additive, quadratic divergences. The Polyakov loop is required to be renormalized to regulate ultraviolet divergences. They can be removed by subtracting the free energy at $T = 0$. This is the usual renormalization procedure for the free energy or pressure. The renormalized Polyakov loop can be calculated as

$$|\langle L_{\text{ren}} \rangle| = |\langle L \rangle| \mathrm{e}^{V(r)/2T} \tag{5.104}$$

where $V(r) = F_{q\bar{q}}(T = 0)$. The scale of $V(r)$ is fixed by the renormalizing condition,

$$V(r_0) = 0 \tag{5.105}$$

where $r_0 = 0.49$ fm is the Sommer scale.

5.5.12 Lattice QCD with finite chemical potential

Lattice QCD with quarks has the fermion problem. At finite chemical potential $\mu_q \neq 0$, lattice QCD faces a severe problem called the sign problem. To understand the sign problem, consider the partition function with finite chemical potential $\mu_q = \mu$,

$$\begin{aligned} Z &= \mathrm{Tr} e^{-\beta(H-\mu N)} \\ &= \int \mathrm{d}[U]\mathrm{d}[\bar{\psi}]\mathrm{d}[\psi] e^{-\int_0^\beta \mathrm{d}\tau \int \mathrm{d}^3x(\mathcal{L}+\mu n)}. \end{aligned} \tag{5.106}$$

The fermion action in equation (5.75) now includes an extra term,

$$\begin{aligned} S_\mathrm{F}(\mu) &= S_\mathrm{F}(\mu = 0) + \mu N \\ &= \int_0^\beta \mathrm{d}\tau \int \mathrm{d}^3x\bar{\psi}\left[\gamma^\mu\left(\partial^\mu - \mathrm{i}gA_\mu\right) + m + \mu\gamma^0\right]\psi \end{aligned} \tag{5.107}$$

where we have introduced the chemical potential as $\mu\bar{\psi}\gamma_0\psi$. For $\mu = \mu_\mathrm{R} + \mathrm{i}\mu_\mathrm{I}$, the chemical potential effectively changes the gauge potential,

$$A_0 \to A_0 - \mathrm{i}\mu = \left(A_0 + \mu_I\right) - \mathrm{i}\mu_\mathrm{R}. \tag{5.108}$$

If the quark fields are integrated out, the partition function can be obtained as

$$Z(\mu) = \int \mathrm{d}U \det M(\mu) e^{-\beta S_\mathrm{G}} \tag{5.109}$$

with the fermion determinant,

$$M(\mu) = \gamma_\nu D_\nu + m + \mu\gamma_0. \tag{5.110}$$

The finite chemical potential introduces some severe problems in lattice simulations. Adopting anti-Hermitian D and Hermitian γ, it is easy to show that

$$M(\mu)^\dagger = -\gamma_\nu D_\nu + m + \mu^*\gamma_0 = \gamma_5 M\left(-\mu^*\right)\gamma_5 \tag{5.111}$$

and then,

$$\left[\det M(\mu)\right]^* = \det M(\mu)^\dagger = \det M\left(-\mu^*\right). \tag{5.112}$$

Thus while for $\mu = 0$, the fermion determinant is real, at finite baryon density, the fermion determinant is complex. It will not be elaborated but the standard technique of MC importance sampling then fails. This is known as the 'fermion sign' problem. Several techniques have been suggested to circumvent the problem: (i) reweighting [64, 65], (ii) analytical continuation of imaginary chemical potential [66, 67] and (iii) Taylor expansion [68, 69]. Below, they are discussed briefly.

Reweighting. In the reweighting method [64, 65], the observable is reweighted with the oscillating part of the partition function and ensemble averaging is done with a smooth partition function. For example, for the partition function,

$$Z(\mu) = \int \mathrm{d}U \det M(\mu) \mathrm{e}^{-\beta S_G} \tag{5.113}$$

in the reweighting method, the expectation value of an observable is calculated as

$$\begin{aligned} \langle O \rangle &= \frac{1}{Z} \int \mathrm{d}U O \det M(\mu) \mathrm{e}^{-\beta S_G} \\ &= \frac{\int \mathrm{d}U O \frac{\det M(\mu)}{\det M(0)} \det M(0) \mathrm{e}^{-\beta S_G}}{\int \mathrm{d}U \frac{\det M(\mu)}{\det M(0)} \det M(0) \mathrm{e}^{-\beta S_G}} = \frac{\left\langle O \frac{\det M(\mu)}{\det M(0)} \right\rangle_{\mu=0}}{\left\langle \frac{\det M(\mu)}{\det M(0)} \right\rangle_{\mu=0}}. \end{aligned} \tag{5.114}$$

In the improved reweighting method, the gauge coupling is also used as a parameter for the reweighting.

Taylor's expansion. Taylor's expansion method [68, 69] is based on an expansion of the partition function or pressure in powers μ/T around the vanishing quark chemical potential,

$$\frac{P(T, \mu)}{T^4} = \sum_{n=0}^{\infty} c_{2n} \left(\frac{\mu}{T} \right)^{2n} \tag{5.115}$$

with

$$c_n = \frac{1}{n!} \frac{\partial^n}{\partial(\mu/T)^n} \left(\frac{p(T, \mu)}{T^4} \right) \Bigg|_{\mu=0}. \tag{5.116}$$

Due to reflection symmetry, $Z(\mu) = Z(-\mu)$, only even powers contribute in the expansion. The evaluation of higher order derivatives with controlled accuracy is difficult.

Imaginary chemical potential method. For complex chemical potential, $\mu = \mathrm{i}\mu_I$, the fermion determinant is real and the standard importance sampling method works. In the imaginary chemical potential method [66, 67], one computes the observables with imaginary chemical potential and analytically continue to real chemical potential

$\mu \to -i\mu$. For finite chemical potential, QCD has an extended Z_3 symmetry, which is realized as Roberge–Weise [70] periodicity of the partition function,

$$Z(\mu_I) = Z\left(\mu_I + \frac{2\pi T}{3}\right). \tag{5.117}$$

The imaginary chemical potential method is then limited to $\mu/T \leqslant 2\pi/3$.

5.6 Chiral phase transition

We have talked about QCD confinement–deconfinement phase transition. However, QCD has a well-known phase transition called 'chiral phase transition'. Chirality means 'handedness'. Handedness can be understood from the helicity concept. Let us define the helicity operator,

$$h = \mathbf{J} \cdot \hat{\mathbf{p}} = (\mathbf{L} + \mathbf{s}) \cdot \hat{\mathbf{p}} = \mathbf{s} \cdot \hat{\mathbf{p}}. \tag{5.118}$$

h is the projection of spin on the momentum direction. For spin half fermions, the helicity operator will have two eigenvalues, $+1/2$ and $-1/2$. A particle with helicity $+1/2$ ($-1/2$) is called the right (left) handed particle.

In figure 5.10, two particles with helicity $+1/2$ and $-1/2$ are shown. It is clear that for massive particles helicity is not a good quantum number. A massive particle will move with finite speed $v < c$ and one can go to a frame from where the particle will move backward and helicity will be reversed. However, massless particles move with speed c and helicity is a good quantum number for massless particles.

The concept of chirality is more abstract. Consider a Dirac field ψ for massless particles. The Lagrangian is

$$\mathcal{L} = i\bar{\psi}\gamma^\mu \partial_\mu \psi \tag{5.119}$$

with $\bar{\psi} = \psi^\dagger \gamma_0$.

Now, consider the following transformation,

$$\Lambda_V: \psi \to e^{-i\frac{\sigma}{2}\Theta}\psi \approx \left(1 - i\frac{\sigma}{2}\Theta\right)\psi. \tag{5.120}$$

σ is the Pauli matrices and Θ is the rotation angle. This is the general structure of a unitary transformation. The conjugate field transforms under Λ_V as

$$\Lambda_V: \bar{\psi} \to e^{+i\frac{\sigma}{2}\Theta}\bar{\psi} \approx \left(1 + i\frac{\sigma}{2}\Theta\right)\bar{\psi}. \tag{5.121}$$

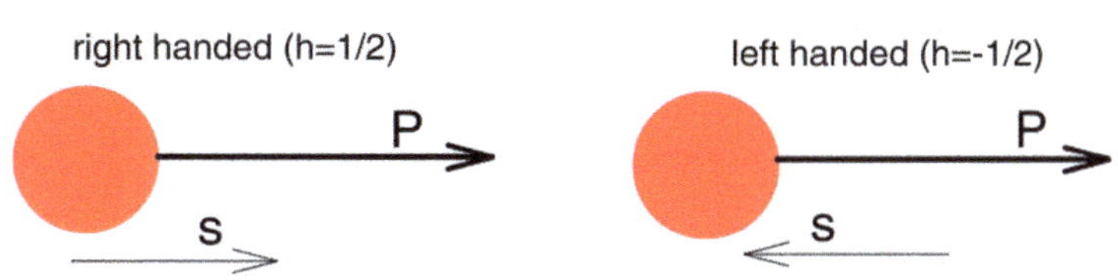

Figure 5.10. Two particles with helicity +1/2 and −1/2 are shown.

The Lagrangian is invariant under the transformation Λ_V.

$$i\bar{\psi}\partial\!\!\!/\psi \rightarrow i\bar{\psi}\partial\!\!\!/\psi - i\Theta\left(\bar{\psi}i\partial\!\!\!/\frac{\sigma}{2}\psi - \bar{\psi}i\partial\!\!\!/\frac{\sigma}{2}\psi\right) \tag{5.122}$$

$$= i\bar{\psi}\partial\!\!\!/\psi \tag{5.123}$$

One can say that the vector current $V_\mu^a = \bar{\psi}\gamma_\mu\frac{\sigma^a}{2}\psi$ is conserved.

Let us now consider the following transformation,

$$\Lambda_A\text{:}\quad \psi \rightarrow e^{-i\gamma_5\frac{\sigma}{2}\Theta}\psi \approx \left(1 - i\gamma_5\frac{\sigma}{2}\Theta\right)\psi \tag{5.124}$$

$$\bar{\psi} \rightarrow e^{-i\gamma_5\frac{\sigma}{2}\Theta}\bar{\psi} \approx \left(1 - i\gamma_5\frac{\sigma}{2}\Theta\right)\bar{\psi} \tag{5.125}$$

where the anti-commutation relation $\gamma_0\gamma_5 = -\gamma_5\gamma_0$ is used. The Lagrangian for the massless Dirac particle transforms as

$$i\bar{\psi}\partial\!\!\!/\psi \rightarrow i\bar{\psi}\partial\!\!\!/\psi - i\Theta\left(\bar{\psi}i\partial_\mu\gamma^\mu\frac{\sigma}{2}\psi - \bar{\psi}\gamma 5\frac{\sigma}{2}i\partial\gamma^\mu\psi\right)$$
$$= i\bar{\psi}\partial\!\!\!/\psi \tag{5.126}$$

and the second term vanishes due to the anti-commutation relation $\{\gamma_5, \gamma_\mu\} = 0$. The Lagrangian for the massless Dirac particle is also invariant under the transformation Λ_A, with conserved 'axial current', $A_\mu^a = \bar{\psi}\gamma_\mu\gamma_5\frac{\sigma}{2}\psi$.

Let us introduce the mass term in the free Dirac Lagrangian,

$$\delta\mathcal{L} = -m\bar{\psi}\psi \tag{5.127}$$

and see how it transforms under Λ_V and Λ_A.

$$\Lambda_V\text{:}\quad m\bar{\psi}\psi \rightarrow e^{+i\frac{\sigma}{2}\Theta}\bar{\psi}e^{-i\frac{\sigma}{2}\Theta}\psi = m\bar{\psi}\psi \tag{5.128}$$

$$\Lambda_A\text{:}\quad m\bar{\psi}\psi \rightarrow m\bar{\psi}\psi - 2im\Theta\left(\bar{\psi}\frac{\sigma}{2}\gamma_5\psi\right). \tag{5.129}$$

For massive Dirac particles only the vector current is conserved. Thus for massless fermions, the Dirac Lagrangian is invariant under the transformation, Λ_A and Λ_V, i.e. vector and axial vector currents are conserved. This symmetry is called chiral symmetry and its group structure is $SU(2)_V \times SU(2)_A$.

Chiral transition is signaled by the quark condensate $\langle\bar{\psi}\psi\rangle$. In a chiral symmetric phase, $\langle\bar{\psi}\psi\rangle = 0$. In the chiral symmetry broken phase $\langle\bar{\psi}\psi\rangle \neq 0$. In QCD, quark masses are small but non-zero. Chiral symmetry is broken and quark condensate $\langle\bar{\psi}\psi\rangle \neq 0$. However, at sufficiently high temperatures, quark mass decreases and condensate $\langle\bar{\psi}\psi\rangle \rightarrow 0$, and one says that chiral symmetry is restored. Some key properties of the chiral transition and confinement–deconfinement transition are listed in table 5.3.

Table 5.3. Some key properties of chiral and deconfinement transitions in QCD

	Chiral phase transition	Deconfinement phase transition
Quark mass	0	∞
Symmetry	chiral symmetry	center group symmetry
Order parameter	quark condensate	Polyakov loop

5.7 The nature of QCD phase transition

In lattice QCD, one calculates the partition function,

$$Z = \int [\mathrm{d}U] \mathrm{e}^{-\beta S_G(U)} \prod_q \det\left[M\left(U, m_q\right)\right] \tag{5.130}$$

where S_G is the gauge action, β is related to the gauge coupling, $\beta = 6/g^2$, M is the Dirac matrix and m_q is the quark mass for flavor q. Once the partition function is known, all the thermodynamic variables can be calculated using the thermodynamic relations. For example, in the thermodynamic limit,

$$\text{Pressure:} \; P = \lim_{V\to\infty} \frac{T}{V} \ln Z. \tag{5.131}$$

From pressure $P(T)$, one generally calculates the trace anomaly $I = \varepsilon - 3P$, which is just the derivative of the normalized pressure,

$$I = \varepsilon - 3P = T^5 \frac{\partial}{\partial T} \frac{P}{T^4}. \tag{5.132}$$

The energy density ε, the entropy density s and the speed of sound c_s then simply follow as

$$\varepsilon = I + 3P \qquad s = \frac{\varepsilon + P}{T} \qquad c_s^2 = \frac{\partial P}{\partial \varepsilon}. \tag{5.133}$$

Several groups worldwide, e.g. the MILC collaboration, HPQCD/UKQCD collaboration, USQCD collaboration, RBC collaboration, HotQCD collaboration and Wuppertal–Budapest collaboration, etc, have simulated QCD on a lattice. Most of the lattice simulations were performed for baryon-free ($\mu_B = 3\mu_q = 0$) QCD matter. In recent years there has been considerable progress in simulations with finite baryon density [71]. A few results of lattice simulations for baryon-free QCD matter are shown below.

In figure 5.11, Wuppertal–Budapest simulations for energy density (ε) and pressure (p), entropy density (s) and speed of sound (c_s), as a function of temperature, are shown. The simulations were for baryon-free ($\mu_q = 0$) $N_f = 2 + 1$ flavor QCD matter, with approximately physical quark mass. Details of the simulations can be found in [59–61]. The renormalized thermodynamic variables ε/T^4, pressure P/T^4 and entropy density s/T^3, rise sharply over a narrow temperature range

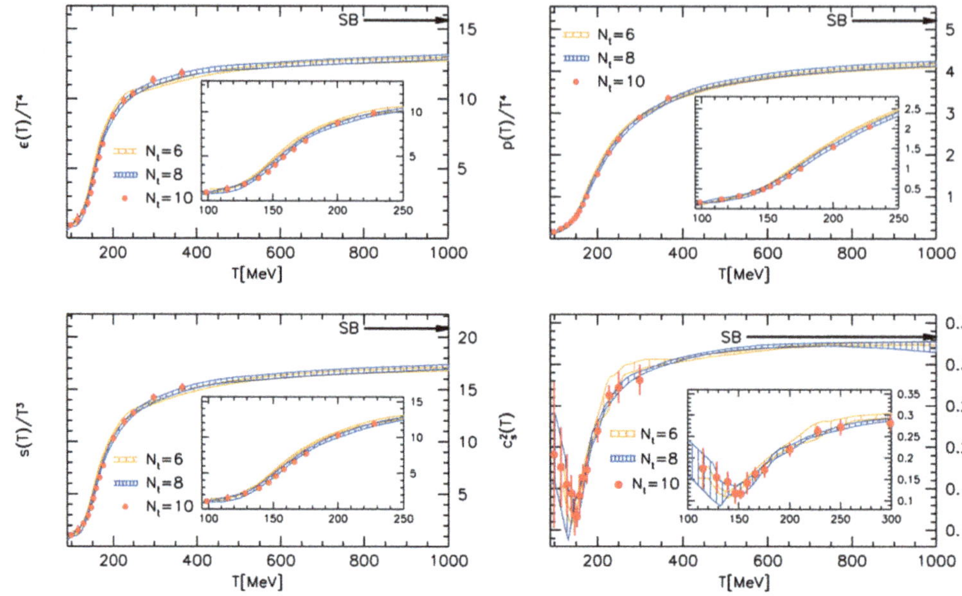

Figure 5.11. Wuppertal-Budapest simulations for temperature dependence of (i) energy density (top left panel), (ii) pressure (top right panel), (iii) entropy density (bottom left panel) and (iv) speed of sound (bottom right panel). The arrows indicate the Stefan–Boltzmann limit. The figures are reproduced with the authors' permission from [60].

150–200 MeV. At high temperatures, they saturate. ε/T^4, p/T^4 or s/T^3 are effectively proportional to the degeneracy of the medium. The temperature dependence of thermodynamic variables thus indicates that effective degrees of freedom rapidly change across the narrow temperature range $T = 150$–200 MeV. In figure 5.11, the Stefan–Boltzmann limits are indicated. Simulated ε/T^4, p/T^4 and s/T^3, though saturates, remain below the Stefan–Boltzmann limit. For example, at the highest temperature $T = 1000$ MeV, lattice simulated P is ~20% below the Stefan–Boltzmann limit. If we believe that at high temperature QCD matter exists as quark-gluon plasma (QGP), its constituents are not free, but they are interacting. This is the reason that QGP is called strongly interacting QGP. In figure 5.11 the variation of the square of the speed of sound (c_s^2) with temperature is also shown. At large temperature c_s^2 is consistent with the Stefan–Boltzmann value of $\frac{1}{3}$. As the temperature is lowered, the speed of sound decreases and shows a dip around temperature $T = 145(5)$ MeV. At still lower temperatures c_s^2 again increases. In a first order phase transition, the speed of sound is zero at the critical temperature. The dip of c_s^2 is then indicative of the phase transition.

Trace anomaly is an important observable in lattice QCD. Trace anomaly is identically zero for point particles, ($p = \frac{1}{3}\varepsilon$), and also in a conformal theory (in conformal field theory, action is invariant under a change of scale). Trace anomaly is also related to bulk viscosity. In figure 5.12, Wuppertal–Budapest simulations for

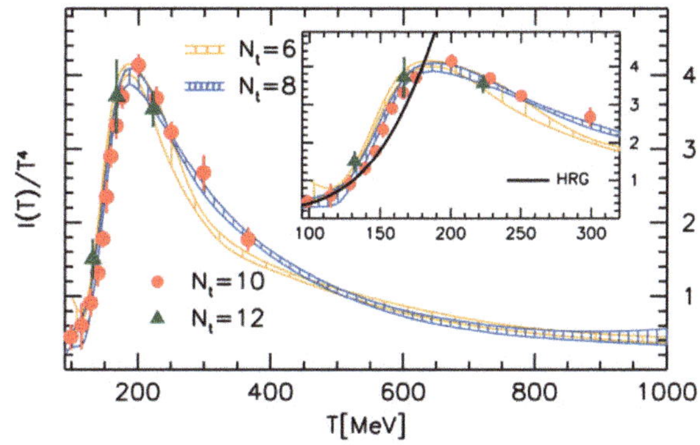

Figure 5.12. Wuppertal–Budapest simulation for trace anomaly $I = \varepsilon - 3P$ normalized by T^4 for various lattice spacings. In the inset lattice simulations are compared with the HRG model. The figure is reproduced with the authors' permission from [60].

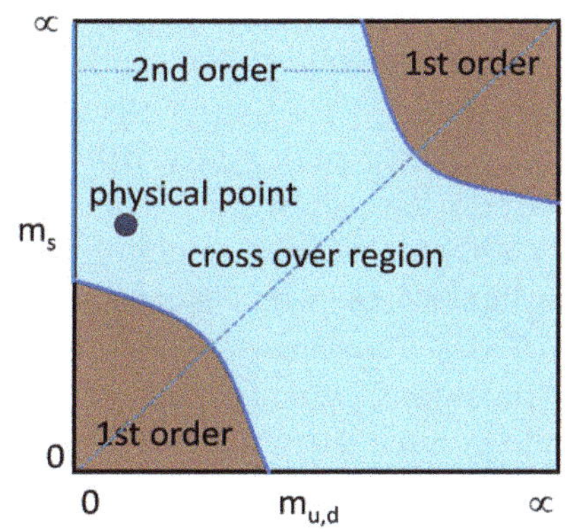

Figure 5.13. Current understanding of the nature of confinement–deconfinement phase transition as a function of quark mass m_u, m_d and m_s.

the trace anomaly are shown. Trace anomaly is non-zero, except at very high temperatures. The non-zero value of the trace anomaly indicates that QCD is not a conformal theory and QCD matter is not an ideal fluid. Wuppertal–Budapest lattice simulations for the trace anomaly have an inflexion point at temperature $T \approx 152(4)$ MeV. It turns out to be the cross-over temperature for confinement–deconfinement transition. In the inset, the trace anomaly of hadronic resonance gas (HRG) is compared with the simulation results. Below the inflexion point, the trace anomaly of HRG is close to lattice simulations.

In figure 5.13, current understanding [72] of the nature of the confinement–deconfinement phase transition, in baryon-free matter as a function of quark mass m_u, m_d and m_s, is shown. The plot is known as a Columbia plot. The results can be summarized as follows.

(i) In a pure gauge theory ($m_q \to \infty$), the transition is first order.

(ii) For $m_q \to 0$, the Lagrangian is chirally symmetric and there is a chiral symmetry restoration phase transition. It is also first order.

(iii) For $0 < m_q < \infty$, there is neither a confinement–deconfinement phase transition nor a chiral symmetry restoring phase transition. The system undergoes a cross-over transition. The order parameter, e.g. the Polyakov loop, or the susceptibility shows a sharp temperature dependence and it is possible to define a pseudo-critical cross-over transition temperature.

That the confinement–deconfinement transition is a cross-over was conclusively proved in [62]. There, finite size scaling of quark susceptibility

$$\chi = \frac{\partial^2}{\partial m_{ud}^2} \frac{T}{V} \log Z, \tag{5.134}$$

was studied. Quark susceptibility diverges at the critical point, which is manifested as a pronounced peak around the critical temperature. In finite volume, for real phase transitions, the singular behavior of susceptibility is reflected on the peak. For example, for a first-order phase transition the height is proportional to V and the width is proportional to $\frac{1}{V}$. For a second-order transition the singular behavior is given by some power of V. However, for analytic cross-over, there will be no singular behavior and the susceptibility peak and width will be volume independent. Explicit lattice simulations do not show any volume dependence, even when the volume is increased by a factor of four or eight. The cross-over nature of the transition is also shown in figure 5.14, where quark susceptibilities $T^4/(m^2\Delta\chi)$ are shown as a function of $1/(T_c^3 V)$. Expected asymptotic behavior for first-order and O(4) (second-order) phase transitions are shown by dotted and dashed lines. For true phase transitions the infinite volume extrapolation should be consistent with zero. Lattice simulations for continuum-extrapolated susceptibilities do not show phase-transition-like volume dependence. The $V \to \infty$ extrapolated value is 22(2) which is 11σ away from zero. The results confirm that for baryon-free matter, QCD confinement–deconfinement transition is not a true phase transition but a cross-over.

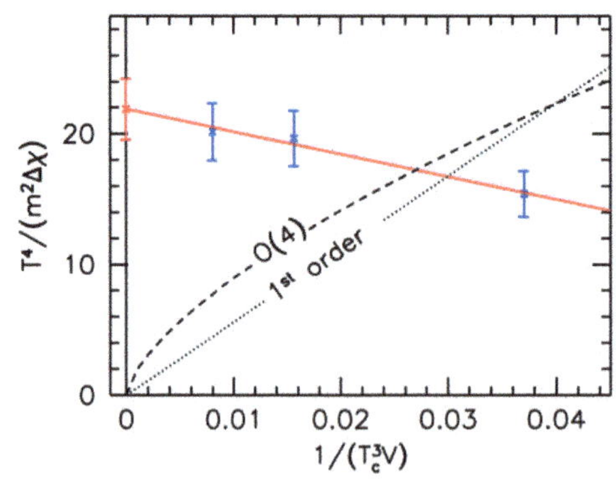

Figure 5.14. Lattice simulations for continuum extrapolated susceptibilities $T^4/(m^2\Delta\chi)$ as a function of $1/(T_c^3 V)$. The asymptotic behavior for first-order and O(4) (second-order) phase transitions are shown by dotted and dashed lines. The figure is reproduced with permission from [62].

5.8 The QCD phase diagram in the $\mu - T$ plane

The QCD phase diagram in the $\mu - T$ plane is of interest. Simulation studies suggest that the curvature parameter κ in the expansion,

$$\frac{T_c(\mu_B)}{T_c(\mu_B = 0)} = 1 - \kappa \left(\frac{\mu_B}{T_c(\mu_B = 0)} \right)^2 \tag{5.135}$$

is small [73]. As an example, in figure 5.15, the QCD phase diagram obtained in the analytical continuation method [64] (the filled circles) and in the Taylor expansion [74] (the filled squares) are shown. Both the methods give nearly identical phase diagrams for $\mu_B/T_c(\mu_B = 0) < 3\,\text{GeV}$, and the curvature parameter is small, $\kappa \approx 0.006$. At larger μ_B, they differ marginally.

From theoretical considerations, QCD phase transition is expected to be first order in baryon-dense matter. Since at $\mu_q \approx 0$ the deconfinement transition is a cross-over, one expects a QCD critical end point (CEP) where the first order transition line ends up at the cross-over. The location of the QCD CEP is of current interest. At the CEP, the first order transition becomes continuous, resulting in long range correlation and fluctuations at all length scales. Mathematically, it is true thermodynamic singularity.

The experimental signature of the QCD CEP is tricky. Since at the CEP, fluctuations exist at all length scales, one expects these fluctuations to percolate in the observables. Event-by-event fluctuations of baryon number or charge number can possibly signal a QCD CEP.

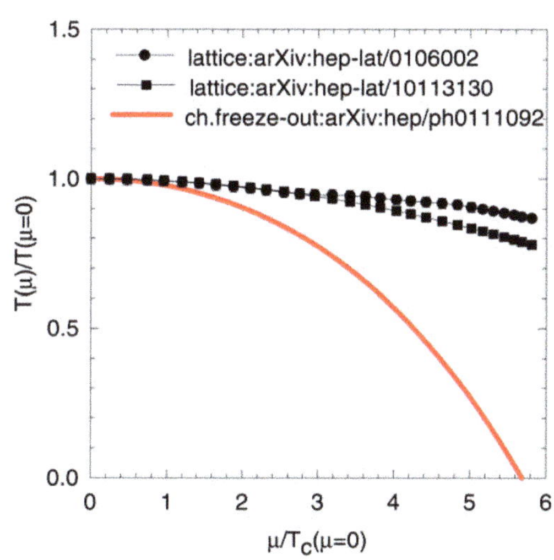

Figure 5.15. Lattice QCD calculations of the QCD phase diagram in the imaginary chemical potential method [64] and Taylor expansion method [74] are shown. The red line is the chemical freeze-out curve obtained in a statistical model [76].

In figure 5.15, the chemical freeze-out curve [75, 76], obtained in statistical model analysis of particle ratios, is shown (the red line). Curvature of the chemical freeze-out curve is a factor of four larger than the curvature in the QCD phase diagram. The small curvature of the freeze-out curve compared to the chemical freeze-out is interesting. The experimental signal of the CEP will get diluted as the deconfined medium produced at the CEP will evolve longer to reach chemical freeze-out. Fluid will have more time to wash out any signature of CEP.

Bibliography

[1] Gell-Mann M 1964 *Phys. Lett.* **8** 214
[2] Zweig G 1964 *CERN preprint* 8182/TH.401
[3] Greenberg O W 1964 *Phys. Rev. Lett.* **13** 598
[4] Han M Y and Nambu Y 1965 *Phys. Rev.* **139** 8100
[5] Yang C-N and Mills R L 1954 *Phys. Rev.* **96** 191
[6] 't Hooft G 1971 *Nucl. Phys.* B **33** 173
't Hooft G and Veltman M 1972 *Nucl. Phys.* B **44** 189
[7] Gross D J and Wilczek F 1973 *Phys. Rev. Lett.* **30** 1343
[8] Politzer H D 1973 *Phys. Rev. Lett.* **30** 1346
[9] Beringer J *et al* (Particle Data Group Collaboration) 2012 *Phys. Rev.* D **86** 010001
[10] Greiner W, Schramm S and Stein E 2007 *Quantum Chromodynamics* (Berlin: Springer)
[11] Muta T 2009 *Foundations of Quantum Chromodynamics: an Introduction to Perturbative Methods in Gauge Theories* (Singapore: World Scientific)
[12] Field R D 1989 *Applications of Perturbative QCD* (Reading, MA: Addison-Wesley)
[13] Lee T D 1981 *Particle Physics and Introduction to Field Theory* (New York: Harwood Academic)
[14] Sterman G 1993 *An Introduction to Quantum Field Theory* (Cambridge: Cambridge University Press)
[15] Weinberg S 1995 *The Quantum Theory of Fields* vol. 1–3 (Cambridge: Cambridge University Press)
[16] Peskin M E and Schroeder D V 1995 *An Introduction to Quantum Field Theory* (New York: Perseus)
[17] Quigg C 1997 *Gauge Theories of the Strong, Weak and Electromagnetic Interactions* (Reading, MA: Addison-Wesley)
[18] Kovchegov Yu V and Levin E 2012 *Quantum Chromodynamics at High Energy* (Cambridge: Cambridge University Press)
[19] Grozin A 2007 *Lectures on QED and QCD* (Singapore: World Scientific)
[20] Gribov L V, Levin E M and Ryskin M G 1983 *Phys. Rep.* **100** 1
[21] Brodsky S J, Pauli H-C and Pinsky S S 1998 *Phys. Rep.* **301** 299
[22] Collins J C, Soper D E and Sterman G F 1988 *Adv. Ser. Direct High Energy Phys.* **5** 1
[23] Marciano W J and Pagels H 1978 *Phys. Rep.* **36** 137
[24] Dokshitzer Y L, Diakonov D and Troian S I 1980 *Phys. Rep.* **58** 269
[25] Mueller A H 1981 *Phys. Rep.* **73** 237
[26] Reya E 1981 *Phys. Rep.* **69** 195
[27] Shuryak E V 1980 *Phys. Rep.* **61** 71
[28] Abers E S and Lee B W 1973 *Phys. Rep.* **9** 1

[29] Kronfeld A S and Quigg C 2010 *Am. J. Phys.* **78** 1081
[30] Gribov V N and Lipatov L N 1972 *Sov. J. Nucl. Phys.* **15** 438
Altarelli G and Parisi G 1977 *Nucl. Phys.* B **126** 298
Dokshitzer Yu L 1977 *Sov. Phys. JETP* **46** 641
[31] Soper D E 1997 *Nucl. Phys. Proc. Suppl.* **53** 69
[32] Pumplin J, Stump D R, Huston J, Lai H L, Nadolsky P M and Tung W K 2002 *JHEP* **0207** 012
[33] Jimenez-Delgado P and Reya E 2009 *Phys. Rev.* D **79** 074023
[34] Martin A D, Stirling W J, Thorne R S and Watt G 2007 *Phys. Lett.* B **652** 292
[35] Brock R *et al* (CTEQ Collaboration) 1995 *Rev. Mod. Phys.* **67** 157
[36] Kniehl B A, Kramer G and Potter B 2000 *Nucl. Phys.* B **582** 514
[37] Greiner W, Schramm S and Stein E 2007 *Quantum Chromodynamics* (Berlin: Springer)
[38] Creutz M 1983 *Quarks, Gluons and Lattices* (Cambridge: Cambrigde University Press)
[39] Rothe H 2005 *Lattice Gauge Theories—An Introduction* (Singapore: World Scientific)
[40] Smit J 2002 *Introduction to Quantum Fields on a Lattice* (*Cambridge Lecture Notes in Physics* vol 15) (Cambridge: Cambrigde University Press)
[41] Kogut J B 1979 *Rev. Mod. Phys.* **51** 659
[42] Kogut J B 1983 A review of the lattice gauge theory approach to quantum chromodynamics *Rev. Mod. Phys.* **55** 775
[43] Creutz M, Jacobs L and Rebbi C 1983 *Phys. Rep.* **95** 201
[44] Davies C 2002 arXiv: hep-ph/0205181
[45] Muroya S, Nakamura A, Nonaka C and Takaishi T 2003 *Prog. Theor. Phys.* **110** 615
[46] Karsch F 2002 *Lattice QCD at High Temperature and Density* (*Lecture Notes in Physics* vol 583) (Berlin: Springer) p 209
[47] Lepage G P 2005 arXiv: hep-lat/0506036
[48] Sommer R 1994 *Nucl. Phys.* B **411** 839
[49] Nielsen H B and Ninomiya M 1981 *Nucl. Phys.* B **185** 20
Nielsen H B and Ninomiya M 1981 *Phys. Lett.* B **105** 219
[50] Wilson K G 1974 *Phys. Rev.* D **10** 2445
[51] Kogut J B and Susskind L 1975 *Phys. Rev.* D **11** 395
[52] Susskind L 1977 *Phys. Rev.* D **16** 3031
[53] Banks T, Susskind L and Kogut J B 1976 *Phys. Rev.* D **13** 1043
[54] Kaplan D B 1992 *Phys. Lett.* B **288** 342
[55] Shamir Y 1993 *Nucl. Phys.* B **406** 90
[56] Narayanan R and Neuberger H 1993 *Phys. Rev. Lett.* **71** 3251
[57] Narayanan R and Neuberger H 1994 *Nucl. Phys.* B **412** 574
[58] Metropolis N, Rosenbluth A W, Rosenbluth M N, Teller A H and Teller E 1953 *J. Chem. Phys.* **21** 1087
[59] Borsanyi S *et al* (Wuppertal–Budapest Collaboration) 2011 *Acta Phys. Polon. Suppl.* **4** 593
[60] Borsanyi S, Endrodi G, Fodor Z, Jakovac A, Katz S D, Krieg S, Ratti C and Szabo K K 2010 *JHEP* **1011** 077
[61] Borsanyi S, Fodor Z, Hoelbling C, Katz S D, Krieg S, Ratti C and Szabo K K 2010 arXiv: 1011.4230 [hep-lat]
[62] Aoki Y, Endrodi G, Fodor Z, Katz S D and Szabo K K 2006 *Nature* **443** 675
[63] Cheng M *et al* 2008 *Phys. Rev.* D **77** 014511
[64] Fodor Z and Katz S D 2002 *JHEP* **0203** 014

[65] Fodor Z and Katz S D 2004 *JHEP* **0404** 050
[66] de Forcrand P and Philipsen O 2002 *Nucl. Phys.* B **642** 290
[67] D'Elia M and Lombardo M-P 2003 *Phys. Rev.* D **67** 014505
[68] Allton C R, Ejiri S, Hands S J, Kaczmarek O, Karsch F, Laermann E and Schmidt C 2003 *Phys. Rev.* D **68** 014507
[69] Gavai R V and Gupta S 2003 *Phys. Rev.* D **68** 034506
[70] Roberge A and Weiss N 1986 *Nucl. Phys.* B **275** 734
[71] Borsanyi S, Endrodi G, Fodor Z, Katz S D, Krieg S, Ratti C and Szabo K K 2012 *JHEP* **1208** 053
[72] Laermann E and Philipsen O 2003 *Annu. Rev. Nucl. Part. Sci.* **53** 163
[73] Philipsen O 2008 *Prog. Theor. Phys. Suppl.* **174** 206
[74] Kaczmarek O *et al* 2011 *Phys. Rev.* D **83** 014504
[75] Becattini F, Manninen J and Gazdzicki M 2006 *Phys. Rev.* C **73** 044905
[76] Cleymans J, Oeschler H, Redlich K and Wheaton S 2006 *J. Phys. G: Nucl. Part. Phys.* **32** S165

IOP Publishing

A Short Course on Relativistic Heavy Ion Collisions

A K Chaudhuri

Chapter 6

Equation of state for QGP and hadronic resonance gas

6.1 Introduction

From theoretical considerations, it is expected that in ultra-relativistic heavy ion collisions, an extended, deconfined medium of quarks and gluons, which is commonly termed quark–gluon plasma (QGP), will be produced. For theoretical treatment, the extended medium is considered as a macroscopic system. A macroscopic system is characterized by a few macroscopic or thermodynamic variables. The relation between the thermodynamic variables is called the equation of state for the system. For example, if pressure (P), volume (V) and temperature (T) are used to characterize the system, the equation

$$f(P,V,T) = 0 \tag{6.1}$$

is the equation of state. An equation of state effectively reduces the number of independent thermodynamic variables. In the preceding chapter, we have shown that if the partition function or the thermodynamic potential for the macroscopic system is known, all the equilibrium properties of the system can be calculated thereof. In other words, knowledge of the thermodynamic potential or the partition function is equivalent to knowledge of the equation of state.

One of the fundamental interests in high-energy nuclear physics is understanding the equation of state of QGP, the quantum chromodynamic (QCD) matter at high temperature or density. As discussed earlier, due to the non-abelian nature of QCD theory, first principles calculation of QCD equations of state is very complex. For finite baryon systems, only limited progress has been made in lattice QCD. An approximate equation of state of QGP can be obtained if the simplifying assumption is made that the quarks and gluons in QGP are interaction-free. In the following, we write down the equation of state for interaction-free QGP. At high temperatures,

doi:10.1088/978-0-750-31060-4ch6

QCD coupling is weak and to a good approximation, quarks and gluons can be treated as interaction-free particles. However, lattice QCD calculations do indicate that even at very high temperatures, QCD equations of state indicate departure from an ideal, interaction-free nature.

QGP is a transient state. Even if it is formed in heavy ion collisions, ultimately, it will convert into a system of hadrons, or more accurately, into a hadronic resonance gas (HRG). HRG is a mixture of non-interacting, all possible hadrons and hadron resonances. Indeed, QGP and HRG can be considered as two phases of QCD matter: QGP, the high temperature or high density phase; and HRG, the low temperature and low density phase. In the following, explicit relations for the equation of state for QGP matter and HRG will be presented.

6.2 Equation of state for QGP with vanishing quark mass

The thermodynamic potential or the partition function for a system of interaction-free quarks and gluons can be written as

$$\Omega_{\text{QGP}} = -T \ln Z = \sum_{i=\text{quark,gluon}} \Omega_i. \tag{6.2}$$

Gluons are massless bosons and equations (4.106), derived for massless bosons, with appropriate degeneracy factors, are applicable. For gluons, there are eight colors and two helicity states and the degeneracy factor is

$$g = g_{\text{gluon}} = \text{color} \times \text{spin} = 8 \times 2 = 16. \tag{6.3}$$

Quarks come in several flavors. If the masses of N_f flavors are assumed to be same, the degeneracy factor for quarks can be obtained as

$$\begin{aligned} g_{\text{quark}} &= (\text{particle} + \text{antiparticle}) \times \text{spin} \times \text{color} \times \text{flavor} \\ &= 2 \times 2 \times 3 \times N_f = 12 \times N_f. \end{aligned} \tag{6.4}$$

In table 5.1, we have summarized the properties of quarks. Current quark mass is the relevant mass here, and enters into the QCD Lagrangian. Constituent quark masses are used in modeling hadrons. In a sense they are dressed current quarks. In RHIC or LHC energy collisions, QGP is expected to be produced at a temperature of $T \sim 300$–500 MeV. As charm, bottom and top quarks are quite heavy (see table 5.1), their contributions to the QGP equation of state can be neglected. The QGP equation of state will be dominated by the up and down quarks. The strange quarks, which are not as heavy as charm or bottom and not as light as up and down, will also contribute. Equations derived in the preceding chapter for fermions can be used directly. Equations of state become simpler in the massless limit. Up and down quarks are approximately of similar mass and are light. For practical purposes, their mass can be neglected and equations for massless fermions can be applied with the degeneracy factor $g_{\text{quark}} = 12\, N_f, \times N_f = 2$. To account for the non-negligible mass of strange quarks, yet using the equations for the massless fermion, in several applications, the number of quark flavors is taken as $N_f = 2.5$.

For completeness, for baryonic chemical potential $\mu_B = 3\mu_q$, where μ_q is the quark chemical potential, equations of state for a system of non-interacting massless gluons, quarks and anti-quarks are given below.

$$\text{Net baryon density: } n = g_{\text{quark}} \frac{1}{3}\left[\frac{\mu_B T^2}{36} + \frac{\mu_B^3}{324\pi^2}\right] \tag{6.5a}$$

$$\text{Pressure: } P = \left(g_{\text{gluon}} + \frac{7}{8} g_{\text{quark}}\right)\frac{\pi^2}{90} T^4 + g_{\text{quark}}\left(\frac{\mu_B^2 T^2}{216} + \frac{\mu_B^4}{324\pi^2}\right) \tag{6.5b}$$

$$\text{Energy density: } \varepsilon = 3P \tag{6.5c}$$

$$\text{Entropy density: } s = \left(g_{\text{gluon}} + \frac{7}{8} g_{\text{quark}}\right)\frac{2\pi^2}{45} T^3 + g_{\text{quark}} \frac{\mu_B^2 T}{108}. \tag{6.5d}$$

Noting that the thermodynamic quantities of the bosonic and fermionic system differ by a factor of only $\frac{7}{8}$ it is convenient to introduce an effective degeneracy factor for a system of baryon-free quarks and gluons,

$$g_{\text{QGP}} = g_{\text{gluon}} + \frac{7}{8} g_{\text{quark}} = 16 + \frac{21}{2} N_f. \tag{6.6}$$

At RHIC and LHC energies, the central rapidity region is essentially baryon-free. Equations of state for a system of baryon-free QGP, in terms of the effective degeneracy factor, look simple,

$$\text{Net baryon density: } n = 0 \tag{6.7a}$$

$$\text{Pressure: } P = g_{\text{QGP}} \frac{\pi^2}{90} T^4 \tag{6.7b}$$

$$\text{Energy density: } \varepsilon = g_{\text{QGP}} \frac{\pi^2}{30} T^4 \tag{6.7c}$$

$$\text{Entropy density: } s = g_{\text{QGP}} \frac{2\pi^2}{45} T^3. \tag{6.7d}$$

While quarks and gluons in QGP are in a deconfined state, they must satisfy the asymptotic properties of QCD, that free quarks and gluons are not observed in isolation. The deconfined quarks and gluons of QGP must be restricted to a certain spatial volume to prevent them from escaping to infinity. Borrowing the idea of the MIT bag model, the quarks and gluons of QGP can be confined within a certain spatial region by introducing bag pressure, which acts inward and balances the outward pressure of the quarks and gluons. The resulting equation of state is called the bag model equation of state and was used extensively in analysis of relativistic heavy ion collision. To appreciate the bag model equation of state, we briefly discuss the MIT bag model.

6.2.1 The MIT bag model

The experimental fact that fractional charges or quarks are not observed in isolation led to the confinement property of QCD that quarks can exist only as a bound state or inside a hadron. As the confinement property was not properly understood (a statement that remains true to-date), it was realized very early that some model was required to confine the quarks inside the hadron. Bogoliubov proposed a simple model whereby quarks of mass m are placed inside a spherical volume within which they feel a scalar field, also of strength m. The model confines the quarks by making them infinitely heavy, $m \to \infty$. In 1974, a group of physicists at MIT developed a model for hadron structure, which is a much improved version of the Bogoliubov model. The model has become very popular and is known as the MIT bag model [12, 13]. In the model, the quarks are forced by a fixed external pressure to move only inside a fixed spatial region called a bag. Inside the bag, they are quasi-free. Appropriate boundary conditions are imposed such that no quark can leave the bag. The MIT bag model predicts fairly accurate hadron masses. Color confinement is built in to the model. Now confinement is essentially a non-perturbative phenomenon. One can interpret the bag pressure in the MIT bag model as the essential effect of non-perturbative QCD. The quantum stress from the quarks inside the bag is counterbalanced by the inward bag pressure. We may note here that chiral symmetry is explicitly broken at the bag surface. A remedy was suggested in the cloudy bag model [14]. In the following, we briefly discuss the bag model.

In the MIT bag model, the Lagrangian density for a massless quark, inside a three-dimensional volume V with a surface S, is written as

$$\mathcal{L} = \left[\frac{i}{2}\left(\bar{\psi}\gamma^{\mu}\partial_{\mu}\psi - \partial_{\mu}\bar{\psi}\gamma^{\mu}\psi\right) - B\right]\theta_V - \frac{1}{2}\bar{\psi}\psi\delta_S \tag{6.8}$$

where θ_V is a step function,

$$\theta_V(x) = 1 \quad \text{inside the bag} \tag{6.9a}$$

$$= 0 \quad \text{outside the bag.} \tag{6.9b}$$

Furthermore,

$$\frac{\partial\theta_V}{\partial x^{\mu}} = n_{\mu}\delta_S \tag{6.10}$$

where n_μ is the interior unit vector normal to the surface of the volume V and δ_S is the surface delta function. For a static, spherical case, e.g. the quark is confined in a spherical cavity of radius R,

$$\theta_V = \theta(R - r) \qquad \delta_S = \delta(R - r). \tag{6.11}$$

The equation of motion can be obtained from the Euler–Lagrangian equation,

$$\frac{\partial \mathcal{L}}{\partial \overline{\psi}} - \partial_\mu \frac{\partial \mathcal{L}}{\partial\left(\partial_\mu \overline{\psi}\right)} = 0. \tag{6.12}$$

For the Lagrangian in equation (6.8),

$$\frac{\partial \mathcal{L}}{\partial \overline{\psi}} = \frac{\mathrm{i}}{2}\gamma^\mu \partial_\mu \psi \theta_V - \frac{1}{2}\psi \delta_S \tag{6.13a}$$

$$\partial_\mu \frac{\partial \mathcal{L}}{\partial\left(\partial_\mu \overline{\psi}\right)} = -\frac{\mathrm{i}}{2}\gamma^\mu \partial_\mu \psi \theta_V - \frac{\mathrm{i}}{2}\gamma^\mu n_\mu \psi. \tag{6.13b}$$

Inserting equations (6.13*a*)–(6.13*b*) into the equation of motion, and separately equating θ_V and δ_S, we obtain the boundary conditions,

$$\text{inside the bag:} \quad \mathrm{i}\gamma^\mu \partial_\mu \psi = 0 \tag{6.14a}$$

$$\text{on the surface of the bag:} \quad \mathrm{i}\gamma^\mu n_\mu \psi = \psi. \tag{6.14b}$$

Taking the Hermitian conjugate of equation (6.14*b*) and multiplying by γ^0 from the right, the boundary condition on the surface can also be written as

$$\begin{aligned} -\mathrm{i}\gamma^\mu n_\mu\left(\psi^\dagger \gamma^0\right) &= \left(\psi^\dagger \gamma^0\right) \\ -\mathrm{i}\gamma^\mu n_\mu \overline{\psi} &= \overline{\psi}. \end{aligned} \tag{6.15}$$

Equation (6.14*b*) can be interpreted as the condition that the quark current $j^\mu = \overline{\psi}\gamma^\mu\psi$ normal to bag surface is zero.

$$\begin{aligned} \mathrm{i}n_\mu j^\mu &= \overline{\psi}\left(\mathrm{i}\gamma^\mu n_\mu \psi\right) = \overline{\psi}\psi \\ &= \left(\mathrm{i}\gamma^\mu n_\mu \overline{\psi}\right)\psi = -\overline{\psi}\psi \\ &= 0. \end{aligned} \tag{6.16}$$

Equation (6.14*a*) is the free Dirac equation for a massless quark. Inside the bag, the quarks obey the free Dirac equation. The Dirac equation for massless quarks in a spherical cavity can be solved easily, see appendix A.8. For a massless quark in a spherical cavity of radius R, energy eigenvalues satisfy the condition

$$j_0(ER) = j_1(ER). \tag{6.17}$$

From the tabulated values of the spherical Bessel functions, the ground state energy can be obtained as

$$E = \frac{2.04}{R}. \tag{6.18}$$

To obtain the implication of the bag term B, we consider the energy–momentum tensor for the bag model Lagrangian,

$$\begin{aligned} T^{\mu\nu} &= -g^{\mu\nu}\mathcal{L} + \frac{\partial\mathcal{L}}{\partial(\partial_\mu\psi)}\partial^\nu\psi + \frac{\partial\mathcal{L}}{\partial(\partial_\mu\overline{\psi})}\partial^\nu\overline{\psi} \\ &= -g^{\mu\nu}\mathcal{L} + \frac{\mathrm{i}}{2}(\overline{\psi}\gamma^\mu\partial^\nu\psi - \gamma^\mu\psi\partial^\nu\overline{\psi})\theta_V \\ &= -g^{\mu\nu}\left(-B\theta_V - \frac{1}{2}\overline{\psi}\psi\delta_S\right) + \frac{\mathrm{i}}{2}(\overline{\psi}\gamma^\mu\partial^\nu\psi - \gamma^\mu\psi\partial^\nu\overline{\psi})\theta_V. \end{aligned} \tag{6.19}$$

Conservation condition $\partial_\mu T^{\mu\nu} = 0$ gives

$$Bn^\nu\delta_S + \frac{\mathrm{i}}{2}(\overline{\psi}\gamma^\mu\partial^\nu\psi - \gamma^\mu\psi\partial^\nu\overline{\psi})n_\mu\delta_S = 0 \tag{6.20a}$$

$$\frac{1}{2}\partial^\nu(\overline{\psi}\psi\delta_S) = 0. \tag{6.20b}$$

Surface boundary condition equations (6.14*b*) and (6.15) can be used in equation (6.20*a*) to obtain

$$Bn^\nu = \frac{1}{2}\partial^\nu(\psi\overline{\psi}). \tag{6.21}$$

We have already found that $\psi\overline{\psi} = 0$ on the surface (see equation (6.16)). The derivative of $\psi\overline{\psi}$ must lie along the normal,

$$Bn^\nu = \frac{1}{2}\partial^\nu(\psi\overline{\psi}) = 2n^\nu P_{\mathrm{Dirac}}. \tag{6.22}$$

P_{Dirac} is recognized as the pressure on the surface, in the instantaneous rest frame of the surface, the momentum flow normal to the surface. Equation (6.22) is nothing but a pressure balance equation; the pressure exerted by the quark is balanced by the quantity B, the bag pressure. Since $n_\nu n^\nu = -1$ (n^ν being space-like),

$$B = -\frac{1}{2}n_\nu\partial^\nu(\psi\overline{\psi}). \tag{6.23}$$

6.2.2 Bag model equation of state

As in the bag model, in high temperature QGP quarks are approximately free and even though it is a deconfined medium, it is confined in a limited region (albeit that the confinement region is of nuclear size rather than of hadronic size). To keep the deconfined quarks and gluons within a certain spatial region, one introduces bag pressure, which acts inward and balances the outward pressure exerted by the quarks and gluons. The thermodynamic potential or the partition function then modifies to

$$\Omega_{\mathrm{QGP}} = \sum_{i=\mathrm{quark,gluon}} \Omega_i + \Omega_{\mathrm{bag}} \tag{6.24}$$

with $\Omega_{bag} = BV$, B being the 'external bag pressure'. Note that the introduction of Ω_{bag} does not change the particle numbers or entropy density. Only the expressions for pressure and energy density need to be augmented by the term $-B$ and $+B$, respectively. For completeness, the bag model equations of state for interaction-free massless quarks and gluons are written below.

$$\text{Energy density: } \varepsilon = g_{QGP}\frac{\pi^2}{30}T^4 + B \tag{6.25a}$$

$$\text{Pressure: } p = g_{QGP}\frac{\pi^2}{90}T^4 - B \tag{6.25b}$$

$$\text{Entropy density: } s = g_{QGP}\frac{4\pi^2}{90}T^3 \tag{6.25c}$$

$$\text{Number density: } n \approx g_{QGP}\frac{3}{4\pi^2}T^3. \tag{6.25d}$$

Equations (6.25*a*)–(6.25*d*) constitute the bag model equation of state for QGP. We note that the bag pressure is still undefined. In the MIT bag model for hadrons, bag pressure $B^{1/4} \sim 200$ MeV. However, in the QGP equation of state, bag pressure is obtained by the consideration that QGP is a transient state and below a critical or pseudo-critical temperature T_c, QGP transforms into hadronic matter or HRG. If the transformation is a first order phase transition, the bag constant is obtained by demanding that at the transition temperature T_c, the pressures of the two phases are equal,

$$p_{QGP}(T_c) = p_{HRG}(T_c). \tag{6.26}$$

It was discussed earlier that explicit simulations of QCD on a lattice indicate that for baryon-free ($\mu_B = 0$) matter, the transformation of QGP to HRG is not a phase transition in the thermodynamic sense, but rather is a smooth cross-over. In that case, thermodynamic variables in two phases can be joined smoothly to obtain the bag pressure.

6.3 Hadronic resonance gas

QGP is a transient state. If formed in heavy ion collisions, it will cool back to hadronic matter at low temperature. It is natural to ask what the composition or constituents of the hadronic matter are at low temperatures. Let us consider, for example, a system of non-interacting particles. In the Maxwell–Boltzmann limit, density (n), energy density (ε) and average energy per particle (ω) can easily be obtained as

$$n \approx \frac{1}{\pi^2}T^3 \tag{6.27a}$$

$$\varepsilon \approx \frac{3}{\pi^2}T^4 \tag{6.27b}$$

$$\omega = \frac{\varepsilon}{n} \approx 3T. \tag{6.27c}$$

Qualitatively, one understands that at sufficiently low temperatures, the thermodynamics of the hadronic matter will be dominated by the pions, the lowest hadronic state. Equations (6.27) indicate that if the system energy is increased, it will lead to (i) higher temperatures, (ii) a greater number of particles and (iii) more energetic particles. Now if we assume that the particles are interacting, and resonance formation is the basic characteristic of the interaction, a fraction of the available energy can go into exciting resonance states. As the energy increases, more and more of the heavier resonances will form. For temperature $T \geqslant 150$ MeV, heavy states dominate the energy density. However, densities of heavy particles are still small, $\rho_i \sim e^{-M_i/T}$. Their mutual interaction, being proportional to $\rho_i\rho_j \sim e^{-(M_i+M_j)/T}$, is suppressed. One can use virial expansion to obtain an effective interaction. Virial expansion together with experimental phase shifts was used by Prakash and Venugopal to study the thermodynamics of low temperature hadronic matter [1]. It was shown that the interplay of attractive interactions (characterized by positive phase shifts) and repulsive interactions (characterized by negative phase shifts) is such that the theory is effectively interaction-free. One can then consider that interaction-free resonances constitute the hadronic matter at low temperatures. The interaction-free combination of all possible hadrons and hadronic resonances is called HRG.

For a non-interacting HRG comprising N hadrons (mesons and baryons), at temperature T and chemical potential μ, the thermodynamic potential can be written as

$$\Omega(T, \mu) = \sum_{i=1}^{N} \Omega_i(T, \mu_i). \tag{6.28}$$

The chemical potential μ_i of the ith hadron is generalized to include baryonic chemical potential μ_B, strangeness chemical potential μ_S and isospin chemical potential μ_I,

$$\mu_i = B_i\mu_B + S_i\mu_S + I_i\mu_I, \tag{6.29}$$

where B_i, S_i and I_i are the baryon number, the strangeness quantum number and the third component, the isospin quantum number of the ith hadron. Chemical potentials μ_B, μ_S and μ_I allow one to satisfy appropriate conservation laws.

The expressions for energy density, pressure, number density and entropy density for HRG, comprising N hadrons, can be obtained by summing over the same for the individual components of HRG,

$$\varepsilon(T, \mu) = \sum_{i=1}^{N} \varepsilon_i(T, \mu_i) \tag{6.30a}$$

$$P(T, \mu) = \sum_{i=1}^{N} P_i(T, \mu_i) \tag{6.30b}$$

$$n(T, \mu) = \sum_{i=1}^{N} n_i(T, \mu_i) \tag{6.30c}$$

$$s(T, \mu) = \sum_{i=1}^{N} s_i(T, \mu_i). \tag{6.30d}$$

In equations (4.92), (4.93), (4.98) and (4.99), the expressions for number density, pressure, energy density and entropy density were developed for massive particles obeying the Fermi–Dirac/Bose–Einstein distribution. They can be used in the above equations. However, in deriving those expressions it was implicitly assumed that the particles are point particles. The expressions need to be corrected to account for the finite size of hadrons. There is a well-established method called the 'excluded volume method' to account for the finite volume of hadrons. It is also called the van der Waals excluded volume method. Johannes Diderik van der Waals modified the ideal gas equation of state to account for the molecular interaction in an approximated manner. It is a remarkable equation of state as, for a particular class of interaction, it can be cast into a form devoid of all the details of the system. For completeness, a short description of the van der Waals equation of state is given below.

6.4 The van der Waals equation of state

van der Waals attempted to modify the ideal gas equation of state,

$$PV = NkT \qquad \text{or} \qquad PV = nRT, \tag{6.31}$$

approximately accounting for the molecular interaction. The general form of molecular interaction is shown in figure 6.1. It has a repulsive part and an attractive part. van der Waals approximated the repulsive part by an infinite hard sphere potential. The presence of a hard sphere potential or hard core forbids any molecule to be within a certain diameter of a test molecule. Thus the volume available to a

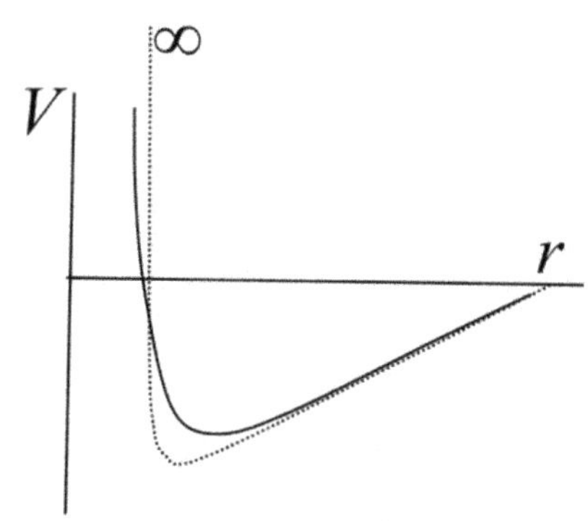

Figure 6.1. The solid line shows the general form of intermolecular interaction. The van der Waals approximation of the potential is shown as the dotted line.

molecule will be reduced by a constant, b, depending upon the molecular diameter and number of molecules present,

$$V \rightarrow V - b. \tag{6.32}$$

Qualitatively, the effect of the attractive part of the potential is to reduce pressure. For sufficiently strong attractive interaction, the molecules will form an N-body composite requiring no container. The decrease in pressure is proportional to the square of the density or inversely to the square of the volume. The effect of attractive interaction is then to change the pressure as

$$P \rightarrow P_{\text{kinetic}} - \frac{a}{V^2} \tag{6.33}$$

with a constant, a, characteristic of the system. van der Waals then modified the ideal gas equation of state as

$$\left(P - \frac{a}{V^2}\right)(V - b) = NT. \tag{6.34}$$

Some isotherms (constant temperature PV curves) corresponding to the van der Waals equation of state are shown in figure 6.2. At low temperatures, the isotherms show 'kinks' which disappear above a critical temperature T_c. The point of inflection C is called the critical point. The van der Waals equation of state is cubic in V. For a given pressure P and temperature T, it has three solutions. The solutions V_1, V_2 and V_3, corresponding to points 1, 2 and 3 on the isotherm for $T < T_c$, are explicitly shown in figure 6.2. Two of the roots (V_1 and V_3) correspond to two stable thermodynamic states for first order transition (the liquid and gas phases). The third root is a thermodynamically unstable solution. One notices that in the vicinity of the root V_2, compressibility $\kappa = -\frac{1}{V}\frac{\partial V}{\partial P}$ is negative, and P increases with V, a characteristic opposite to the other two roots. The unphysical behavior of the van der Waals equation of state is understood. The system is assumed to be homogeneous, with no allowance for a coexistence of two phases, i.e. a mixed phase. The coexistence of two phases requires

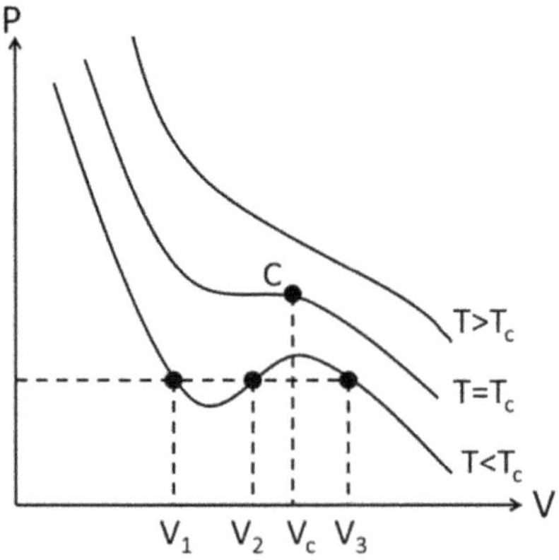

Figure 6.2. van der Waals isotherms in the PV plane.

that the pressure and temperature of the two phases are equal, which is satisfied by the roots V_1 and V_3. In Maxwell's construction, the unphysical behavior of the van der Waals equation of state is corrected by eliminating the second root V_2 altogether and joining roots V_1 and V_3 by a straight line. The position of the straight line is obtained from the free energy minimization condition; the system is either in a single homogeneous phase or in a mixed phase, depending on the free energy.

Before we end this section, we note that the van der Waals equation of state is a remarkable one. As noted, for a given P and T, the equation has three roots. As the temperature increases, the roots move together and at $T = T_c$ merge into one root V_c at the critical point C. In the neighborhood of the critical point, the van der Waals equation of state can be written as

$$
\begin{aligned}
&(V - V_c)^3 = 0 \\
\text{or} \quad & V^3 - 3V^2V_c + 3VV_c^2 - V_c^3 = 0.
\end{aligned}
\tag{6.35}
$$

Equation (6.35) can be compared with van der Waals equation (6.34) at $T = T_c$, $P = P_c$, yielding

$$NT_c = \frac{8a}{27b} \tag{6.36a}$$

$$P_c = \frac{a}{27b^2} \tag{6.36b}$$

$$V_c = 3b. \tag{6.36c}$$

If we define reduced variables,

$$\hat{P} = \frac{P}{P_c} \qquad \hat{V} = \frac{V}{V_c} \qquad \hat{T} = \frac{T}{T_c} \tag{6.37}$$

the van der Waals equation can be cast into a universal form (independent of the details of the potential),

$$\left(\hat{P} + \frac{3}{\hat{V}^2}\right)\left(\hat{V} - \frac{1}{3}\right) = \frac{8}{3}\hat{T}. \tag{6.38}$$

6.5 HRG in the excluded volume method

Let us return to the equation of state for HRG. Several procedures are in vogue to include the finite volume effect in HRG [2–11]. For example, in [3] the excluded volume effect is taken into account by reducing all the thermodynamic quantities including pressure by the reduction factor

$$r = \left[1 + \sum_j v_j n_j^{pt}\left(T, \mu_j\right)\right]^{-1},$$

where v_j and n_j^{pt} are, respectively, the proper volume and the point particle density of the jth hadron. In [2], the excluded volume effect is taken into account by reducing all the thermodynamic quantities including pressure by the reduction factor $r = [1 + \varepsilon^{pt}/4B]$, with ε^{pt} being the point particle energy density, and B the bag constant of the bag model. However those procedures are not thermodynamically consistent. Several authors have obtained thermodynamically consistent excluded volume correction to thermodynamic variables. To obtain thermodynamically consistent excluded volume correction, we follow [4, 5]. We write the grand canonical partition $\mathcal{Z}^{pt}(T, \mu, V)$ function for a system of point particles as a sum over canonical partition functions,

$$\mathcal{Z}^{pt}(T, \mu, V) = \sum_{N=0}^{\infty} \exp\left(\frac{\mu N}{T}\right) Z^{pt}(T, N, V). \tag{6.39}$$

In equation (6.39), $Z^{pt}(T, N, V)$ is the canonical partition function for a system of N point particles. In the van der Waals excluded volume approach, for a fixed N number of particles, the system volume V is replaced by $V - vN$, where v is the excluded volume per particle or the proper volume of the particles. For hadron radius $r_{\rm h}$, proper volume can be easily calculated. Consider figure 6.3 where the two small circles represent a pair of hadrons in touching configuration. The outer circle is of radius $d = 2r_{\rm h}$. The minimum volume or excluded volume for a pair of hadrons is then calculated as

$$\frac{4\pi}{3} d^3 \tag{6.40}$$

or excluded volume per particle is

$$v = \frac{1}{2}\frac{4\pi}{3}(2r_{\rm h})^3 = \frac{16\pi}{3} r_{\rm h}^3. \tag{6.41}$$

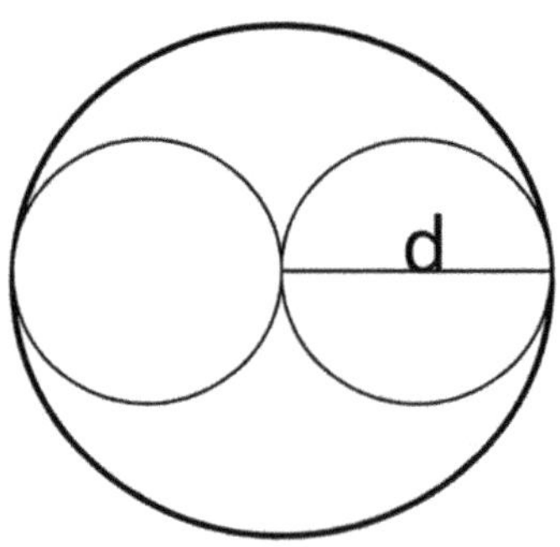

Figure 6.3. Schematic diagram for a pair of hadrons in touching configuration.

Accounting for the finite size of the particles, the grand canonical partition function in equation (6.39) can be re-written,

$$\mathcal{Z}^{xv}(T, \mu, V) = \sum_{N=0}^{\infty} \exp\left(\frac{\mu N}{T}\right) Z^{pt}(T, N, V - vN)\theta(V - vN). \tag{6.42}$$

There is a difficulty in evaluation of the sum over N in equation (6.42) as the available volume depends upon N. To overcome it, one performs a Laplace transform,

$$\begin{aligned}
\hat{\mathcal{Z}}^{xv}(T, \mu, x) &= \int_0^{\infty} \mathrm{d}V \mathrm{e}^{-xV} \mathcal{Z}^{xv}(T, \mu, V) \\
&= \int_0^{\infty} \mathrm{d}V \mathrm{e}^{-xV} \sum_{N=0}^{\infty} \mathrm{e}^{\mu N/T} Z^{pt}(T, N, V - vN)\theta(V - vN) \\
&= \int_0^{\infty} \mathrm{d}\hat{V} \mathrm{e}^{-x\hat{V}} \sum_{N=0}^{\infty} \mathrm{e}^{\hat{\mu} N/T} Z^{pt}\left(T, N, \hat{V}\right)\theta(\hat{V}) \\
&= \int_0^{\infty} \mathrm{d}\hat{V} \mathrm{e}^{-x\hat{V}} \mathcal{Z}^{pt}\left(T, \hat{\mu}, \hat{V}\right)
\end{aligned} \tag{6.43}$$

where $\hat{\mu} = \mu - xvT$, $\hat{V} = V - vN$. One understands that the partition function accounting for the finite particle volume at chemical potential μ is equal to the partition function for point particles with shifted chemical potential $\hat{\mu} = \mu - vTx$. To obtain the shifted chemical potential we note the definition of pressure,

$$P(T, \mu) = T \lim_{V\to\infty} \frac{\ln \mathcal{Z}(T, \mu, V)}{V}. \tag{6.44}$$

Alternatively, in the thermodynamic limit, the partition function $\mathcal{Z}(T, \mu, V)$ approaches

$$\lim_{V\to\infty} \mathcal{Z}(T, \mu, V) \sim \exp\left(\frac{p(T, \mu)}{T} V\right). \tag{6.45}$$

In the thermodynamic limit, the Laplace transform in equation (6.43) can be re-written as

$$\begin{aligned}
\hat{\mathcal{Z}}^{xv}(T, \mu, x) &= \int_0^{\infty} \mathrm{d}V \mathrm{e}^{-xV} \mathcal{Z}^{xv}(T, \mu, V) \\
&= \int_0^{\infty} \mathrm{d}V \exp\left[\left(\frac{P^{xv}(T, \mu)}{T} - x\right)V\right].
\end{aligned} \tag{6.46}$$

For $x < P^{xv}(T, \mu)/T$ the integration over V diverges at the upper limit. The extreme right singularity of $\hat{\mathcal{Z}}^{xv}$ at some $x = x^*(T, \mu)$ gives the system pressure,

$$P^{xv}(T, \mu) = Tx^*(T\mu) \qquad \text{or} \qquad x^*(T, \mu) = \frac{P^{xv}(T, \mu)}{T} \tag{6.47}$$

which corresponds to shifted chemical potential,

$$\tilde{\mu} = \mu - vTx^*(T, \mu) = \mu - vP^{xv}(T, \mu). \tag{6.48}$$

For a single species hadron gas, excluded volume corrected thermodynamic parameters can be obtained as

$$\text{Pressure: } P^{xv}(T, \mu) = P^{pt}(T, \tilde{\mu}) \qquad \tilde{\mu} = \mu - vP^{xv}(T, \mu) \tag{6.49a}$$

$$\text{Number density: } n^{xv}(T, \mu) = \frac{\partial P^{xv}(T, \mu)}{\partial \mu} = \frac{n^{pt}(T, \tilde{\mu})}{1 + vn^{pt}(T, \tilde{\mu})} \tag{6.49b}$$

$$\text{Entropy density: } s^{xv}(T, \mu) = \frac{\partial P^{xv}(T, \mu)}{\partial T} = \frac{s^{pt}(T, \tilde{\mu})}{1 + vn^{pt}(T, \tilde{\mu})} \tag{6.49c}$$

$$\text{Energy density: } \varepsilon^{xv}(T, \mu) = Ts + p + \mu n = \frac{\varepsilon^{pt}(T, \tilde{\mu})}{1 + vn^{pt}(T, \tilde{\mu})}. \tag{6.49d}$$

The excluded volume approach can be easily extended to a mixture of several particle species. If the proper volume of particle species i is v_i, when excluded volume corrected the grand canonical partition function can be written as

$$\begin{aligned}\mathcal{Z}^{xv}(T, \mu_1, \ldots \mu_n, V) &= \prod_{i=1}^{h} \mathcal{Z}_i^{xv}(T, \mu_i, V) \\ &= \prod_{i=1}^{h} \left[\sum_{N_i=0}^{\infty} \exp(\mu_i N_i/T) \mathcal{Z}_i^{pt}\left(T, N_i, V - \sum_{i=1}^{h} v_i N_i\right)\right].\end{aligned} \tag{6.50}$$

Laplace transform of the above equation will give the excluded volume corrected pressure as

$$P^{xv}(T, \mu_1, \ldots \mu_h) = T \lim_{V \to \infty} \frac{T \ln Z^{xv}(T, \mu_1, \ldots \mu_h)}{V} = \sum_{i=1}^{h} P^{pt}(T, \tilde{\mu}_i), \tag{6.51}$$

with shifted chemical potential,

$$\tilde{\mu}_i = \mu_i - v_i P^{xv}(T, \mu_1, \ldots \mu_h) \qquad i = 1, 2, \ldots h. \tag{6.52}$$

The particle number density of the ith species can be calculated as

$$n_i^{xv}(T,\mu_1,\ldots\mu_h) = \left(\frac{\partial P^{xv}}{\partial \mu_i}\right)_{T,\mu_1,\ldots\mu_{i-1},\mu_{i+1},\ldots\mu_h}$$
$$= \frac{n_i^{pt}(T,\tilde{\mu}_i)}{1+\sum_{j=1}^{h} v_j n_j^{pt}(T,\tilde{\mu}_j)}. \tag{6.53}$$

The total particle density will be the sum over the partial densities n_i. The entropy density and energy density can also be calculated,

$$\text{Entropy density: } s^{xv}(T,\mu_1,\ldots\mu_h) = \frac{\sum_{i=1}^{h} s_i^{pt}(T,\tilde{\mu}_i)}{1+\sum_{j=1}^{h} v_j n_j^{pt}(T,\tilde{\mu}_j)} \tag{6.54}$$

$$\text{Energy density: } \varepsilon^{xv}(T,\mu_1,\ldots\mu_h) = \frac{\sum_{i=1}^{h} \varepsilon_i^{pt}(T,\tilde{\mu}_i)}{1+\sum_{j=1}^{h} v_j n_j^{pt}(T,\tilde{\mu}_j)}. \tag{6.55}$$

According to the bag model, hadrons occupy a finite volume proportional to their own mass, $m/4B$. Kapusta and Olive [10] advocated a procedure where the spatial volume is then replaced by

$$V \rightarrow V - \sum_{i=1}^{N} \varepsilon_i/4B. \tag{6.56}$$

Details of their method will not be discussed; we will just mention the results. The finite or excluded volume corrected pressure, energy density, temperature and entropy densities are

$$\text{Pressure: } P^{xv} = \frac{P^{pt}(T^*,\mu^*)}{1-\frac{P^{pt}(T^*,\mu^*)}{4B}} \tag{6.57a}$$

$$\text{Temperature: } T_{xv} = \frac{T^*}{1-\frac{P^{pt}(T^*,\mu^*)}{4B}} \tag{6.57b}$$

$$\text{Chemical potential: } \mu_{xv} = \frac{\mu^*}{1-\frac{P^{pt}(T^*,\mu^*)}{4B}} \tag{6.57c}$$

$$\text{Energy density: } \varepsilon^{xv} = \frac{\varepsilon^{pt}(T^*, \mu^*)}{1 + \frac{\varepsilon^{pt}(T^*, \mu^*)}{4B}} \tag{6.57d}$$

$$\text{Entropy density: } s^{xv} = \frac{s^{pt}(T^*, \mu^*)}{1 + \frac{\varepsilon^{pt}(T^*, \mu^*)}{4B}} \tag{6.57e}$$

$$\text{Number density: } n^{xv} = \frac{n^{pt}(T^*, \mu^*)}{1 + \frac{\varepsilon^{pt}(T^*, \mu^*)}{4B}}. \tag{6.57f}$$

B is the bag pressure, $B^{1/4} = 340$ MeV. In practical calculations, one chooses a T^* and μ^* and using point particle pressure can obtain real temperature T_{xv}, chemical potential μ_{xv} and all the thermodynamic variables.

6.6 The statistical bootstrap model

In the 1960s, particle physicists were confronted with two challenges: (i) hadronic collision produces a large number of particles (mostly pions) with exponential spectra, with a fixed slope $T \approx 160$ MeV, independent of collision energy; and (ii) proliferation of so-called elementary particles, hadrons and hadron resonances. Rolf Hagedorn met both these challenges with his statistical bootstrap model [15, 16]. The statistical bootstrap model was introduced in 1965 but several experimental facts and theoretical concepts of the preceding years, which were recounted by Hagedorn in [17], were important and instrumental for its development. At that time the concept of the fireball was introduced. In the fireball approach, all the collision energy is released within a small space–time volume, from which particles are emitted. Hagedorn took the fireball approach to its limit. His observation that heavier resonances decay into smaller ones led him to the bootstrap model, where a hadronic resonant state is itself a constituent of a still heavier resonance while also being composed of lighter ones. Identifying resonant states with fireballs, he describes them as fireballs going into fireballs living on the fireballs forever, in a self-consistent manner. The model led to exponentially increasing density of states for hadron resonances,

$$\rho(m) \sim m^{\alpha} \mathrm{e}^{m/T_{\mathrm{H}}} \tag{6.58}$$

with $-\frac{7}{2} \leqslant \alpha \leqslant -\frac{5}{2}$ and $135\,\text{MeV} \leqslant T_{\mathrm{H}} \leqslant 185$. Hagedorn originally interpreted T_{H} as the highest possible temperature of strongly interacting matter; hadronic matter cannot exist for $T > T_{\mathrm{H}}$. Somewhat later, when the concept of quarks as fundamental constituents of matter was firmly established, Cabibbo and Parisi [18] argued that the exponentially increasing mass spectrum in the statistical bootstrap model did not necessarily indicate a limiting temperature, but rather were generic to a system undergoing second order phase transition. It is now clear that $T_{\rm H}$ defines the transition from hadronic matter to a QGP. Hadron physics alone can only specify its inherent limit; to go beyond this limit, we need QCD.

Hagedorn's original model was refined by several authors. In an elegant formulation Frautschi [19] gave a phase space formulation for the bootstrap model. It was also more restrictive than the original model. An analytical solution of the model was obtained by Nahm [21]. A manifestly Lorentz invariant form was proposed and solved by Yellin [20].

The Hagedorn bootstrap model was introduced before QCD. Post QCD, its relevance has been lost to some extent. In recent years, the model has found its use in constructing equations of state of HRG [11, 22, 23] for use in hydrodynamic evolution. Following [17], we consider the toy model to establish the limiting temperature in HRG. Let $\rho(m)\mathrm{d}m$ be the density of states between mass m and $m + \mathrm{d}m$. In the limit of vanishing kinetic energy, the bootstrap equation can be written as

$$\rho(m) = \delta(m - m_0) + \sum_{n=2}^{\infty}\frac{1}{n!}\int \delta\left(m - \sum_{i=1}^{n} m_i\right)\prod_{i=1}^{n}\rho(m_i)\mathrm{d}m_i. \tag{6.59}$$

The equation can be understood as follows: a hadronic resonance with mass m is either the 'input particle' with mass m_0 or else it is composed of any number of resonances m_i such that $\sum_i m_i = m$.

The Laplace transform of equation (6.59) reads

$$\int \rho(m)\mathrm{e}^{-\beta m}\mathrm{d}m = \mathrm{e}^{-\beta m_0} + \sum_{n=2}^{\infty}\frac{1}{n!}\prod_{i=1}^{n}\int \rho(m_i)\mathrm{e}^{-\beta m_i}\mathrm{d}m_i. \tag{6.60}$$

Define,

$$z(\beta) = \mathrm{e}^{-\beta m_0} \qquad G(z) = \int \rho(m)\mathrm{e}^{-\beta m}\mathrm{d}m \tag{6.61}$$

and equation (6.60) becomes

$$G(z) = z(\beta) + \sum_{n=2}^{\infty}\frac{1}{n!}[G(z)]^n = z(\beta) + \mathrm{e}^{G(z)} - 1 - G(z) \tag{6.62}$$

or

$$z(\beta) = 2G(z) - \mathrm{e}^{G(z)} + 1. \tag{6.63}$$

Equation (6.63) is the bootstrap equation for Laplace transformed mass spectrum $\rho(m)$. The equation is universal and not restricted to the toy model considered here but appears in all realistic statistical bootstrap models without cut off. Moreover, it is independent of (i) the number of space–time dimensions, (ii) the number of input particles and (iii) abelian or non-abelian symmetry.

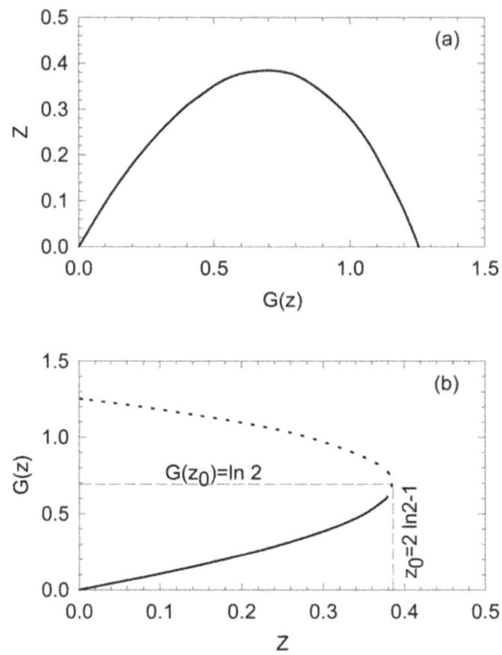

Figure 6.4. Graphical solution of bootstrap equation (6.63). (a) z as a function of $G(z)$ and (b) $G(z)$ as a function of z.

Equation (6.63) can be solved for $G(z)$, the Laplace transformed mass spectrum. It can easily be solved graphically: plot $z(G)$ against G and exchange the axis. In figure 6.4(a) and (b), the graphical solution is shown. The parabola-like maximum of $z(G)$ at z_0 implies that

$$z = e^{-\beta m_0} < z_0$$
$$\text{or} \quad \frac{1}{T} \geqslant -\frac{\ln z_0}{m_0}. \tag{6.64}$$

The toy model then predicts a limiting temperature for the system of hadrons. Locations of maxima $G(z_0)$ and z_0 can be found from equation (6.63),

$$0 = \left.\frac{dz}{dG}\right|_{z=z_0} = 2 - e^G$$
$$\Rightarrow G(z_0) = \ln 2, \qquad z_0 = 2\ln 2 - 1. \tag{6.65}$$

For $m_0 = m_\pi$, the toy model then predicts the limiting temperature (also called the Hagedorn temperature) as

$$T_H \approx \frac{m_0}{0.95} \approx 145 \text{ MeV}. \tag{6.66}$$

Nahm [21] first noted that the parabola-like maximum of $z(G)$ at z_0 implies that there is a square root singularity of $G(z)$ at z_0 and inverse Laplace transformation of $G(z)$ leads to an exponentially increasing mass spectrum,

$$\rho(m) \sim \frac{1}{m^3} \exp\left(\frac{m}{T_\mathrm{H}}\right). \tag{6.67}$$

The bootstrap model prediction for the exponential mass spectrum is supported experimentally. In [23], for hadrons with masses up to 11.019 GeV, the densities of states were fitted with the Hagedorn mass spectrum,

$$\rho(m) = \frac{c}{(m^2 + m_a^2)^a} \exp\left(\frac{m}{T_\mathrm{H}}\right) \tag{6.68}$$

with $c = 0.16 \pm 0.02$, $a = 5/4$, $m_a = 0.5$ GeV and $T_H = 0.174 \pm 0.11$ GeV. It must be mentioned here that equivalent fit could be obtained with changed parameter values.

6.7 Extended HRG

The HRG model described in section 6.3 and in section 6.5 can be extended to include the exponentially increasing hadron mass spectrum. The thermodynamic potential or the partition function can then be written as

$$\ln Z(T, \mu) = \int \mathrm{d}m\rho(m) \ln Z(T, \mu, m) \tag{6.69}$$

where $Z(T, \mu, m)$ is the partition function for a species of mass m. One generally recognizes that low mass hadrons have a definitely discrete spectrum. This is accounted for by decomposing the mass spectrum as

$$\rho(m) = \sum_{i=1}^{m_i \leqslant M} g_i \delta(m - m_i) + \frac{c}{(m^2 + m_0^2)^{5/4}} \theta(m - M) \exp\left(\frac{m}{T_\mathrm{H}}\right). \tag{6.70}$$

The assumed mass spectrum is discrete up to mass M and continuous thereafter. All the thermodynamic parameters can now be calculated from equation (6.69).

Bibliography

[1] Venugopalan R and Prakash M 1992 *Nucl. Phys.* A **546** 718

[2] Hagedorn R and Rafelski J 1980 *Phys. Lett.* B **97** 136

[3] Cleymans J, Redlich K, Satz H and Suhonen E 1986 *Z. Phys.* C **33** 151
Cleymans J and Satz H 1993 *Z. Phys.* C **57** 135

[4] Rischke D H, Gorenstein M I, Stoecker H and Greiner W 1991 *Z. Phys.* C **51** 485

[5] Cleymans J, Gorenstein M I, Stalnacke J and Suhonen E 1993 *Phys. Scr.* **48** 277

[6] Singh C P, Patra B K and Singh K K 1996 *Phys. Lett.* B **387** 680

[7] Yen G D, Gorenstein M I, Greiner W and Yang S-N 1997 *Phys. Rev.* C **56** 2210

[8] Gorenstein M I 2012 arXiv: 1205.1762

[9] Andronic A, Braun-Munzinger P, Stachel J and Winn M 2012 *Phys. Lett.* B **718** 80
[10] Kapusta J I and Olive K A 1983 *Nucl. Phys.* A **408** 478
[11] De S, Srivastava D K and Chatterjee R 2010 *J. Phys. G: Nucl. Part. Phys.* **37** 115004
[12] Chodos A, Jaffe R L, Johnson K, Thorn C B and Weisskopf V F 1974 *Phys. Rev.* D **9** 3471
[13] Chodos A, Jaffe R L, Johnson K and Thorn C B 1974 *Phys. Rev.* D **10** 2599
[14] Theberge S, Thomas A W and Miller G A 1980 *Phys. Rev.* D **22** 2838
Theberge S, Thomas A W and Miller G A 1981 *Phys. Rev.* D **23** 2106 (erratum)
[15] Hagedorn R 1965 *Nuovo Cimento Suppl.* **3** 147
[16] Hagedorn R and Ranft J 1968 *Nuovo Cimento Suppl.* **6** 169
[17] Hagedorn R 1995 The long way to the statistical bootstrap model *NATO ASI Series* vol 346 (Berlin: Springer) pp 13–46
[18] Cabibbo N and Parisi G 1975 *Phys. Lett.* B **59** 67
[19] Frautschi S C 1971 *Phys. Rev.* D **3** 2821
[20] Yellin J 1973 *Nucl. Phys.* B **52** 583
[21] Nahm W 1972 *Nucl. Phys.* B **45** 525
[22] Noronha-Hostler J, Beitel M, Greiner C and Shovkovy I 2010 *Phys. Rev.* C **81** 054909
[23] Cleymans J and Worku D 2011 *Mod. Phys. Lett.* A **26** 1197
[24] Ericson T E O and Rafelski J 2003 The tale of the Hagedorn temperature *CERN Courier* **43**(7) 30

IOP Publishing

A Short Course on Relativistic Heavy Ion Collisions

A K Chaudhuri

Chapter 7

Modeling relativistic heavy ion collisions

7.1 Introduction

Nucleus–nucleus collision is a well established tool to study nuclear matter. With increasing collision energy, nuclear matter is probed at finer and finer resolution and several facets of nuclear matter are revealed. Let us consider what happens when the energy of the colliding system increases.

In very low-energy collisions, the nucleus as a whole interacts. Indications are obtained from giant dipole resonances[1], where the compound nucleus[2] undergoes dipole oscillation. One can also excite the nucleus and populate various excited states. As the energy is increased, nucleons in the nucleus start to interact, and one can see the production of new fragments or new particles, e.g. π. At still higher energy, the quarks inside the nucleons will interact. Here also, the production of different particle species will be observed. However, in contrast to low/medium-energy nuclear reactions, where one can describe pA/AA collisions entirely in terms of nucleon–nucleon (NN) collisions, in relativistic energy, such a description will fail.

A nucleus–nucleus collision at relativistic energy passes through different stages. A schematic picture of different stages of the collision is shown in figure 7.1. One can broadly classify the following stages.

(i) *Pre-equilibrium stage*. At relativistic energy, the initial collisions are expected to be at the partonic level. Initial partonic collisions produce a fireball in a highly excited state. In all possibility, the fireball is not in equilibrium. Constituents of the fireball collide frequently to establish a

[1] Giant dipole resonances are collective excitations of a nucleus. An example is the collective oscillations of all the protons against all the neutrons in a nucleus.

[2] Niels Bohr proposed the compound nucleus model for low-energy nuclear reactions. It is a two-stage process. In the first step, the bombarding particle loses all its energy in the target nucleus and becomes an integral part of an unstable nucleus called a 'compound nucleus'. In the second stage, the compound nucleus decays to reaction products. The decay of the compound nucleus is independent of the reactants.

doi:10.1088/978-0-750-31060-4ch7

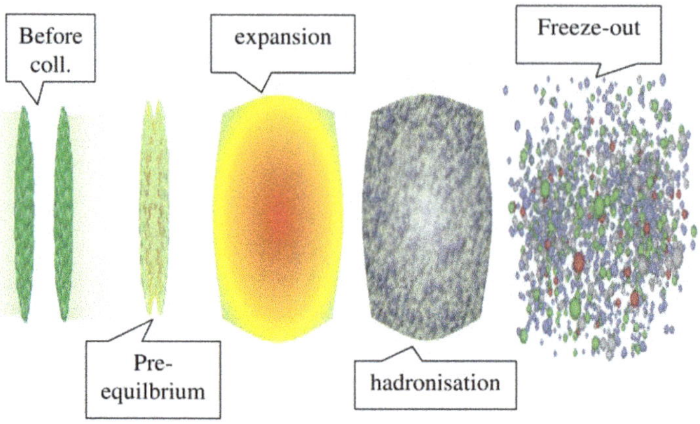

Figure 7.1. Different stages of nuclear collisions at ultra-relativistic energy. Reproduced with permission from Steffen Bass, Quark Matter 2001.

'local' equilibrium state. The time taken to establish the local equilibrium is called the 'thermalization' or 'equilibration' time.

(ii) *Expansion stage.* In the equilibrium or the thermalized state, the constituents of the fireball, the partons or more specifically, the quarks and gluons, are in the deconfined state. The system has thermal pressure which acts against the surrounding vacuum. The fireball then undergoes collective (hydrodynamic) expansion. As it expands, the density (energy density) decreases and the system cools. If the system supports phase transition, then below a critical temperature, the deconfined quarks and gluons will hadronize. In the hadronization stage, over a small temperature interval, the entropy density will decrease very fast. Since total entropy cannot decrease, it implies that the fireball will expand rapidly, while temperature remains approximately constant.

(iii) *Freeze-out.* Even after the hadronization, the matter can be in thermal equilibrium. Constituent hadrons will collide to maintain local equilibrium. The system will expand and cool. A stage will come when inelastic collisions, in which hadrons change identity, become too small to keep up with expansion. The stage is called 'chemical freeze-out'. Hadron abundances will remain fixed after the chemical freeze-out. However, due to elastic collisions, local equilibrium can still be maintained and the system will further expand and cool, with fixed hadron abundances. Eventually a stage will come when the average distance between the constituents will be larger than the strong interaction range. Collisions between the constituents will be so infrequent that 'local' thermal equilibrium cannot be maintained. The hydrodynamic description will break down. The hadrons decouple or freeze out. This is called 'kinetic freeze-out'. Hadrons from the kinetic freeze-out surface will be detected in the detector.

There are several models in vogue for relativistic heavy ion collisions. They can be categorized into two types: (i) static models and (ii) dynamic models. The static models aim to explain experimental outcomes by modeling the freeze-out condition. They do not try to answer the question of how the freeze-out condition is arrived at. Dynamical models try to answer that question. Currently two types of dynamical models are in vogue: (i) models based on the transport approach and (ii) models based on the hydrodynamic approach. In the transport approach, the evolution of the system from the pre-equilibrium stage to freeze-out can be modeled within certain approximations. Hydrodynamics based models, on the other hand, are limited to expansion to the freeze-out stage only. In the following, short descriptions of some of the prevailing models are given.

7.2 Statistical or thermal models

In statistical or thermal models, it is assumed that the hadrons are emitted from regions of thermal equilibrium, called fireballs or clusters. In a single collision event, there might be several fireballs or clusters with different collective momenta, different overall charges and volumes. Lorentz-invariant quantities like final multiplicities depend on the distribution of initial conserved charges (baryon number, strangeness and electric charge) among the produced clusters but are independent of cluster momenta. This distribution is determined by the dynamics of the collision, which is beyond the statistical model. Most of the analyses then assume that a single fireball is produced. The assumption is reasonable if either one of the two conditions below is fulfilled.

(i) All clusters are large enough to allow a grand canonical description and all of them have the same values of relevant intensive parameters, i.e. temperature and chemical potential.

(ii) The clusters are small with the same temperature, allowing for canonical treatment. The distribution governing fluctuations of charges is the same as that obtained by splitting one large cluster (equivalent global cluster), having as its volume the sum of all cluster rest frame volumes, and as its charge the sum of all cluster charges.

Thermal models are most successful in explaining hadron multiplicities, or more specifically ratios of hadron multiplicities, with a limited number of parameters. Hadron multiplicities are frozen at the chemical freeze-out, and provide a snapshot of fireballs/clusters at that particular stage of the evolution. Thermal models thus give information about fluid temperature and chemical potential at the chemical equilibration stage. Resonances play an important role in hadron multiplicities. For example, at any temperature, a large proportion of pions are from the decay of resonances η, ρ, ω, etc. In the thermal model resonances are treated in the spirit of hadronic resonance gas (HRG; described earlier), i.e. treating them as free. This amounts to taking into account a considerable number of interactions among the stable hadrons. It is also possible to include van der Waals type interactions via the excluded volume correction.

In the statistical or thermal model, the basic quantity required to compute the thermal composition of hadron yields measured in heavy ion collisions is the thermodynamic potential (Ω) or the partition function $Z(T, V)$. The partition function is discussed in detail in chapter 4. For a single cluster, the thermodynamic potential or the grand canonical partition function for a hadron species i is

$$\Omega_i = -T \ln Z_i = -\frac{g_i V}{6\pi^2} \int_0^\infty \frac{p^4 \mathrm{d}p}{\sqrt{p^2 + m_i^2}} \frac{1}{\left[\mathrm{e}^{\beta\left(\sqrt{p^2+m_i^2} - \mu_i\right)} + \eta \right]}. \tag{7.1}$$

Average particle number is then calculated as,

$$\langle N_i \rangle = T \frac{\partial \ln Z_i}{\partial \mu_i} = g_i V \frac{T m_i^2}{2\pi^2} \sum_{n=1}^{\infty} \frac{(-\eta)^{n+1}}{n} K_2\left(\frac{n m_i}{T}\right) \mathrm{e}^{\frac{n\mu_i}{T}} \tag{7.2}$$

where $g_i = (2J_i + 1)$ is the spin degeneracy factor and m_i is the mass of the hadron. V is the fireball volume and $\beta = 1/T$ is the inverse temperature. The factor η takes into account the bosonic/fermionic nature of the particle. $\eta = +1$ for fermions and $\eta = -1$ for bosons.

For hadron i of baryon number B_i, the third component of the isospin I_{3i}, strangeness S_i and charmness C_i, the chemical potential is $\mu_i = \mu_B B_i + \mu_{I3} I_{3i} + \mu_S S_i + \mu_C C_i$. The baryon chemical potential μ_B, isospin chemical potential μ_{I3}, strangeness chemical potential μ_S and charm chemical potential μ_C ensure the conservation (on average) of the respective quantum numbers:

$$\text{(i) baryon number:} \quad \sum_i N_i B_i = N_B, \tag{7.3a}$$

$$\text{(ii) isospin:} \quad \sum_i N_i I_{3i} = I_3^{\text{tot}}, \tag{7.3b}$$

$$\text{(iii) strangeness:} \quad \sum_i N_i S_i = 0, \tag{7.3c}$$

and

$$\text{(iv) charmness:} \quad \sum_i N_i C_i = 0. \tag{7.3d}$$

The (net) baryon number N_B and the total isospin I_3^{tot} of the system are input values which need to be specified according to the colliding nuclei studied. The degree of stopping of the colliding nuclei is energy dependent and cannot be precisely determined experimentally. Thus the inputs N_B and I_3^{tot} are uncertain to some extent. The volume V of the system is also uncertain. In thermal models, one generally fits the particle ratios, whence the volume gets cancelled and the ratios remain largely unaffected by the uncertainty in inputs N_B and I_3^{tot}. Experimental data can be largely reproduced with two parameters, the temperature T and baryon chemical potential μ_B.

The finite widths of resonances are taken into account in the calculation by an additional integration, over the particle mass, with a Breit–Wigner distribution as a weight:

$$\langle N_i^R \rangle = \frac{g_i V}{(2\pi)^3} \frac{1}{N_{\rm BW}} \int_{M_0}^{\infty} {\rm d}m \int_0^{\infty} {\rm d}^3p \frac{\Gamma_i^2}{(m - m_i)^2 + \frac{\Gamma_i^2}{4}} \frac{1}{\left[{\rm e}^{\beta\left(\sqrt{p^2+m^2}-\mu_i\right)} + \eta \right]} \tag{7.4}$$

where m_i is the nominal mass and Γ_i is the width of the resonance particle i. M_0 is the threshold for the dominant decay channel. $N_{\rm BW}$ is the normalization for the Breit–Wigner distribution. Finally, the overall multiplicity to be compared with the data is calculated as the sum of primary multiplicity and the contribution from the decay of resonances,

$$\langle N_i \rangle = \langle N_i^{\rm primary} \rangle + \sum_j {\rm Br}(j \to i) \langle N_j^R \rangle. \tag{7.5}$$

$\mathrm{Br}(j \to i)$ is the branching ratio for the resonance j to decay into hadron i. Branching ratios for resonances are listed in various issues of *Review of Particle Physics.*

In elementary collisions or in peripheral nucleus–nucleus collisions (as well as for light nuclei), complete equilibration in the strangeness sector is not expected. To account for the possible violation of the strangeness equilibrium, in some of the thermal models an additional parameter, $\gamma_S < 1$, is used and the distribution function is written as,

$$\frac{1}{{\rm e}^{\beta\left(\sqrt{p^2+m^2}-\mu_i\right)} + \eta} \to \frac{1}{\gamma_S^{-1}{\rm e}^{\beta\left(\sqrt{p^2+m^2}-\mu_i\right)} + \eta}. \tag{7.6}$$

γ_S is called the strangeness suppression factor and it is established that $\gamma_S \sim 1$ in RHIC energy collisions and decreases at lower energy.

The thermal model is simple and various authors have analyzed experimental data in the model [1–16]. Astonishingly, the simple model does give a very good description of the ratios of particle yield over a wide range of energy. An example of fit to particle yield ratios in $\sqrt{s}_{\rm NN} = 200$ GeV central Au+Au collision data is shown in figure 7.2. The agreement between experimental data and thermal model predictions is excellent. Similarly, a good quality fit is obtained at other collision energies. In figure 7.3, the energy dependence of hadron yield relative to pion yield is shown over a large energy range. Here also, the thermal model describes the experimental data rather satisfactorily. The differences in energy dependence of various hadrons are qualitatively and quantitatively understood. The steep decrease of the p/π^+ ratio is due to the combination of two effects: (i) decreased proton stopping with energy and (ii) increased pion yield with energy. The ratio flattens beyond $\sqrt{s}_{\rm NN} = 100$ GeV as a consequence of the dominance of newly created baryons. The energy dependence of the kaon yield relative to pion is interesting. The ratio K^-/π^- monotonically increases followed by saturation beyond 100 GeV. The K^+/π^+ ratio on the other hand shows a broad maximum around $\sqrt{s}_{\rm NN} \approx 8$ GeV. K^+ consist of a u and $\bar{s}$ quark. The u quark may come from the initial

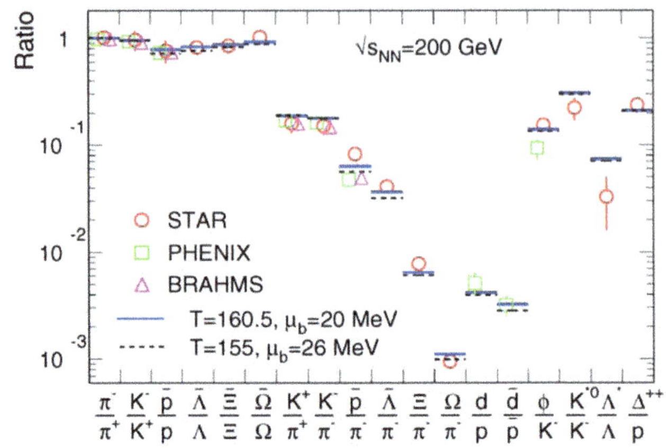

Figure 7.2. Thermal model fits to hadron yield ratios in $\sqrt{s}_{\rm NN} = 200$ GeV Au+Au collisions. The figure is reproduced with permission from [13].

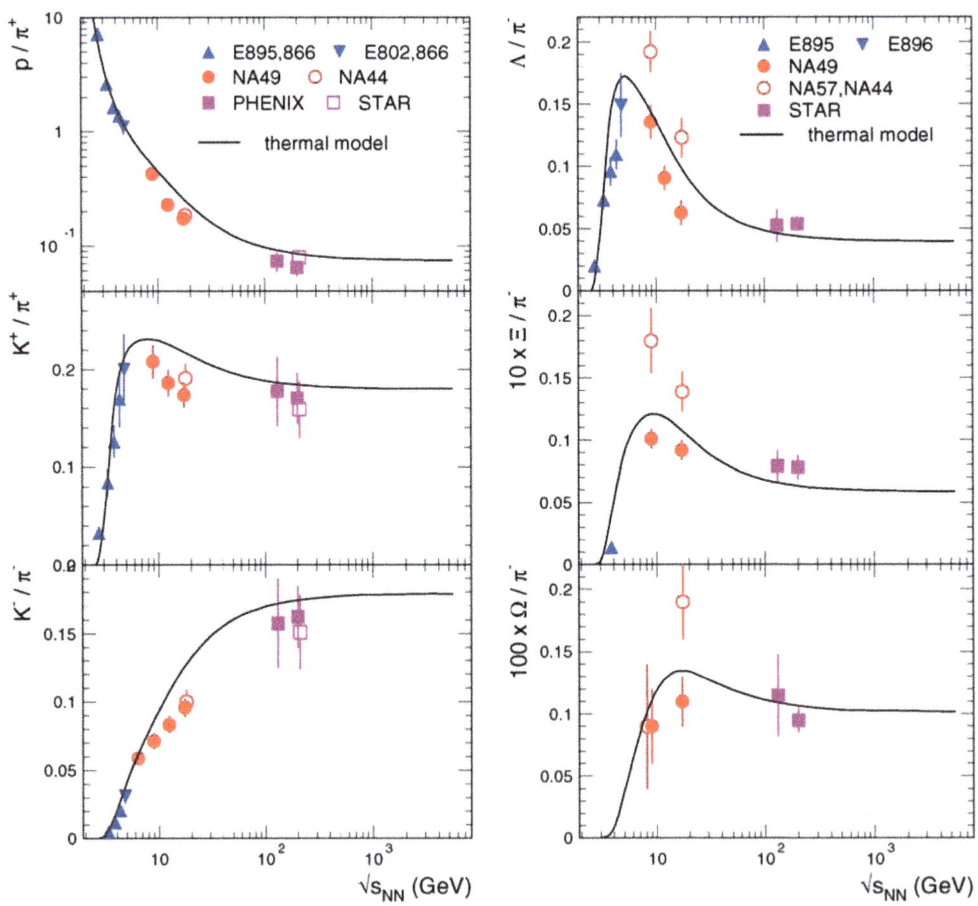

Figure 7.3. The energy dependence of hadron yields relative to pions. The figures are reproduced with permission from [13].

colliding system as well as from newly created partons. K^+ yield then depends on two competing mechanisms, an increase in u from newly created partons with energy and a decrease in initial u with increasing energy. The 'horn'-like structure in the K^+/π^- ratio was predicted to be a signature of quark–gluon plasma (QGP) formation [3]. The statistical model naturally reproduces the 'horn'-like structure but the predicted width is rather broad.

The fit parameters, T and μ_B, in the thermal model measure the temperature and baryonic chemical potential at the chemical freeze-out (particle ratios are fixed at the chemical freeze-out). They depend strongly on NN center-of-mass (cm) energy ($\sqrt{s}_{\rm NN}$). The energy dependence however is rather smooth and it is possible to obtain an 'empirical' chemical freeze-out curve in the $T - \mu_B$ plane. In [10] the chemical freeze-out curve was parameterized as

$$T = 0.1675 - 0.1583\mu_B^2. \tag{7.7}$$

The energy dependence of μ_B was obtained as

$$\mu_B(\sqrt{s}) = 2.06\frac{\log\sqrt{s}_{\rm NN}}{\left(\sqrt{s}_{\rm NN}\right)^{1.13}}. \tag{7.8}$$

In equations (7.7) and (7.8), the variables $\sqrt{s}_{\rm NN}$, T and μ_B are expressed in GeV. Fitted parameters and the freeze-out curve are shown in figure 7.4. An alternative and nearly equivalent parameterization was given in [12],

$$T = 0.166 - 0.139\mu_B^2 - 0.053\mu_B^4, \tag{7.9}$$

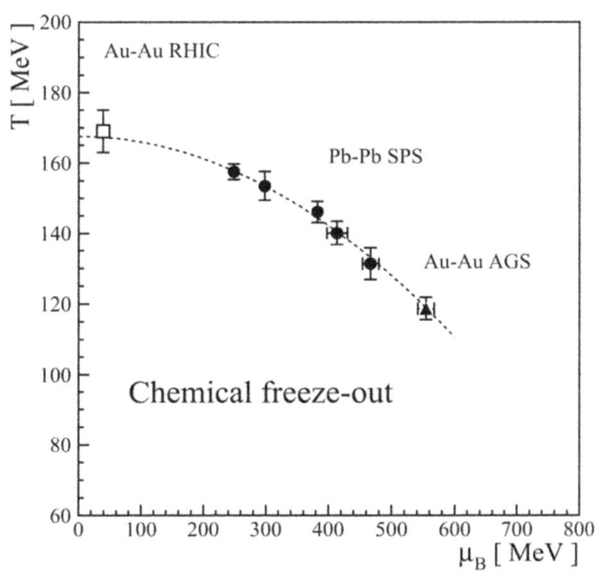

Figure 7.4. Chemical freeze-out points in the $\mu_B - T$ plane for central Pb–Pb and Au–Au collisions. The dashed line shows the parabolic interpolation equation (7.7). The figure is reproduced with the author's permission from [10].

$$\mu_B(\sqrt{s}) = \frac{1.308\ \text{GeV}}{1 + 0.273\ \text{GeV}^{-1}\sqrt{s}_{\text{NN}}}. \tag{7.10}$$

7.2.1 Rapidity spectra in the thermal model

Allowing for the rapidity dependent chemical potential and temperature, the thermal model has been extended to describe the rapidity distribution of the produced hadrons in ultra-relativistic nuclear collisions [17–21]. It is assumed that the rapidity axis is populated with fireballs or clusters, moving along the beam axis with increasing rapidity, y_{FB}. The emitted particles leave these regions (fireballs) at the freeze-out following a (local) thermal distribution. The resulting rapidity distribution of any given particle species is then obtained by a superposition of the contributions of these regions (fireballs) as follows:

$$\frac{\mathrm{d}N^i}{\mathrm{d}y} = A\int \rho(y_{\text{FB}})\frac{\mathrm{d}N_1^i(y - y_{\text{FB}})}{\mathrm{d}y} \tag{7.11}$$

where y is the particle's rapidity in the rest frame of the colliding nuclei and A is an overall normalization factor. $\rho(y_{\text{FB}})$ is the probability distribution for the fireballs. A convenient choice is the Gaussian distribution,

$$\rho(y_{\text{FB}}) = \frac{1}{\sqrt{2\pi}\sigma}\mathrm{e}^{-\frac{y_{\text{FB}}^2}{2\sigma^2}}. \tag{7.12}$$

$\frac{\mathrm{d}N_1^i}{\mathrm{d}y}$ in equation (7.11) is the thermal distribution of the ith hadron species from a single fireball and can easily be calculated. The distribution function

$$f(x, p) = \frac{1}{\left[\mathrm{e}^{\beta\left(\sqrt{p^2+m_i^2}-\mu_i\right)} + \eta\right]}$$

gives the average number of particles of the ith hadron in a small volume element d^3x around $\mathbf{x}$ with momenta between $\mathbf{p}$ and $\mathbf{p} + \mathrm{d}\mathbf{p}$. The average number of particles can be obtained by integrating the distribution function over the phase space,

$$N_1^i = \frac{g_i}{(2\pi)^3}\int\frac{\mathrm{d}^3x\mathrm{d}^3p}{\left[\mathrm{e}^{\beta\left(\sqrt{p^2+m_i^2}-\mu_i\right)} + \eta\right]} = \frac{g_iV}{(2\pi)^3}\int\frac{\mathrm{d}^3p}{\left[\mathrm{e}^{\beta\left(\sqrt{p^2+m_i^2}-\mu_i\right)} + \eta\right]}, \tag{7.13}$$

where g_i is the degeneracy factor. The invariant distribution for a hadron species can be obtained from the above as (omitting the subscript i)

$$E\frac{\mathrm{d}N_1}{\mathrm{d}^3p} = \frac{\mathrm{d}N_1}{\mathrm{d}y\mathrm{d}^2p_{\text{T}}} = \frac{gV}{(2\pi)^3}\frac{m_{\text{T}}\cosh y}{\left[\mathrm{e}^{\beta(m_{\text{T}}\cosh y-\mu)} + \eta\right]} \tag{7.14}$$

where we have expressed the energy in terms of the rapidity variable, $y = \frac{1}{2}\ln\frac{E+p_L}{E-p_L}$ and transverse mass, $m_T = \sqrt{m^2 + p_T^2}$. The equation can be integrated to obtain the rapidity distribution,

$$\frac{dN_1}{dy} = \frac{gV}{(2\pi)^3}\int d^2p_T \frac{m_T \cosh y}{\left[e^{\beta(m_T \cosh y - \mu)} + \eta\right]}. \tag{7.15}$$

In the Boltzmann limit, rapidity density in the thermal model can be expressed analytically

$$\frac{dN_1}{dy} = \frac{gV}{(2\pi)^2}\left[\frac{2T^3}{\cosh^2 y} + \frac{2mT^2}{\cosh y} + m^2 T\right]e^{\mu/T}. \tag{7.16}$$

Equation (7.14) in the Boltzmann limit can be integrated to obtain thermal model particle spectra as

$$\frac{dN_1}{m_T dm_T} = \frac{gV}{2\pi^2} m_T K_1(\beta m_T) \xrightarrow{m_T \gg T} V'\sqrt{m_T} e^{-\beta m_T}. \tag{7.17}$$

The thermal model extended for rapidity density introduces a new, energy dependent parameter, the Gaussian width σ of the fireball distribution. One generally fixes it by fitting experimental rapidity density for pions. At RHIC energy $\sqrt{s}_{NN} =$ 200 GeV, $\sigma = 2.18$ and it decreases at lower energy. Energy dependence is approximately given by $\sigma^2 = \ln\frac{\sqrt{s}_{NN}}{2m_p}$. In the analysis one generally assumes universality of the chemical freeze-out condition, i.e. the temperature and chemical potential are related by a freeze-out curve as in equation (7.7) or equation (7.9). If the temperature varies along the rapidity axis, then the chemical potential will also vary. Thus a decrease in the temperature of the fireball will be accompanied by an increase in the baryon chemical potential. Variation of μ_B as a function of energy was studied in [18].

$$\mu_B = 0.025 + 0.011 y_{FB}^2 \quad \text{at RHIC} \tag{7.18}$$

$$\mu_B = 0.237 + 0.011 y_{FB}^2 \quad \text{at SPS.} \tag{7.19}$$

7.3 The blast wave model

If a large amount of energy (E) is released in an infinitesimally small volume of ambient medium of density ρ_1, a shock wave or blast wave is created[3]. The spherical shock front will expand into the ambient medium. Early in the course of expansion,

[3] Technically, blast waves and shock waves are distinguished. In a shock wave thermodynamic variables, e.g. pressure, density, etc, vary discontinuously across the shock front. In a blast wave, there is a peak structure.

the pressure within the shock, $P_2 \sim \rho u_{sh}^2$, is much larger than the ambient pressure P_1 and any radiated energy is much smaller than the explosion energy E. This regime, during which the energy E remains constant, is known as the blast wave regime. In a blast wave, the expansion velocity $u_{sh}(r, t)$, density $\rho(r, t)$, pressure $P(r, t)$ and other properties are determined solely by the two initial parameters of the system, initial energy E and ambient density ρ_1, and do not depend on the details of hydrodynamic evolution. This can be seen from dimensional analysis. The dimensions of the principal quantities are

$$[E] = ML^2T^{-2} \tag{7.20a}$$

$$[\rho_1] = ML^{-3} \tag{7.20b}$$

$$[t] = T. \tag{7.20c}$$

The only quantity of dimension length that can be constructed is

$$\left(\frac{Et^2}{\rho_1}\right)^{\frac{1}{5}} = L, \tag{7.21}$$

and any radius relevant to the problem must depend on these variables through this combination. At any time, the shock position can be obtained as

$$R(t) = \eta_s\left(\frac{Et^2}{\rho}\right)^{1/5} \tag{7.22}$$

and shock velocity,

$$v_{sh}(t) = \frac{\mathrm{d}R(t)}{\mathrm{d}t} = \frac{2}{5}\frac{R(t)}{t} \propto t^{-3/5}. \tag{7.23}$$

The blast wave model has been used by several authors to analyze experimental data in relativistic energy heavy ion collisions [22–26]. The model assumes that in a collision a fireball is created, which expands suddenly to the freeze-out condition. The model thus assumes that all the particles, pions, nucleons and kaons decouple quite suddenly and for all of them the freeze-out happens at the same time, measured in a frame that co-moves longitudinally with the fluid element of the expanding fireball. Most importantly, in the blast wave model the final fluid parameters do not depend upon the details of the evolution and one can treat them as parameters. Using the well known Cooper–Frye prescription [27], the invariant distribution for the particles can be written in terms of those parameters. The parameters are then fitted from experimental observables such as particle spectra, flow, etc. Before we continue, we discuss the Cooper–Frye prescription for particle invariant distribution.

7.3.1 The Cooper–Frye prescription for particle distribution

Consider a three-dimensional hypersurface $\Sigma(x)$ in four-dimensional Minkowski space–time and count the number of particles crossing the hypersurface. Let $\mathrm{d}\sigma_\mu$ be an infinitesimal element perpendicular to $\Sigma(x)$ and directed outward. If j^μ is the current of particles, then the scalar product $\mathrm{d}\sigma_\mu j^\mu$ gives the number of particles crossing the infinitesimal surface $\mathrm{d}\sigma$. The total number crossing the hyper surface Σ is

$$N = \int_\Sigma \mathrm{d}\sigma_\mu j^\mu = \int_\Sigma \mathrm{d}\sigma_\mu \left(\frac{1}{(2\pi)^3} \int \frac{\mathrm{d}^2 p}{E} p^\mu f(x, p) \right) \tag{7.24}$$

where we have used the kinetic theory definition for the current with $f(x, p)$ as the one-body distribution function.

Equation (7.24) can be cast into a differential form,

$$E\frac{\mathrm{d}N}{\mathrm{d}^3 p} = \frac{\mathrm{d}N}{\mathrm{d}y\mathrm{d}^2 p_T} = \frac{1}{(2\pi)^3} \int_\Sigma \mathrm{d}\sigma_\mu p^\mu f(x, p). \tag{7.25}$$

This is the Cooper–Frye prescription [27] for the invariant distribution of particles.

In general, a three-dimensional hypersurface is parameterized by three locally orthogonal coordinates (ζ, η, Φ) and the normal vector on the hypersurface is determined by

$$\mathrm{d}\Sigma_\mu = \epsilon_{\mu\alpha\beta\gamma} \frac{\mathrm{d}\Sigma^\alpha}{\mathrm{d}\zeta} \frac{\mathrm{d}\Sigma^\beta}{\mathrm{d}\eta} \frac{\mathrm{d}\Sigma^\gamma}{\mathrm{d}\Phi} \mathrm{d}\zeta \mathrm{d}\eta \mathrm{d}\Phi \tag{7.26}$$

where $\epsilon_{\mu\alpha\beta\gamma}$ is the Levi–Civita antisymmetric tensor. $\epsilon_{\mu\alpha\beta\gamma} = -1$ for $(\mu\alpha\beta\gamma)$, an even permutation of (0123). Different components can be obtained as,

$$\mathrm{d}\Sigma^0 = \mathrm{d}\Sigma_0 = \epsilon_{0\alpha\beta\gamma} \frac{\mathrm{d}\Sigma^\alpha}{\mathrm{d}\zeta} \frac{\mathrm{d}\Sigma^\beta}{\mathrm{d}\eta} \frac{\mathrm{d}\Sigma^\gamma}{\mathrm{d}\Phi} \mathrm{d}\zeta \mathrm{d}\eta \mathrm{d}\Phi = \frac{\partial(\Sigma^1, \Sigma^2, \Sigma^3)}{\partial(\zeta, \eta, \Phi)} \mathrm{d}\zeta \mathrm{d}\eta \mathrm{d}\Phi \tag{7.27a}$$

$$\mathrm{d}\Sigma^1 = -\mathrm{d}\Sigma_1 = -\epsilon_{1\alpha\beta\gamma} \frac{\mathrm{d}\Sigma^\alpha}{\mathrm{d}\zeta} \frac{\mathrm{d}\Sigma^\beta}{\mathrm{d}\eta} \frac{\mathrm{d}\Sigma^\gamma}{\mathrm{d}\Phi} \mathrm{d}\zeta \mathrm{d}\eta \mathrm{d}\Phi = \frac{\partial(\Sigma^2, \Sigma^3, \Sigma^0)}{\partial(\zeta, \eta, \Phi)} \mathrm{d}\zeta \mathrm{d}\eta \mathrm{d}\Phi \tag{7.27b}$$

$$\mathrm{d}\Sigma^2 = -\mathrm{d}\Sigma_2 = -\epsilon_{2\alpha\beta\gamma} \frac{\mathrm{d}\Sigma^\alpha}{\mathrm{d}\zeta} \frac{\mathrm{d}\Sigma^\beta}{\mathrm{d}\eta} \frac{\mathrm{d}\Sigma^\gamma}{\mathrm{d}\Phi} \mathrm{d}\zeta \mathrm{d}\eta \mathrm{d}\Phi = \frac{\partial(\Sigma^3, \Sigma^0, \Sigma^1)}{\partial(\zeta, \eta, \Phi)} \mathrm{d}\zeta \mathrm{d}\eta \mathrm{d}\Phi \tag{7.27c}$$

$$\mathrm{d}\Sigma^3 = -\mathrm{d}\Sigma_3 = -\epsilon_{3\alpha\beta\gamma} \frac{\mathrm{d}\Sigma^\alpha}{\mathrm{d}\zeta} \frac{\mathrm{d}\Sigma^\beta}{\mathrm{d}\eta} \frac{\mathrm{d}\Sigma^\gamma}{\mathrm{d}\Phi} \mathrm{d}\zeta \mathrm{d}\eta \mathrm{d}\Phi = \frac{\partial(\Sigma^0, \Sigma^1, \Sigma^2)}{\partial(\zeta, \eta, \Phi)} \mathrm{d}\zeta \mathrm{d}\eta \mathrm{d}\Phi \tag{7.27d}$$

where

$$\frac{\partial(x, y, z)}{\partial(\alpha, \beta, \gamma)} = \begin{vmatrix} \frac{\partial x}{\partial \alpha} & \frac{\partial x}{\partial \beta} & \frac{\partial x}{\partial \gamma} \\ \frac{\partial y}{\partial \alpha} & \frac{\partial y}{\partial \beta} & \frac{\partial y}{\partial \gamma} \\ \frac{\partial z}{\partial \alpha} & \frac{\partial z}{\partial \beta} & \frac{\partial z}{\partial \gamma} \end{vmatrix}. \tag{7.28}$$

If the freeze-out surface is properly parameterized, the invariant distribution of particles can be calculated from equation (7.25) if the distribution function $f(x, p)$ is known. The simplest freeze-out surface can be constructed as

$$\mathrm{d}\Sigma^{\mu} = \left(\mathrm{d}^3x, 0, 0, 0\right) \tag{7.29}$$

whence the Cooper–Frye invariant distribution and thermal model spectra become identical.

The Cooper–Frye prescription for a boost invariant system. If we assume boost-invariance, the freeze-out surface, in

$$\left(\tau = \sqrt{t^2 - z^2}, x, y, \eta = \frac{1}{2}\ln\frac{t+z}{t-z}\right)$$

coordinates, can be characterized by a longitudinal proper time $\tau_f(x, y) \equiv \tau_f(r_\perp)$,

$$\Sigma^{\mu}(r_\perp, \eta) = \left(t_f, x_f, y_f, z_f\right) = \left(\tau_f(x, y)\cosh\eta, x, y, \tau_f(x, y)\sinh\eta\right), \tag{7.30}$$

and from equation (7.26), one readily obtains,

$$\mathrm{d}\Sigma_{\mu} = \left(\cosh\eta, -\frac{\partial\tau_f}{\partial x}, -\frac{\partial\tau_f}{\partial y}, -\sinh\eta\right)\tau_f(x, y)\mathrm{d}x\mathrm{d}y\mathrm{d}\eta. \tag{7.31}$$

If the particle four-momentum p^{μ} is parameterized as,

$$p^{\mu} = \left(m_{\mathrm{T}}\cosh y, p_x, p_y, m_T\sinh y\right), \tag{7.32}$$

it is easy to calculate the volume element $p^{\mu}\mathrm{d}\Sigma_{\mu}$.

$$p^{\mu}\mathrm{d}\Sigma_{\mu} = \left(m_{\mathrm{T}}\cosh(y - \eta) - p_x\frac{\partial\tau_f}{\partial x} - p_y\frac{\partial\tau_f}{\partial y}\right)\tau_f\mathrm{d}x\mathrm{d}y\mathrm{d}\eta. \tag{7.33}$$

For a boost-invariant system, the fluid four-velocity can be parameterized as,

$$u^{\mu} = \gamma_{\mathrm{T}}\left(\cosh\eta, v_x, v_y, \sinh\eta\right), \tag{7.34}$$

$$\gamma_{\mathrm{T}} = \frac{1}{\sqrt{1 - v_T^2}} = \frac{1}{\sqrt{1 - v_x^2 - v_y^2}}, \tag{7.35}$$

and one can calculate the scalar product,

$$p \cdot u = p^\mu u_\mu = \gamma_T \left(m_T \cosh(y - \eta) - v_x p_x - v_y p_y \right). \tag{7.36}$$

One generally uses the Lorentz-covariant form for the distribution function,

$$f(x, p) = \frac{g}{e^{\left[p^\mu u_\mu(x) - \mu(x)\right]/T(x)} \pm 1} = g \sum_{n=1}^{\infty} (\mp)^{n+1} e^{-n\left[p^\mu u_\mu(x) - \mu(x)\right]/T(x)} \tag{7.37}$$

where g is the degeneracy of the particle. The plus and minus sign in the denominator accounts for proper quantum statistics of the particle species, (+) for fermions and (−) for bosons. The factor $p \cdot u$ in the exponent is the energy of the particle in the local rest frame ($p^\mu u_\mu \to p^0 = E$ when $u^\mu \to (1, 0)$).

The Cooper–Frye invariant distribution can now be computed as

$$E\frac{dN}{d^3p} = \frac{g_i}{(2\pi)^3} \sum_{n=1}^{\infty} (\mp)^{n+1} \int \tau_f \, dx dy d\eta \left(m_T \cosh(y - \eta) - p_x \frac{\partial \tau_f}{\partial x} - p_y \frac{\partial \tau_f}{\partial y} \right)$$
$$\times \exp\left[\frac{n\mu}{T}\right] \exp\left[\frac{-n\gamma_T \left\{ m_T \cosh(y - \eta_s) - v_x p_x - v_y p_y \right\}}{T}\right]. \tag{7.38}$$

Using the standard relation for modified Bessel functions,

$$K_\nu(z) = \int_0^\infty e^{-z \cosh t} \cosh(\nu t) dt, \tag{7.39}$$

the integration over the rapidity variable η can be performed analytically,

$$E\frac{dN}{d^3p} = \frac{2g_i}{(2\pi)^3} \sum_{n=1}^{\infty} (\mp)^{n+1} \int \tau_f \, dx dy \left[m_T K_1\left(\frac{n\gamma_T m_T}{T}\right) - \left(p_x \frac{\partial \tau_f}{\partial x} + p_y \frac{\partial \tau_f}{\partial y} \right) K_0\left(\frac{n\gamma_T m_T}{T}\right) \right]$$
$$\times \exp\left[\frac{n\mu}{T}\right] \exp\left[\frac{n\gamma_T \left\{ v_x p_x + v_y p_y \right\}}{T}\right]. \tag{7.40}$$

If the system possesses azimuthal symmetry, then

$$dx dy = r_T dr_T d\phi, \tag{7.41}$$

$$v_x p_x + v_y p_y = v_T p_T \cos\left(\phi - \phi_p\right), \tag{7.42}$$

$$p_x \frac{\partial \tau_f}{\partial x} + p_y \frac{\partial \tau_f}{\partial y} = p_T \frac{\partial \tau_f}{\partial r_T} \cos\left(\phi - \phi_p\right) \tag{7.43}$$

and the integration over ϕ can be formally completed using the relation for the modified Bessel function,

$$I_\nu(z) = \frac{1}{\pi}\int_0^\pi e^{z\cos\theta}\cos(\nu\theta)d\theta, \tag{7.44}$$

$$E\frac{dN}{d^3p} = \frac{g_i}{2\pi^2}\sum_{n=1}^{\infty}(\mp)^{n+1}\int \tau_f r_T dr_T\left[m_T K_1\left(\frac{n\gamma_T m_T}{T}\right)I_0\left(\frac{n\gamma_T v_T p_T}{T}\right)\right. \tag{7.45}$$

$$\left. - p_T\frac{\partial\tau_f}{\partial r_T}K_0\left(\frac{n\gamma_T m_T}{T}\right)I_1\left(\frac{n\gamma_T v_T p_T}{T}\right)\right]\exp\left[\frac{n\mu}{T}\right]. \tag{7.46}$$

7.3.2 A blast wave model with cylindrical symmetry and boost invariance

When applied to a blast wave, several simplifications can be made to the Cooper–Frye prescription for invariant distribution. For example, in the blast wave model, it is assumed that the freeze-out occurs at a fixed time τ_f. Then terms involving $\frac{\partial}{\partial\tau_f}$ do not contribute. In the blast wave model, the fluid parameters do not depend on the details of the evolution and they may be assumed to be constant and come out of the integration. Furthermore, one generally neglects the quantum statistics. When these simplifications are made, invariant distribution in a boost-invariant, cylindrical symmetric blast wave model reduces to

$$E\frac{dN}{d^3p} = \frac{g_i}{2\pi^2}\tau_f R_f^2 m_T K_1\left(\frac{\gamma_T m_T}{T}\right)I_0\left(\frac{\gamma_T v_T p_T}{T}\right)\exp\left[\frac{\mu}{T}\right]. \tag{7.47}$$

This is the most widely used form of blast wave model. The variables, τ_f, R_f, T, v_T and μ are treated as parameters. They can be obtained by fitting experimental data.

7.4 Color glass condensate

Theoretical considerations indicate that prior to QGP, a new form of matter, 'color glass condensate' (CGC) may be formed. We briefly describe here the beautiful concept behind the CGC. A more elaborate description may be found in [28–32]. According to the theory, CGC controls the high-energy limit of the strong interaction and should describe (i) high-energy cross sections, (ii) distribution of produced particles in high-energy collisions, (iii) distribution of small x particles in a hadron and (iv) initial conditions for heavy ion collisions.

As we know hadrons consist of gluons, quarks and anti-quarks. The constituents of hadrons, e.g. quarks and gluons, are generically called partons (the parton name was given by Richard Feynman). At very high energy, the hadron wave function has contributions from partons, e.g. gluons, quarks and anti-quarks. A convenient variable to measure the contribution of constituents to the hadron wave function is the fraction of the momentum carried by the constituents (Bjorken x variable),

$$x = E_{\text{constituent}}/E_{\text{hadron}}. \tag{7.48}$$

The probability $G(x)$ of obtaining a parton with momentum fraction x and $x + \mathrm{d}x$ is generally called the parton distribution function. The parton distribution function depends weakly on the resolution scale Q^2. One can write the density of small x partons as

$$\frac{\mathrm{d}N}{\mathrm{d}y} = xG\left(x,\, Q^2\right). \tag{7.49}$$

In figure 7.5, the gluon distribution function as measured in the Hadron Electron Ring Accelerator (HERA) is shown. One observes that gluon density rapidly increases at small x. It is also an increasing function of the resolution scale (Q^2). Increase in gluon density at small x is commonly referred to as the small x problem. It means that if we view the proton head-on with increasing energy, gluon density grows. QCD is an asymptotically free theory, and coupling constant decreases at short distances. As the density increases, typical separation between the gluons decreases and the strong coupling constant gets weaker. The higher the density, the more weakly the gluons interact. However, density cannot be increased indefinitely. Indefinite increase will then lead to infinite scattering amplitude and violate the unitary bound[4]. One then argues that as the gluon density increases, repulsive gluon interaction becomes important and in the balance, gluon density saturates. The saturation density will correspond to a saturation momentum scale, Q_{sat}.

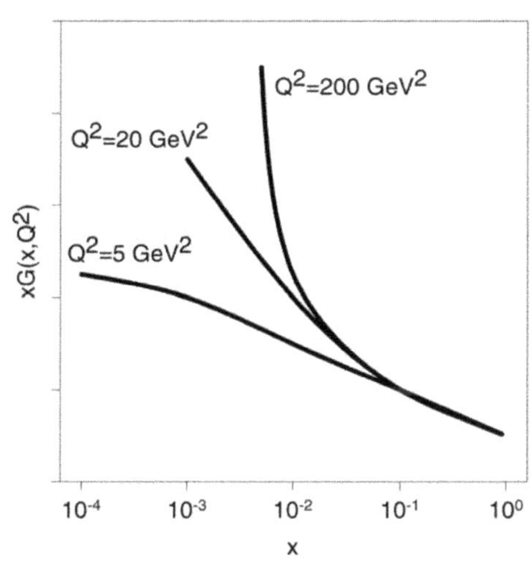

Figure 7.5. Schematic diagram for the gluon density measured in HERA in the three-momentum scale.

[4] The unitary bound is the constraint on any quantum system the that sum of all possible outcomes of evolution is unity.

Qualitatively, one can argue as follows: imagine a proton is being packed with fixed-size gluons. Then after a certain saturation density or close-pack density, repulsive interaction will take over and no more gluons can be added to the proton. Naturally, the saturation density depends on the gluon size; for a smaller sized gluon, the saturation density will increase. Then there is a characteristic momentum scale $Q_{\rm sat}$ which corresponds to the inverse of the smallest sized gluons which are close packed. Note that the saturation scale only tells us that gluons of size $1/Q_{\rm sat}$ have stopped growing. It does not mean that the number of gluons has stopped growing.

It is very reasonable to assume that some effective potential V describes the system of gluons. If the phase space density of gluons is denoted by ρ,

$$\rho = \frac{1}{\pi R^2}\frac{dN}{dyd^2p_T}, \tag{7.50}$$

at low density, the system will want to increase the density and $V \sim -\rho$. On the other hand, repulsive interaction balances the inclination to condensate, $V_{\rm repulsion} \sim \alpha_s \rho^2$. These contributions balance each other when $\rho \sim 1/\alpha_s$. Density scaling as the inverse of interaction strength is characteristic of condensate phenomena such as superconductivity.

Phase space density $\rho = \frac{1}{\pi R^2}\frac{dN}{dyd^2p_T} \sim 1/\alpha_s$ can be integrated to obtain saturation momentum scale ($Q_{\rm sat}$),

$$\frac{1}{\pi R^2}\frac{dN}{dy} \sim \frac{1}{\alpha_s} Q_{\rm sat}^2. \tag{7.51}$$

The origin of the name 'color glass condensate' is now clear. The word color refers to gluons which are colored. The system is at very high density, hence the word condensate. The matter is of glassy nature. Glasses are disordered systems, which behave like liquids on long time scales and like solids on short time scales. The word 'glass' arises because the gluons evolve on a longer time scale than their natural one $1/Q_{\rm sat}$. The small x gluons are produced from gluons at larger values of x. The time scale of the fast gluons is Lorentz diluted and can be approximated as a static field. This scale is transferred to the small x gluons. The small x gluons can then be approximated as static classical fields.

CGC acts as an infrared cut off when computing total multiplicity. For momentum scale $p_T > Q_{\rm sat}$, the particles produced are incoherent and ordinary perturbation applies. For momentum scale $p_T \leqslant Q_{\rm sat}$, the particles produced are in a coherent state, which is color neutral on the length scale $1/Q_{\rm sat}$.

One may wonder about the quark's degrees of freedom. At high energy, gluon density grows faster than quark density and distribution is overwhelmingly gluonic. Fields associated with CGC can be treated as classical fields. Since they arise from fast moving partons, they are plane polarized, with mutually orthogonal color magnetic and electric fields perpendicular to the direction of motion of the hadron. They are also random in two dimensions (see figure 7.6).

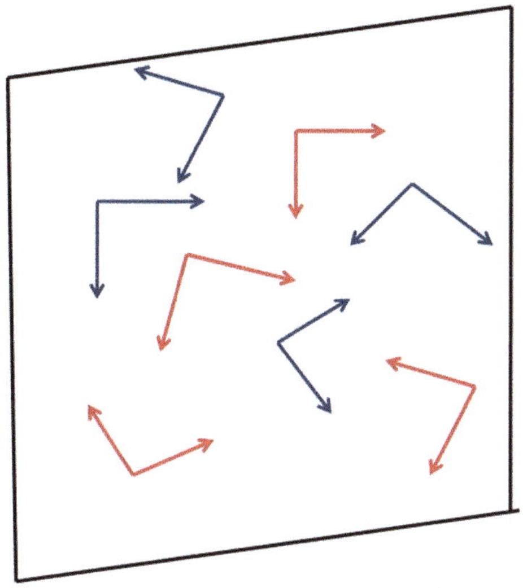

Figure 7.6. CGC representation of a hadron traveling near the speed of light. The hadron is contracted to a colored sheet with color electric and magnetic fields orthogonal to the direction of motion.

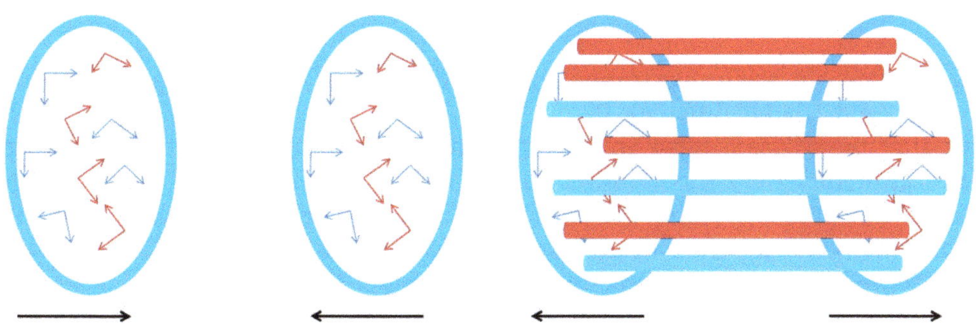

Figure 7.7. Left panel: collision of two sheets of color glass. Right panel: The longitudinal color electric and magnetic fields made in hadron collisions as the colored sheets pass each other.

High-energy hadronic or nuclear collisions can be considered as collisions of two color glass sheets as shown in figure 7.7. They can be viewed as classical fields of one hadron/nucleus propagating along the forward light cone $z = t$ and that of the other along the light-cone $z = -t$. The initial classical field can be obtained by adding them until the hadrons/nuclei collide. After the collision, classical Yang–Mills equations in the forward light cone can be solved to obtain the field configuration. Explicit solutions indicate that after the hadrons pass through one another, they develop a surface color charge (both electric and magnetic) density. This charge density is equal and opposite on each hadron. Because the hadrons have become charged, longitudinal color electric and color magnetic fields will form between them. A schematic diagram of this is shown in figure 7.7. The typical scale of transverse variation of these fields is the inverse saturation momentum $r \sim 1/Q_{\text{sat}}$.

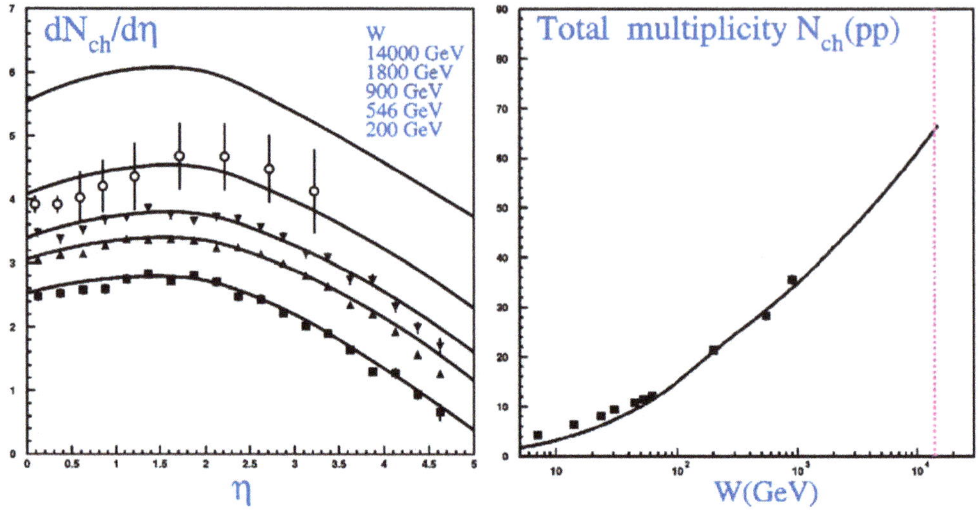

Figure 7.8. Left panel: CGC model predictions for rapidity dependence $dN/d\eta$ of charged hadron multiplicities in proton–proton (antiproton) collisions as a function of the pseudo-rapidity at different energies. Right panel: Energy dependence of total multiplicity in proton–proton (antiproton) collisions. The vertical doted line marks the LHC energies for proton–proton collisions ($W = 14000$ GeV). The figures are reproduced from [33] with permission.

These fields will evolve classically. Field configurations before and after the collisions are remarkably different. Before the collisions there are only transverse fields, but after the collision there are, in addition, longitudinal fields. The longitudinal fields grow on a very short time scale $t \ll 1/Q_{\text{sat}}$. The new matter formed in the collision is produced from the CGC, and is called 'glasma'. It has properties remarkably different from the CGC. The glasma fields eventually dilute to a distribution of gluons and later thermalize to form QGP.

There are many successful applications of the CGC model in explaining various experimental results. For the purpose of completeness, we will show two results obtained in the CGC model [33]. In figure 7.8, in two panels, the rapidity density of charged particles in pp collisions and the energy dependence of charge multiplicity are shown. The solid lines in the figure were obtained in a CGC based model. It is no small wonder that a CGC based model can explain the data. Such a description of the data from a first principle, model was not available earlier.

7.5 Hydrodynamic model

One of the most successful 'dynamical' models for relativistic energy heavy ion collision is the hydrodynamic model. In figure 7.1, we have depicted the different stages of collision. From the expansion stage to freeze-out, the collision process can be modeled by relativistic hydrodynamic equations. If the macroscopic properties of the fluid, e.g. local energy density, pressure, fluid velocity, etc, are known at the initial time τ_i, hydrodynamic equations can be solved to obtain the space–time evolution of the fireball until the freeze-out. At the freeze-out a suitable algorithm (e.g. Cooper–Frye [27]) can be used to convert the fluid information at the freeze-out into a particle's invariant yield and be compared with experimental data.

Hydrodynamic models are unique in that it is imperative to use an equation of state, a thermodynamic relation between the energy density, pressure and number density of the fluid. By explicitly incorporating phase transition into the equation of state, one can study dynamically the effect of phase transition on the fluid evolution and associated particle production. Ideal hydrodynamic models have been largely successful in explaining a variety of experimental data, e.g. transverse momentum spectra and elliptic flow (to be explained later) of identified particles in $\sqrt{s}_{\rm NN} = 200$ GeV Au+Au collisions [34]. Indeed, the success of *ideal* fluid dynamics in explaining several experimental data, together with the string theory motivated lower limit of shear viscosity over entropy ratio $\frac{\eta}{s} \geqslant$ [35–37] leads to a paradigm that in Au+Au collisions, a nearly 'perfect' fluid is created.

However, the paradigm of 'perfect fluid' produced in Au+Au collisions at RHIC needs to be clarified. It so happens that the ideal fluid dynamic models do have their limitations [38, 39]. For example, experimentally, elliptic flow tends to saturate at large transverse momentum. The ideal fluid dynamics on the other hand predicts a continually increasing elliptic flow. The transverse momentum spectra of identified particles also starts to deviate from ideal fluid dynamics predictions beyond $p_T \approx$ 1.5 GeV. Experimentally determined Hanbury Brown–Twiss (HBT) radii (also to be explained later) are not reproduced in the ideal fluid dynamic models, the famous 'HBT puzzle' [40]. Ideal fluid dynamics also works best in central collisions and gets poorer in more peripheral collisions. The shortcomings of ideal fluid dynamics possibly indicate the greater importance of dissipative effects in p_T ranges greater than 1.5 GeV or in more peripheral collisions. Indeed, ideal fluid is a concept which is never realized in nature. As suggested in string theory motivated models [35–37], QGP viscosity could be small, $\eta/s \geqslant 1/4\pi$, but nevertheless it is non-zero. Indeed, much earlier, Danielewicz and Gyulassy [41], using the uncertainty principle, estimated the lower bound of the viscosity to entropy ratio, $\eta/s \geqslant 1/12$, a value very close to the AdS/CFT estimate. It is thus important to study the effect of viscosity, even if small, on the space–time evolution of QGP fluid and quantify its effect. Furthermore, QGP fluid has to be characterized by its transport coefficients, e.g. heat conductivity, bulk and shear viscosity. Theoretically, it is possible to obtain these transport coefficients in a kinetic theory model. However, in the current status of the theory, the goal cannot be achieved immediately, even more so for a strongly interacting QGP (sQGP). Alternatively, one can compare viscous hydrodynamic simulations to experimental data and obtain a 'phenomenological' limit to the transport coefficients of sQGP.

There is another incentive to study dissipative hydrodynamics. Ideal hydrodynamics depends on the assumption of local equilibrium. In dissipative hydrodynamics, the strict assumption of local thermal equilibrium is relaxed to the assumption of 'near' local thermal equilibrium, extending the range of validity of hydrodynamic description. The term 'near' in the previous sentence has certain implications. Indeed, theoretical modeling of a strongly non-equilibrium system has not yet been developed. With the qualification of 'near' equilibrium, dissipative hydrodynamics then explores a time earlier that than in ideal hydrodynamics. It may be mentioned here that the success of hydrodynamical modeling of relativistic heavy ion collisions does not necessarily imply the realization of local thermodynamic equilibrium, although the inverse is true. Whether or not local equilibrium is achieved in high-energy nuclear collisions is still in debate.

A theory of dissipative relativistic fluid was formulated quite early. The original dissipative relativistic fluid equations were given by Eckart [42] and Landau and Lifshitz [43]. They are called first order theories. Formally, relativistic dissipative hydrodynamic equations are obtained from an expansion of entropy four-current, in terms of the dissipative fluxes. In first order theories, entropy four-current contains terms linear in dissipative quantities. The first order theory of dissipative hydrodynamics suffers from the problem of causality violation and instabilities [44, 45]. Causality violation is unwarranted in any theory, even more so in a relativistic theory. The problem of causality violation is removed in Israel–Stewart's second order theory of dissipative fluid [46, 47]. In second order theory, the expansion of entropy four-current contains second order terms in dissipative fluxes. However, these lead to complications that dissipative fluxes are no longer functions of the state variables only. They become dynamic. The space of thermodynamic variables has to be extended to include the dissipative fluxes (e.g. heat conductivity, bulk and shear viscosity). In the following section Israel–Stewart's phenomenological theory of dissipative hydrodynamics is briefly discussed. More detailed exposition can be found in [47, 48].

7.5.1 Dissipative fluid dynamics

In a previous chapter, it was shown that hydrodynamic equations follow from the Boltzmann transport equation. A simple fluid[5], in an arbitrary state, is fully specified by primary variables: particle current (N^μ), energy–momentum tensor ($T^{\mu\nu}$) and entropy current (S^μ) and a number of additional variables[6]. Primary variables satisfy the conservation laws

$$\partial_\mu N^\mu = 0, \tag{7.52a}$$

$$\partial_\mu T^{\mu\nu} = 0, \tag{7.52b}$$

and the second law of thermodynamics,

$$\partial_\mu S^\mu \geqslant 0. \tag{7.53}$$

In relativistic fluid dynamics, one defines a time-like hydrodynamic four-velocity, u^μ (normalized as $u^2 = 1$). One also defines a projector, $\Delta^{\mu\nu} = g^{\mu\nu} - u^\mu u^\nu$,

[5] A simple fluid is composed of either a single element or molecules involving no more than two types of atoms. Limitation in microscopic structure allows for more complete determination of the macroscopic dynamics.

[6] In principle, the additional variables are infinite in number. The fluid condition is determined by the distribution function f or alternatively by all moments of the distribution function. N^μ and $T^{\mu\nu}$ are only the first and second moment. All the higher moments are needed to fully specify the fluid.

orthogonal to the four-velocity ($\Delta^{\mu\nu}u_\nu = 0$). In equilibrium, a unique four-velocity (u^μ) exists such that the particle density (n), energy density (ε) and entropy density (s) can be obtained from

$$N^\mu_{\text{eq}} = nu^\mu \tag{7.54a}$$

$$T^{\mu\nu}_{\text{eq}} = \varepsilon u^\mu u^\nu - p\Delta^{\mu\nu} \tag{7.54b}$$

$$S^\mu_{\text{eq}} = su^\mu. \tag{7.54c}$$

An equilibrium state is assumed to be fully specified by five-parameters $(n,\ \varepsilon,\ u^\mu)$ or equivalently by the thermal potential $\alpha = \mu/T$ (μ being the chemical potential) and inverse four-temperature $\beta^\mu = u^\mu/T$. Given an equation of state, $s = s(\varepsilon, n)$, pressure p can be obtained from the generalized thermodynamic relation,

$$S^\mu_{\text{eq}} = p\beta^\mu - \alpha N^\mu_{\text{eq}} + \beta_\lambda T^{\lambda\mu}_{\text{eq}}. \tag{7.55}$$

Using the Gibbs–Duhem relation,

$$\mathrm{d}\left(p\beta^\mu\right) = N^\mu_{\text{eq}}\mathrm{d}\alpha - T^{\lambda\mu}_{\text{eq}}\mathrm{d}\beta_\lambda, \tag{7.56}$$

the following relations can be established on the equilibrium hyper-surface $\Sigma_{\text{eq}}\left(\alpha,\ \beta^\mu\right)$,

$$\mathrm{d}S^\mu_{\text{eq}} = -\alpha\mathrm{d}N^\mu_{\text{eq}} + \beta_\lambda\mathrm{d}T^{\lambda\mu}_{\text{eq}}. \tag{7.57}$$

In a non-equilibrium system, no four-velocity can be found such that equations (7.54*a*), (7.54*b*) and (7.54*c*) remain valid. Tensor decomposition leads to additional terms,

$$N^\mu = N^\mu_{\text{eq}} + \delta N^\mu = nu^\mu + V^\mu \tag{7.58a}$$

$$\begin{aligned} T^{\mu\nu} &= T^{\mu\nu}_{\text{eq}} + \delta T^{\mu\nu} \\ &= [\varepsilon u^\mu u^\nu - p\Delta^{\mu\nu}] + \Pi\Delta^{\mu\nu} + \pi^{\mu\nu} + (W^\mu u^\nu + W^\nu u^\mu) \end{aligned} \tag{7.58b}$$

$$S^\mu = S^\mu_{\text{eq}} + \delta S^\mu = su^\mu + \Phi^\mu. \tag{7.58c}$$

The new terms describe a net flow of charge $V^\mu = \Delta^{\mu\nu}N_\nu$, energy flow $W^\mu = (\varepsilon + p)/nV^\mu + q^\mu$ (where q^μ is the heat flow vector), and entropy flow Φ^μ. $\Pi = -\frac{1}{3}\Delta_{\mu\nu}T^{\mu\nu} - p$ is the bulk viscous pressure and $\pi^{\mu\nu} = \left[\frac{1}{2}\left(\Delta^{\mu\sigma}\Delta^{\nu\tau} + \Delta^{\nu\sigma}\Delta^{\mu\tau} - \frac{1}{3}\Delta^{\mu\nu}\Delta^{\sigma\tau}T_{\sigma\tau}\right]\right.$ is the shear stress tensor. The hydrodynamic four-velocity can be chosen to eliminate either V^μ (the Eckart frame u^μ is parallel to particle flow) or the energy flow W^μ (the Landau frame, u^μ is eigenvector of the energy–momentum tensor). In relativistic heavy ion collisions, the central rapidity region is nearly baryon-free and Landau's frame is more appropriate than Eckart's frame. Dissipative flows are transverse to u^μ and additionally, the shear stress tensor is traceless. Thus a non-equilibrium state requires $1 + 3 + 5 = 9$ additional quantities, the dissipative flows Π, q^μ (or V^μ) and $\pi^{\mu\nu}$. In kinetic theory, N^μ and $T^{\mu\nu}$ are the first and second moment of the distribution function. Unless the function is known *a priori*, two moments do not furnish enough information to enumerate the microscopic states required to determine S^μ, and in an arbitrary non-equilibrium state, no relation exists between N^ν, $T^{\mu\nu}$ and S^μ. *Only in a state close to equilibrium can such a relation be established.* Assuming that the equilibrium relation equation (7.57) also remains valid in a 'near equilibrium state', the entropy current can be generalized as

$$S^\mu = S^\mu_{\rm eq} + {\rm d}S^\mu = p\beta^\mu - \alpha N^\mu + \beta_\lambda T^{\lambda\mu} + Q^\mu \tag{7.59}$$

where Q^μ is an undetermined quantity in the second order in deviations, $\delta N^\mu = N^\mu - N^\mu_{\rm eq}$ and $\delta T^{\mu\nu} = T^{\mu\nu} - T^{\mu\nu}_{\rm eq}$. First order deviations do not contribute due to the Gibbs–Duhem relation. A detailed form of Q^μ is constrained by the second law $\partial_\mu S^\mu \geqslant 0$. With the help of conservation laws and the Gibbs–Duhem relation, the entropy production rate can be written as,

$$\partial_\mu S^\mu = -\delta N^\mu \partial_\mu \alpha + \delta T^{\mu\nu} \partial_\mu \beta_\nu + \partial_\mu Q^\mu. \tag{7.60}$$

The choice of Q^μ leads to first order or second order theories of dissipative hydrodynamics. In first order theories the simplest choice is made, $Q^\mu = 0$, and the entropy current contains terms up to the first order in deviations, δN^μ and $\delta T^{\mu\nu}$. The entropy production rate can be written as,

$$T\partial_\mu S^\mu = \Pi X - q^\mu X_\mu + \pi^{\mu\nu} X_{\mu\nu} \tag{7.61}$$

where, $X = -\nabla . u$; $X^\mu = \frac{\nabla^\mu}{T} - u^\nu \partial_\nu u^\mu$ and $X^{\mu\nu} = \nabla^{<\mu} u^{\nu>}$.

The second law, $\partial_\mu S^\mu \geqslant 0$, can be satisfied by postulating a linear relation between the dissipative flows and thermodynamic forces,

$$\Pi = -\zeta\theta, \tag{7.62}$$

$$q^\mu = -\lambda \frac{nT^2}{\varepsilon + p} \nabla^\mu(\mu/T), \tag{7.63}$$

$$\pi^{\mu\nu} = 2\eta \nabla^{<\mu} u^{\nu>} \tag{7.64}$$

where ζ, λ and η are the positive transport coefficients, bulk viscosity, heat conductivity and shear viscosity, respectively.

In first order theories, causality is violated [44, 45]. Causality violation is corrected in second order theories [47]. In second order theories, the entropy current contains terms up to second order in the deviations, $Q^\mu \neq 0$. The most general Q^μ containing terms up to second order in deviations can be written as,

$$Q^\mu = -\left(\beta_0\Pi^2 - \beta_1 q^\nu q_\nu + \beta_2 \pi_{\nu\lambda}\pi^{\nu\lambda}\right)\frac{u^\mu}{2T} - \frac{\alpha_0 \Pi q^\mu}{T} + \frac{\alpha_1 \pi^{\mu\nu} q_\nu}{T} \tag{7.65}$$

where β_0, β_1, and β_2 are thermodynamic coefficients for the bulk stress (Π), heat flow (q^μ) and shear viscous stress ($\pi^{\mu\nu}$), respectively. α_0 and α_1 are thermodynamic coefficients for the coupling between the heat flow and bulk and shear stress, respectively. As before, one can cast the entropy production rate ($T\partial_\mu S^\mu$) in the form of equation (7.61). Neglecting the terms involving dissipative flows with gradients of equilibrium thermodynamic quantities (both are assumed to be small) and demanding that a linear relation exists between the dissipative flows and thermodynamic forces, the following *relaxation* equations for the dissipative flows can be obtained,

$$\tau_\Pi \dot{\Pi} + \Pi = -\zeta\theta - \left[\frac{1}{2}\zeta T \partial_\mu\left(\frac{\tau_0}{\zeta T}u^\mu\right)\Pi\right] + l_{\Pi q}\nabla_\mu q^\mu, \tag{7.66a}$$

$$\tau_\pi \Delta^\alpha_\mu \Delta^\beta_\nu \dot{\pi}_{\alpha\beta} + \pi_{\mu\nu} = 2\eta\sigma_{\mu\nu} - \left[\eta T \partial_\lambda\left(\frac{\tau_2}{2\eta T}u^\lambda\right)\pi_{\mu\nu}\right] + l_{\pi q}\nabla_{\langle \mu}q_{\nu\rangle}, \tag{7.66b}$$

$$\begin{aligned}\tau_q \Delta^\nu_\mu \dot{q}_\nu + q_\mu = {} & \lambda\left(\nabla_\mu T - T\dot{u}_\mu\right) + \left[\frac{1}{2}\lambda T^2 \partial_\nu\left(\frac{\tau_1}{\lambda T^2}u^\nu\right)q_\mu\right] \\ & - l_{q\Pi}\nabla_\mu - l_{q\pi}\nabla_\mu \pi^\nu_\mu.\end{aligned} \tag{7.66c}$$

The relaxation times are,

$$\tau_\Pi = \zeta\beta_0 \qquad \tau_\pi = 2\eta\beta_2 \qquad \text{and} \qquad \tau_q = \lambda T\beta_1. \tag{7.67}$$

l_{ab} are coupling coefficients between different dissipative flows, specifically,

$$l_{\Pi q} = \zeta\alpha_0 \qquad l_{q\Pi} = \lambda T\alpha_0 \qquad l_{\pi q} = 2\eta\alpha_1 \qquad l_{q\pi} = \lambda T\alpha_1. \tag{7.68}$$

Unlike in the first order theories, in second order theories dynamical equations control the dissipative flows. Even if thermodynamic forces vanish, dissipative flows do not vanish instantly. It is important to mention that the parameters $\alpha = \frac{\mu}{T}$ and $\beta_\lambda = \frac{u_\lambda}{T}$ are not connected to the actual state (N^μ, $T^{\mu\nu}$). The pressure p in equation (7.59) is also not the 'actual' thermodynamic pressure, i.e. not the work done in an isentropic expansion. Chemical potential α and inverse four-temperature β_λ have meaning only for the equilibrium state. Their meaning need not be extended to non-equilibrium states also. However, it is possible to fit a fictitious 'local equilibrium' state, point by point, such that pressure p in equation (7.59) can be identified with the thermodynamic pressure, at least up to the first order. The conditions of fit fix the underlying non-equilibrium phase space distribution.

It may be mentioned here that relaxation equations for the dissipative fluxes can also be derived in kinetic theory [47, 49–52]. In kinetic theory, relaxation equations as well as explicit expressions for transport coefficients, relaxation times, etc, can be obtained. There is no unique method to obtain the relaxation equations from kinetic theory. For example, Israel–Stewart obtained the relaxation equations from the second moment of the kinetic equation. Betz *et al* [49], on the other hand, obtained them from the kinetic equation itself. Both the methods gave identical relaxation equations. However, while the relaxation equations remain unchanged, the relaxation time and coupling coefficients do depend on whether the zeroth moment or the second moment of the kinetic equation is used to derive the relaxation equations.

Among the three dissipative coefficients, shear viscosity appears to be most important in heavy ion collisions. In a collision, initial momentum is predominantly longitudinal. Some shear force must act on it to isotropize the momentum distribution (as required by the assumption of thermal equilibrium). Most of the dissipative hydrodynamic studies for heavy ion collisions are thus concerned with the effect of shear viscosity on fluid evolution and subsequent particle emission. Bulk viscosity in general is an order of magnitude less than shear viscosity. However, there are indications that in QGP, near the transition point, bulk viscosity can be large [53, 54]. Recently, the effect of bulk viscosity on particle production has been investigated [55–60]. There is some uncertainty about the correct form of the non-equilibrium correction to the equilibrium distribution function in the presence of bulk viscosity. Grad's 14-moment method for non-equilibrium correction appears to give very large correction even for small bulk viscosity [60]. However, the authors of [59] concluded differently. Using Grad's 14-moment method in orthogonal basis form [50] along with the bulk viscosity relaxation time of [52] does not lead to such problems. The conductivity of QGP fluid is the least studied dissipative coefficient. The central rapidity region in ultra-relativistic heavy ion collisions is essentially baryon-free and the effect of conductivity is minimal. However, in future Facility for Antiproton and Ion Research (FAIR) energy collisions, the effect of conductivity may be important.

7.5.2 Ideal hydrodynamic equations

Ideal hydrodynamic equations are used extensively to model relativistic heavy ion collisions. In the following, the general procedure followed in hydrodynamic modeling of heavy ion collisions is discussed. The approach can be generalized to the more complex problem of dissipative fluid.

For simplicity, we assume an ideal fluid with a single conserved charge (e.g. baryon density). The five conservation equations,

$$\partial_\mu N^\mu = 0, \tag{7.69}$$

$$\partial_\mu T^{\mu\nu} = 0 \tag{7.70}$$

govern the motion of the fluid. Given an initial configuration of the fluid and an equation of state $p = p(\varepsilon, n_B)$, the equations can be solved numerically to obtain the space–time evolution of the fluid.

In heavy ion collisions, appropriate coordinates for solving hydrodynamic equations are (τ, x, y, η) rather than (t, x, y, z).

$$x^\mu = (t, x, y, z) \to x^m = (\tau, x, y, \eta) \tag{7.71}$$

$$t = \tau \cosh \eta \qquad \tau = \sqrt{t^2 - z^2} \tag{7.72}$$

$$z = \tau \sinh \eta \qquad \eta = \frac{1}{2} \ln\left(\frac{t+z}{t-z}\right). \tag{7.73}$$

In the (τ, x, y, η) coordinate system, the metric is

$$\begin{aligned} ds^2 = g_{\mu\nu} dx^\mu dx^\nu &= dt^2 - dx^2 - dy^2 - dz^2 \\ &= d\tau^2 - dx^2 - dy^2 - \tau^2 d\eta^2 \end{aligned} \tag{7.74}$$

and

$$g^{\mu\nu} = \text{diag}\left(1, -1, -1, -1/\tau^2\right). \tag{7.75}$$

One notes that the space–time is not flat anymore, it is curved. Accordingly, one needs 'affine connections' or the Christoffel symbols,

$$\Gamma^i_{jk} = \frac{1}{2} g^{im}\left(\frac{\partial g_{mj}}{\partial x^k} + \frac{\partial g_{mk}}{\partial x^l} - \frac{\partial g_{jk}}{\partial x^m}\right). \tag{7.76}$$

In the (τ, x, y, η) coordinate the only non-zero Christoffel symbols are,

$$\Gamma^\tau_{\eta\eta} = \tau \qquad \Gamma^\eta_{\tau\eta} = 1/\tau. \tag{7.77}$$

The covariant (semicolon) derivative of a contravariant tensor is given by

$$A^i_{;p} = \frac{\partial A^i}{\partial x^p} + \Gamma^i_{pm} A^m \tag{7.78}$$

$$A^{ik}_{;p} = \frac{\partial A^{ik}}{\partial x^p} + \Gamma^i_{pm} A^{mk} + \Gamma^k_{pm} A^{mi}. \tag{7.79}$$

Five conservation equations in the (τ, x, y, η) coordinate system can be easily derived. For 'ideal fluid', they are

$$\begin{aligned} \text{(i)}\ N^\mu_{;\mu} = 0 &= N^\tau_{;\tau} + N^x_{;x} + N^y_{;y} + N^\eta_{;\eta} \\ &= \left(\partial_\tau N^\tau + \Gamma^\tau_{\tau m} N^m\right) + \partial_x N^x + \partial_y N^y + \left(\partial_\eta N^\eta + \Gamma^\eta_{\eta m} N^m\right) \\ &= \partial_\tau N^\tau + +\partial_x N^x + \partial_y N^y + \partial_\eta N^\eta + \frac{1}{\tau} N^\tau \end{aligned} \tag{7.80}$$

$$
\begin{aligned}
\text{(ii)}\ T^{\mu\tau}_{;\mu} = 0 &= T^{\tau\tau}_{;\tau} + T^{\tau x}_{;x} + T^{\tau y}_{;y} + T^{\tau\eta}_{;\eta} \\
&= \left(\partial_\tau T^{\tau\tau} + 2\Gamma^{\tau}_{\tau m} T^{m\tau}\right) + \partial_x T^{\tau x} + \partial_y T^{\tau y} \\
&\quad + \left(\partial_\eta T^{\tau\eta} + \Gamma^{\tau}_{\eta m} T^{m\eta} + \Gamma^{\eta}_{\eta m} T^{m\tau}\right) \\
&= \partial_\tau T^{\tau\tau} + \partial_x T^{\tau x} + \partial_y T^{\tau y} + \partial_\eta T^{\tau\eta} + \tau T^{\eta\eta} + \frac{1}{\tau} T^{\tau\tau}
\end{aligned}
\tag{7.81}
$$

$$
\begin{aligned}
\text{(iii)}\ T^{\mu x}_{;\mu} = 0 &= T^{\tau x}_{;\tau} + T^{xx}_{;x} + T^{xy}_{;y} + T^{\eta x}_{;\eta} \\
&= \partial_\tau T^{\tau y} + \partial_x T^{xy} + \partial_y T^{yy} + \partial_\eta T^{y\eta} + \frac{1}{\tau} T^{\tau x}
\end{aligned}
\tag{7.82}
$$

$$
\begin{aligned}
\text{(iv)}\ T^{\mu y}_{;\mu} = 0 &= T^{\tau y}_{;\tau} + T^{xy}_{;x} + T^{yy}_{;y} + T^{\eta y}_{;\eta} \\
&= \partial_\tau T^{\tau y} + \partial_x T^{xy} + \partial_y T^{yy} + \partial_\eta T^{y\eta} + \frac{1}{\tau} T^{\tau y}
\end{aligned}
\tag{7.83}
$$

$$
\begin{aligned}
\text{(v)}\ T^{\mu\eta}_{;\mu} = 0 &= T^{\tau\eta}_{;\tau} + T^{x\eta}_{;x} + T^{y\eta}_{;y} + T^{\eta\eta}_{;\eta} \\
&= \partial_\tau T^{\tau\eta} + \partial_x T^{x\eta} + \partial_y T^{y\eta} + \partial_\eta T^{\eta\eta} + \frac{3}{\tau} T^{\eta\tau}.
\end{aligned}
\tag{7.84}
$$

In ideal hydrodynamics, or in first order hydrodynamics with dissipation, the five partial equations noted above are to be solved simultaneously to obtain the space–time evolution of the fluid. One widely used algorithm to solve the non-linear generalized continuity equation of the type which occurs in fluid dynamics is a smooth and sharp transport algorithm (SHASTA) followed by flux corrected transport (FCT) [61]. SHASTA–FCT is essentially a three step process. In the first step, called transport, the velocities of the fluid and the source term are first calculated at half time steps $t_{n+1/2}$. In the second step, using the new velocity and the new source term in the nth step, one calculates the quantities at the $n + 1$th step. The third step is called the anti-diffusive step and is designed to remove the numerical diffusion inherent to the transport scheme. This is done by calculating an anti-diffusive flux, which is subtracted from the time-advanced quantities at the $n + 1$th step to get the final result at the $n + 1$th step. The calculation of the anti-diffusion is carried out by a method called 'flux correction'. For details of the method see [61, 62]. In second order dissipative hydrodynamics, however, ten additional relaxation equations have to be solved simultaneously.

For illustrative purposes, let us specialize in one-dimensional, Bjorken scaling flow [63]. Noting that, in high-energy collisions, the rapidity density $\frac{dN}{dy}$ is flat, Bjorken hypothesized that physics is rapidity independent, i.e. boost-invariant. If transverse expansion is neglected, for a boost-invariant system, hydrodynamic four-velocity can be written as

$$
u^\mu = \left(\frac{t}{\tau}, 0, 0, \frac{z}{\tau}\right) = (\cosh\eta, 0, 0, \sinh\eta). \tag{7.85}
$$

Since the system is boost-invariant, one can also solve the hydrodynamic equations at rapidity $\eta = 0$ and boost to any finite rapidity. At $\eta = 0$ hydrodynamic four-velocity can be written as $u^\mu = (1, 0, 0, 0)$ and relevant energy–momentum components as,

$$T^{\tau\tau} = \varepsilon \qquad T^{\tau x} = T^{\tau y} = T^{\tau\eta} = 0 \qquad T^{\eta\eta} = p/\tau^2. \tag{7.86}$$

Inserting the values into equation (7.81) we obtain,

$$\begin{aligned} 0 &= \partial_\tau T^{\tau\tau} + \partial_x T^{\tau x} + \partial_y T^{\tau y} + \partial_\eta T^{\tau\eta} + \tau T^{\eta\eta} + \frac{1}{\tau} T^{\tau\tau} \\ &= \partial_\tau T^{\tau\tau} + \tau T^{\eta\eta} + \frac{1}{\tau} T^{\tau\tau} = \frac{\partial \varepsilon}{\partial \tau} + \frac{\varepsilon + p}{\tau}. \end{aligned} \tag{7.87}$$

If the fluid is assumed to be baryon-free, the entropy density will be given as $s = \frac{\varepsilon + p}{T}$ and equation (7.87) can be recast as,

$$\frac{\mathrm{d}\tau s}{\mathrm{d}\tau} = 0. \tag{7.88}$$

Equation (7.88) can be solved as $s\tau = \text{constant}$. One-dimensional flow is isentropic. In ideal gas, $s \propto T^3$ and we get the well known T^3 law for Bjorken scaling expansion,

$$T_i^3 \tau_i = T_f^3 \tau_f, \tag{7.89}$$

where the subscripts i and f refer to some initial and final state of the fluid. Given the fluid temperature at initial time τ_i, its value at a subsequent time is easily obtained. Similarly, one can also solve for the number conservation equation. For fluid velocity $u^\mu = (1, 0, 0, 0)$, $N^\mu = (n, 0, 0, 0)$. From equation (7.80) we obtain,

$$\begin{aligned} 0 &= \partial_\tau N^\tau + +\partial_x N^x + \partial_y N^y + \partial_\eta N^\eta + \frac{1}{\tau} N^\tau \\ &= \partial_\tau n + \frac{1}{\tau} n. \end{aligned} \tag{7.90}$$

The equation can be solved to give,

$$n_f = n_i \frac{\tau_i}{\tau_f}. \tag{7.91}$$

Noting that density is inversely proportional to volume, one finds the scaling law for volume expansion,

$$V_f = V_i \frac{\tau_f}{\tau_i}. \tag{7.92}$$

7.5.3 Equations of state

Hydrodynamic equations are closed with an equation of state $p = p(\varepsilon, n_B)$. As mentioned earlier, hydrodynamic models are unique in that the effect of phase transition can be studied dynamically, by including the phase transition in the equation of state. Earlier, hydrodynamical simulations of relativistic heavy ion collisions used the equation of state with first order confinement–deconfinement phase transition. The simple bag model equation of state was used extensively to model the equation of state of the QGP phase. The confined phase is generally modeled as interaction-free hadron resonance gas. At sufficiently low temperatures, the thermodynamics of strongly interacting matter is dominated by pions. As the temperature increases, a larger and larger fraction of the available energy goes into the excitation of more and more, heavier resonances. For temperature $T \geqslant 150$ MeV, heavy states dominate the energy density. However, the densities of heavy particles are still small, $\rho_i \sim \mathrm{e}^{-M_i/T}$. There mutual interaction, being proportional to $\rho_i\rho_j \sim \mathrm{e}^{-(M_i+M_j)/T}$, is suppressed. One can use virial expansion to obtain an effective interaction. Virial expansion together with experimental phase shifts were used by Prakash and Venugopal to study the thermodynamics of low temperature hadronic matter [64]. It was shown that the interplay of attractive interactions (characterized by positive phase shifts) and repulsive interactions (characterized by negative phase shifts) is such that effectively, the theory is interaction-free. One can then consider that interaction-free resonances constitute the hadronic matter at low temperatures.

In recent years, there has been much progress in the lattice simulation of QCD. In particular, the thermodynamic properties of baryon-free QCD have been studied accurately and extensively. Currently, there is consensus that the confinement–deconfinement transition is a cross-over and the cross-over or the pseudo-critical temperature for the transition is $T_c \approx 170$ MeV [65–68]. Accurate lattice simulations also show that at low temperatures, interaction-free hadron resonance gas correctly reproduces lattice simulation results. Recent hydrodynamical simulations generally use equations of state where the lattice simulation results for the deconfined phase are smoothly joined at $T = T_c$ MeV, with hadronic resonance gas equations of state [69–71]. At high temperatures the lattice QCD simulations for trace anomalies can be well-parameterized by the inverse polynomial form,

$$\frac{\varepsilon - 3p}{T^4} = \frac{\mathrm{d}_2}{T^2} + \frac{\mathrm{d}_4}{T^4} + \frac{c_1}{T^{n_1}} + \frac{c_2}{T^{n_2}}. \tag{7.93}$$

The form can be matched to the HRG model at temperature T_0 by requiring that the trace anomaly as well as its first and second derivatives are continuous. The resulting equation of state can mimic a cross-over transition. In table 7.1, the parameter values as obtained in [69] are listed. The HRG result for the trace anomaly can also be parameterized by the simple form [69],

$$\frac{\varepsilon - 3p}{T^4} = a_1 T + a_2 T^3 + a_3 T^4 + a_4 T^{10} \tag{7.94}$$

Table 7.1. Fitted values of the parameters for three different fits of the trace anomaly in [69]. The labels s95 and s90 indicate that for the parameterization entropy, density at $T = 800$ MeV is below the ideal gas limit by 5% and 10% respectively. The labels *p*, *n* and *f* indicate three different treatments of the peak structure in the trace anomaly. For details see [69].

	$d_2(\mathrm{GeV}^2)$	$d_4(\mathrm{GeV}^4)$	$c_1(\mathrm{GeV}^{n_1})$	$c_2(\mathrm{GeV}^{n_2})$	n_1	n_2	T_0(MeV)
s95p	0.2660	$2.403 \cdot 10^{-3}$	$-2.809 \cdot 10^{-7}$	$6.073 \cdot 10^{-23}$	10	30	183.8
s95n	0.2654	$6.563 \cdot 10^{-3}$	$-4.370 \cdot 10^{-5}$	$5.774 \cdot 10^{-6}$	8	9	171.8
s90f	0.2495	$1.355 \cdot 10^{-2}$	$-3.237 \cdot 10^{-3}$	$1.439 \cdot 10^{-14}$	5	18	170.0

with $a_1 = 4.654$ GeV^{-1}, $a_2 = -879$ GeV^{-3}, $a_3 = 8081$ GeV^{-4} and $a4 = -7\,039\,000$ GeV^{-10} .

In [70], analytic parameterization for the equation of state for the baryon-free medium was given. The fitting procedure was slightly different from that of [69]. It is called s95p-PCE. For completeness, the parameterization is given below.

1. Pressure:

$$p(e) = \begin{cases} 0.3299[\exp(0.4346e) - 1]; & e < e_1 \\ 1.024 \cdot 10^{-7} \cdot \exp(6.041e) + 0.007\,273 + 0.145\,78e; & e_1 < e < e_2 \\ 0.301\,95 \exp(0.313\,08e) - 0.256\,232; & e_2 < e < e_3 \\ \left. \begin{array}{l} 0.332 - 0.3223e^{0.4585} - 0.003\,906e \cdot \exp(-0.056\,97e) \\ \quad +0.1167e^{-1.233} + 0.1436e \cdot \exp(-0.9131e) \end{array} \right\}; & e_3 < e < e_4 \\ 0.3327e - 0.3223e^{0.4585} - 0.003\,906e \cdot \exp(-0.056\,97e); & e > e_4 \end{cases}$$

where $e_1 = 0.502\,856\,330\,544\,127\,0$ GeV fm^{-3}, $e_2 = 1.62$ GeV fm^{-3}, $e_3 =$ 1.86 GeV fm^{-3} and $e_4 = 9.987\,835\,578\,627\,354\,5$ GeV fm^{-3}.

2. Entropy density:

$$s^{4/3}(e) = \begin{cases} 12.2304e^{1.168\,49}; & e < e_1 \\ 11.9279e^{1.156\,35}; & e_1 < e < e_2 \\ 0.058\,057\,8 + 11.833e^{1.161\,87}; & e_2 < e < e_3 \\ \left. \begin{array}{l} 18.202e - 62.021\,814 - 4.854\,79 \\ \quad \times \exp\left(-2.724\,07 \cdot 10^{-11} e^{4.548\,86}\right) + 65.1272e^{-0.128\,012} \\ \quad \times \exp\left(-0.003\,696\,24 \cdot e^{1.187\,35}\right) - 4.752\,53e^{-1.184\,23} \end{array} \right\}; & e_3 < e < e_4 \\ \left. \begin{array}{l} 18.202e - 63.0218 - 4.854\,79 \exp\left(-2.724\,07 \cdot 10^{-11} e^{4.548\,86}\right) \\ \quad +65.1272e^{-0.128\,012} \exp\left(-0.003\,696\,24 \cdot e^{1.187\,35}\right) \end{array} \right\}; & e > e_4 \end{cases}$$

where, $e_1 = 0.127\,076\,902\,142\,744\,9$ GeV fm^{-3}, $e_2 = 0.446\,707\,952\,467\,404\,0$ GeV fm^{-3}, $e_3 = 1.940\,283\,253\,419\,378\,8$ GeV fm^{-3}, and $e_4 =$ 3.729 247 457 097 728 5 GeV fm^{-3}.

3. Temperature:

$$T(e) = \begin{cases} 0.203054e^{0.30679}; & e < 0.514\,393\,984\,623\,640\,9 \text{ GeV fm}^{-3} \\ (e+p)/s; & e > 0.514\,393\,984\,623\,640\,9 \text{ GeV fm}^{-3}. \end{cases}$$

An alternative procedure to obtain the equation of state with cross-over transition is to join the entropy density of the two phases as

$$s_{\text{QGP+had.}}(T) = 0.5[1 + \tanh x]s_{\text{had.}} + 0.5[1 - \tanh x]s_{\text{QGP}}, \tag{7.95}$$

where $x = \frac{T - T_c}{\Delta T}$, and to calculate the energy density and pressure from the thermodynamical relations,

$$P(T) = \int_0^\infty \mathrm{d}T' s(T') \tag{7.96a}$$

$$\varepsilon(T) = sT - P. \tag{7.96b}$$

The parameter ΔT controls the transition width between the two phases. An equation of state thus constructed in [71] is shown in figure 7.9. In [71], Wuppertal–Budapest simulations for entropy density were parameterized as

$$\frac{s_{\text{QGP}}}{T^3} = \alpha + [\beta + \gamma T]\left[1 + \tanh \frac{T - T_c}{\Delta T}\right]. \tag{7.97}$$

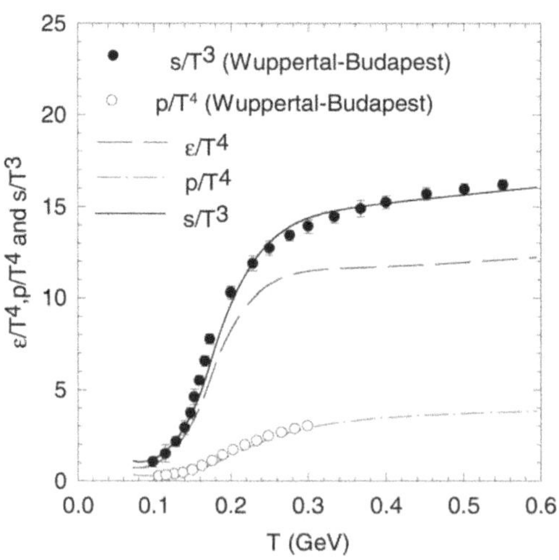

Figure 7.9. The filled and open circles are Wuppertal–Budapest lattice simulations [68] for energy density and entropy density. The lines are parameterized equations of state. For details of the parametric equation of state see [71].

The best fitted values are, $\alpha = 0.64$, $\beta = 6.93$, $\gamma = 0.55$, with fixed $T_c = 170$ MeV, $\Delta T = 0.1T_c$. The calculated pressure from equation (7.96*a*) reproduces the lattice simulations nicely.

7.5.4 Initial conditions for hydrodynamic analysis

It is understood that hydrodynamics is an initial value problem. For example, in Bjorken one-dimensional hydrodynamics, given the density/temperature at some initial time τ_i (the time beyond which hydrodynamics is applicable, itself a parameter of the model), density/temperature evolution of the fluid can be obtained. A kinetic freeze-out condition is also required to define the freeze-out surface such that the Cooper–Frye prescription can yield the particle's invariant distribution. One simple procedure to implement kinetic freeze-out is to assume a fixed freeze-out temperature T_F. Depending on the model, experimental data are fitted with $T_F = 100$–140 MeV.

In a more general system, one has to initialize the (baryon) number density $n(x, y, \eta)$, energy density $e(x, y, \eta)$ and velocity $\mathbf{u}(x, y, \eta) = \gamma\mathbf{v}(x, y, \eta)$ distributions at the initial time τ_i. Indeed, one of the aims of hydrodynamic analysis of heavy ion collisions in ultra-relativistic collisions is to obtain the initial conditions of the produced fluid, by comparing hydrodynamic simulations with experimental data. As discussed earlier, experimental results are given in terms of collision centrality. One then tries to parameterize the initial condition in terms of impact parameter, such that once the parameters are fixed at some particular collision centrality, it can predict for other collision centralities.

In the (τ, x, y, η) coordinate, for the initial energy density, a common practice is to assume a factorized form,

$$e(x, y, \eta) = \varepsilon(x, y)H(\eta), \tag{7.98}$$

$\varepsilon(x, y)$ being the initial energy density in the transverse plane and $H(\eta)$ in the direction of (spatial) rapidity. One can use a Gaussian distribution for $H(\eta)$. Transverse energy distribution $\varepsilon(x, y)$ can be conveniently parameterized in a Glauber model or in a CGC model. The number density distribution or the velocity distribution at the initial time can be similarly parameterized. In general, one assumes zero initial fluid velocity at the initial time, though it is possible that the fluid has non-zero velocity, especially near the surface. The reasoning is simple. Fluid constituents can have random velocity. In the interior of the fluid, the random velocities will balance to produce net zero velocity. However, near the surface random velocities will not be balanced.

Glauber model initial condition. In a previous chapter, we have discussed the Glauber model. Expressions for the number of participant nucleons and number of binary collisions, in AB collisions at impact parameter **b** collisions, were obtained.

$$N_{\text{coll}}(\mathbf{b}) = AB\sigma_{\text{NN}} \int d^2 s T_A(\mathbf{b}) T_B(\mathbf{b} - \mathbf{s}) \tag{7.99}$$

$$N_{\text{part}}(\mathbf{b}) = A\int d^2sT_A(\mathbf{s})\left(1-\left[1-\sigma_{\text{NN}}T_B(\mathbf{b}-\mathbf{s})\right]^B\right)$$
$$+ B\int d^2sT_B(\mathbf{b}-\mathbf{s})\left(1-\left[1-\sigma_{\text{NN}}T_A(\mathbf{s})\right]^A\right). \qquad (7.100)$$

From the above equations, a transverse profile of binary collision and participant numbers in impact parameter **b** collisions can be easily obtained as

$$N_{\text{coll}}(x, y, \mathbf{b}) = AB\sigma_{\text{NN}}T_A(x+\mathbf{b}/2, y)T_B(x-\mathbf{b}/2, y) \qquad (7.101)$$
$$N_{\text{part}}(x, y, \mathbf{b}) = AT_A(x+\mathbf{b}/2, y)\left(1-\left[1-\sigma_{\text{NN}}T_B(x-\mathbf{b}/2, y)\right]^B\right)$$
$$+ BT_B(x-\mathbf{b}/2, y)\left(1-\left[1-\sigma_{\text{NN}}T_A(x+\mathbf{b}/2, y)\right]^A\right). \qquad (7.102)$$

Comparison of hydrodynamic simulations with experimental data indicate that a combined profile

$$e(x, y, \mathbf{b}) \propto \left[(1-f)N_{\text{part}}(x, y, \mathbf{b}) + fN_{\text{coll}}(x, y, \mathbf{b})\right]$$
$$= \varepsilon_0\left[(1-f)N_{\text{part}}(x, y) + fN_{\text{coll}}(x, y)\right] \qquad (7.103)$$

with $f \approx 0.1 - 0.2$ best explains the data. Once the proportionality factor ε_0 is fixed in a given collision centrality, the impact parameter dependence of the model allows one to predict for the energy density distribution at other collision centralities.

CGC initial condition. In a previous section, the CGC model was discussed. CGC is a highly coherent, extremely high energy density ensemble of gluon states. Glasma is matter produced in the collision of CGCs of two hadrons. It has properties very different from those of the CGC and is produced in a very short time after the collision. It eventually evolves from the CGC initial conditions into a QGP. CGC models have been used extensively to model the transverse energy distribution of the initial QGP fluid in hydrodynamic models [72–74]. In the following, we briefly describe the procedure to obtain the initial condition in high-energy nuclear collisions in the Kharzeev–Levin–Nardi (KLN) [33, 75] approach to CGC.

The number of gluons produced in the k_T factorization formula is given by,

$$\frac{dN_g}{d^2r_T dY} = \frac{4\pi^2 N_c}{N_c^2-1}\int\frac{d^2p_T}{p_T^2}\int d^2k_T\alpha_s(k_T)$$
$$\times\,\phi_A(x_1, p_T^2, r_T)\phi_B\left(x_2, (p_T-k_T)^2, r_T\right) \qquad (7.104)$$

where p_T and Y are the transverse momentum and rapidity of the produced gluon. $x_{1,2} = \frac{p_T}{\sqrt{s}}e^{\pm Y}$ is the momentum fraction of colliding gluon ladders at cm energy $\sqrt{s}$.

$\alpha_s(k_T)$ is the strong coupling constant at the momentum scale k_T. The unintegrated gluon distribution functions ϕ_A in equation (7.104) are related to the gluon density in a nucleus at the transverse position $r_T(=x, y)$,

$$xG(x, Q^2) = \int^{Q^2} d^2k_T d^2r_T \phi_A(x, k_T^2; r_T). \tag{7.105}$$

In principle, an unintegrated gluon distribution function should be a solution of a non-linear quantum evolution equation, e.g. the Jalilian-Marian, Iancu, McLearran, Weigert, Leonidov and Kovner (JIMWLK) equation [76–79]. In the KLN approach [75] (which captures the essential features of gluon saturation), the unintegrated gluon distribution functions are taken as [75, 80]

$$\phi_A(x, k_T^2, r_T) \sim \frac{1}{\alpha_s(Q_{sat}^2)} \frac{Q_{sat}^2}{\max(Q_{sat}^2, k_T)} \tag{7.106}$$

where Q_{sat} is the saturation momentum at the given momentum fraction x and at the transverse position r_T.

In the KLN approach, the saturation scale in AB collision is parameterized as [75, 80]

$$Q_{sat,A(B)}(x, r_T) = 2\ \text{GeV}^2 \left(\frac{N_{part}^{A(B)}(r_T)}{1.53} \right) \cdot \left(\frac{0.01}{x} \right)^{\lambda}. \tag{7.107}$$

The form $Q_{sat}(x) \sim x^{-\lambda}$, with $\lambda \approx 0.2 - 0.3$ is motivated from deep inelastic scattering (DIS) experiments. $N_{part}^{A(B)}$ in the above equation is the transverse density of participant nucleons, which can be calculated in a Glauber model (e.g. see equation (7.102))

$$N_{part}^{A}(r_T) = AT_A(x + \mathbf{b}/2, y)\left(1 - \left[1 - \sigma_{NN} T_B(x - \mathbf{b}/2, y)\right]^B\right). \tag{7.108}$$

In the CGC model, the transverse energy density should follow equation (7.104). However, equation (7.104) is valid in the time scale $\tau_s \sim \frac{1}{Q_{sat}}$, when the medium may not be in thermal equilibrium. One assumes that the medium undergoes one-dimensional Bjorken (longitudinal, isentropic) expansion during the period τ_s to τ_i. The density at the time τ_i, when hydrodynamics become applicable, is easily obtained as $n(\tau_i) = \frac{\tau_s}{\tau_i} n(\tau_s)$. The transverse energy density profile at the initial time τ_i is then

$$\varepsilon(x, y, \mathbf{b}) = \varepsilon_0 \left[\frac{dN_g}{dx dy dY} \right]^{4/3} \tag{7.109}$$

with ε_0 a normalizing factor, which is to be fixed from experimental data.

7.5.5 Collective flow

In relativistic heavy ion collisions, one of the important observables is the azimuthal distribution of produced particles. In figure 7.10, the geometry of a collision at non-zero impact parameter collision is shown. The overlap region of the two nuclei is the participant region, where most of the collisions occur. The target and projectile remnants on the periphery act as spectators. It is obvious from figure 7.10 that in non-zero impact parameter collisions, the participant or the reaction zone in coordinate space do not possess azimuthal symmetry. Multiple collisions among the constituent particles translate this spatial anisotropy into momentum anisotropy of the produced particles. The observed momentum anisotropy is called collective flow and has a natural explanation in a hydrodynamic model [34]. In the following we briefly discuss collective flow phenomena. More detailed expositions can be found in [81–83].

Momentum anisotropy is best studied by decomposing the invariant distribution in a Fourier series. For example, the momentum integrated invariant distribution of a particle can be expanded as,

$$\frac{dN}{d\phi} = \frac{N}{2\pi}\left[1 + 2\sum_n v_n \cos\left[n\left(\phi - \psi_n\right)\right]\right] \qquad n = 1, 2, 3, \ldots \tag{7.110}$$

ϕ is the azimuthal angle of the detected particle and ψ_n is the plane of the symmetry of the initial collision zone. For smooth initial matter distribution, the plane of symmetry of the collision zone coincides with the reaction plane Ψ_{RP} (the plane containing the impact parameter and the beam axis). The sine terms are not present in the expansion due to symmetry with respect to the reaction plane.

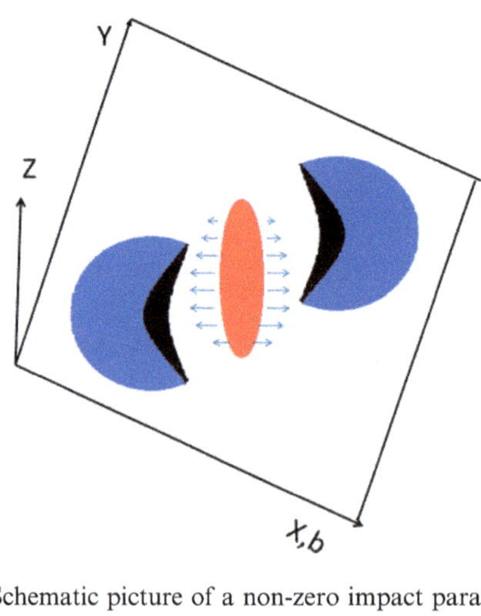

Figure 7.10. Schematic picture of a non-zero impact parameter collision.

Flow coefficients v_n are easily obtained,

$$v_n = \langle\cos(n\phi - n\psi)\rangle = \frac{\int d\phi \frac{dN}{d\phi}\cos(n\phi - n\psi)}{\int d\phi \frac{dN}{d\phi}} \qquad n = 1, 2, 3, \ldots \tag{7.111}$$

v_1 is called (integrated) directed flow, v_2 (integrated) elliptic flow, v_3 (integrated) triangular flow, v_4 (integrated) hexadecapole flow, etc.

Similar to equation (7.110), one can Fourier expand the invariant distribution:

$$E\frac{d^3N}{d^3p} = \frac{1}{2\pi}\frac{d^2N}{p_T dp_T dy}\left[1 + 2\sum_n v_n(p_T, y)\cos[n(\phi - \psi)]\right] \qquad n = 1, 2, 3, \ldots \tag{7.112}$$

and differential flow coefficients are obtained as

$$v_n(p_T, y) = \langle\cos(n\phi - n\psi)\rangle = \frac{\int d\phi \frac{d^3N}{p_T dp_T d\phi dy}\cos(n\phi - n\psi)}{\int d\phi \frac{d^3N}{p_T dp_T d\phi dy}} \qquad n = 1, 2, 3, \ldots \tag{7.113}$$

The second flow coefficient or the elliptic flow has been studied extensively in RHIC and LHC energy collisions. The finite non-zero value of v_2 is thought to be a direct signature of the production of thermalized medium. It is also best understood in a hydrodynamic model [34]. Elliptic flow measures the momentum anisotropy. In non-zero impact parameter collisions, the reaction zone is spatially asymmetric. Spatial asymmetry of the initial reaction zone can be quantified in terms of eccentricity, defined as,

$$\varepsilon = \frac{\langle y^2 - x^2\rangle}{\langle y^2 + x^2\rangle} \tag{7.114}$$

where $\langle\ldots\rangle$ indicate energy/entropy density weighted averaging. In a non-zero impact parameter collision, initial eccentricity is non-zero, positive. If a thermalized medium is produced in the reaction zone, due to thermodynamic pressure, the medium will expand against the outside vacuum. One can immediately see that the pressure gradient will be greater along the minor axis than along the major axis. Due to the differential pressure gradient, as the system evolves with time, eccentricity will reduce. The momentum distribution of particles is isotropic initially. If momentum anisotropy is measured as

$$\varepsilon_p = \frac{\int dxdy\left[T^{xx} - T^{yy}\right]}{\int dxdy\left[T^{xx} + T^{yy}\right]} \tag{7.115}$$

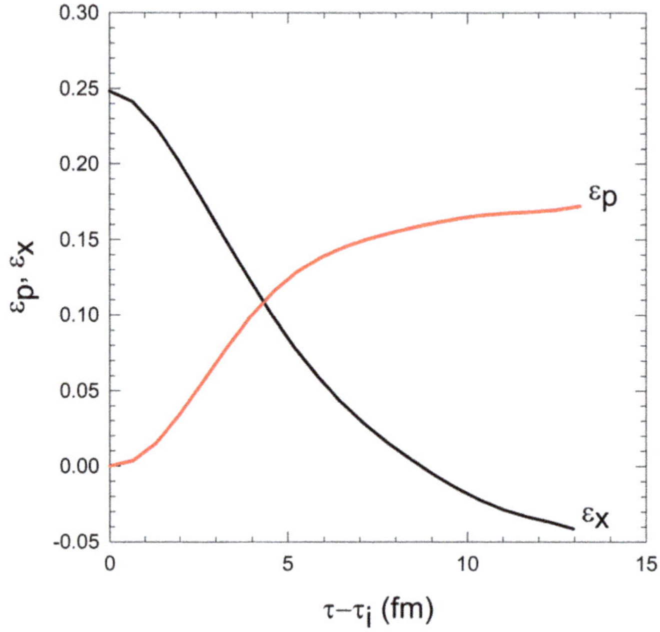

Figure 7.11. Temporal evolution spatial eccentricity (ε_x) and momentum anisotropy (ε_p) with time [34].

initially ε_p will be zero. However, as the fluid evolves, rescattering of particles will introduce asymmetry and ε_p will grow. It is expected to saturate beyond a certain time, when the reaction zone attains azimuthal symmetry. In that sense, elliptic flow is a self-quenching phenomenon, and the driving force of the flow (the reaction zone asymmetry) continuously reduces as the flow grows. In figure 7.11, ideal hydrodynamic model simulations for temporal evolution of spatial eccentricity and momentum anisotropy are shown. They follow our expectations.

The second harmonic coefficient or the elliptic flow (v_2) has been studied extensively in $\sqrt{s}_{\rm NN} = 200$ GeV Au+Au collisions at RHIC [84, 85] as well as in $\sqrt{s}_{\rm NN} =$ 2.76 TeV Pb+Pb collisions at LHC [87, 86]. Large elliptic flow has provided compelling evidence that nearly perfect fluid is produced at RHIC and LHC. Deviation from the ideal fluid behavior is controlled by the shear viscosity to entropy density ratio (η/s). The effect of shear viscosity is to dampen the flow coefficients. Elliptic flow has sensitive dependence on η/s. The sensitivity of elliptic flow has been utilized to obtain phenomenological estimates of η/s [73, 88–94]. The results will not be discussed in detail, but it appears that the QGP viscosity over entropy ratio is close to $\eta/s \approx 1/4\pi$, the value advocated in string theory motivated models.

Flow coefficients measure azimuthal correlations with respect to the reaction plane. However, there are several other sources for azimuthal correlation, not connected with the reaction plane, e.g. transverse momentum conservation, which produces back-to-back correlations, resonance decays such as $\Delta \to p\pi$, HBT correlations and final state interactions. They are termed as non-flow correlations. These correlations increase with impact parameter: most of them vary with the multiplicity N like $1/N$. For accurate measurements of flow or correlations related to the reaction plane, non-flow effects must be eliminated.

Flow coefficients as defined in equation (7.110) or in equation (7.112) require the event plane or the reaction plane angle $\psi = \psi_{RP}$. The reaction plane is not a directly experimental observable. Methods have been devised to calculate the reaction plane angle. Methods for calculating flow not requiring the reaction plane angle have also been devised. In the following, a brief description of the methods is given.

Event plane method. The event plane method [81, 95] uses the anisotropic flow itself to determine the event plane. For the nth flow harmonic, a flow vector $\mathbf{Q}_n$ is defined,

$$Q_n \cos(n\psi_n) = X_n = \sum_i w_i \cos(n\phi) \tag{7.116a}$$

$$Q_n \sin(n\psi_n) = Y_n = \sum_i w_i \sin(n\phi), \tag{7.116b}$$

and the reaction or the event plane angle, for the nth flow coefficient is determined as

$$\psi_n = \frac{1}{n}\tan^{-1}\frac{\sum_i w_i \sin(n\phi)}{\sum_i w_i \cos(n\phi)}. \tag{7.117}$$

The sums in equation (7.116) are over the M particles used in the event plane determination and the w_i are weights. Weights are used to make the flow vector a better defined observable and increase the resolution of the event plane. Experimentally elliptic flow increases with the transverse momentum and $w_i \propto p_{\mathrm{T}}$ is a good choice. In some of the analysis, weight w_i is defined as

$$\begin{aligned} w_i &= \frac{p_T}{p_T^{\mathrm{sat}}} && p_T < p_T^{\mathrm{sat}} \\ &= 1 && p_T \geqslant p_T^{\mathrm{sat}}, \end{aligned} \tag{7.118}$$

with $p_T^{\mathrm{sat}} = 2$ GeV.

In symmetric collisions, due to reflection symmetry, particle distributions in the backward hemisphere of the cm should be the same as in the forward hemisphere if the azimuthal angles of all particles are shifted by π. Signs of the weights for the odd harmonics are reversed in the backward hemisphere while for the even harmonics the signs of the weights are not reversed.

The cumulant method. Accurate reconstruction of the reaction plane in the event plane method requires a large number of particles in the final state. In situations where the number of final state particles is small (e.g. in peripheral collisions), the event plane method has large uncertainty. In the cumulant method [96–98], flow coefficients are calculated from experimental cumulants and thus eliminate the event plane reconstruction altogether.

For example, the two-particle azimuthal correlation can be written as

$$\begin{aligned} c_2\{2\} &= \left\langle\left\langle e^{i2(\phi_1-\phi_2)}\right\rangle\right\rangle = \left\langle\left\langle e^{i2((\phi_1-\psi)-(\phi_2-\psi))}\right\rangle\right\rangle \\ &= \left\langle\left\langle e^{i2((\phi_1-\psi)}\right\rangle\left\langle e^{-i2(\phi_2-\psi)}\right\rangle + \delta_2\right\rangle \\ &= \left\langle v_2^2 + \delta_2\right\rangle \end{aligned} \tag{7.119}$$

where the double brackets $\langle\langle ... \rangle\rangle$ denote an average over all particles within an event, followed by averaging over all events. The two-particle azimuthal correlation then consists of reaction plane correlation v_2 and a non-flow (independent of reaction plane) correlation δ_2. It can be shown that in the event $v_2 \gg 1/\sqrt{M}$, where M is the number of particles used in the analysis, the non-flow correlation can be neglected and v_2 can be obtained from the two-particle azimuthal correlation. The above analysis can be generalized for the nth flow harmonic,

$$v_n\{2\} = \sqrt{\left\langle\left\langle e^{in(\phi_1-\phi_2)}\right\rangle\right\rangle}. \tag{7.120}$$

The sensitivity of the method increases with higher order cumulants. For example, for a four-particle cumulant,

$$\begin{aligned} c_2\{4\} &= \left\langle\left\langle e^{i2(\phi_1+\phi_2-\phi_3-\phi_4)}\right\rangle\right\rangle - 2\left\langle\left\langle e^{i2(\phi_1-\phi_2)}\right\rangle\right\rangle^2 \\ &= \left\langle v_2^4 + \delta_4 + 4v_2^2\delta_2 + 2\delta_2^2\right\rangle - 2\left\langle v_2^2 + \delta_2\right\rangle^2 \\ &= \left\langle -v_2^4 + \delta_4\right\rangle. \end{aligned} \tag{7.121}$$

The last equality in the above equation follows from the assumption that v_2 and δ_2 are uncorrelated and $\langle\delta_2^2\rangle = \langle\delta_2\rangle^2$ and $\langle v_2^4\rangle = \langle v_2^2\rangle^2$. $\delta_4 \propto 1/M^3$ and if $v_2\{4\} \gg 1/M^{3/4}$, the four-particle cumulant will be a good measure of the flow, $v_2\{4\} = \sqrt[4]{c_2\{4\}}$.

In practice, cumulants are calculated from the generating function,

$$G_n(z) = \left\langle \prod_{j=1}^{M} \left[1 + w_j\left(ze^{-in\phi_j} + z^* e^{in\phi_j}\right]\right\rangle \tag{7.122}$$

where z is a complex number and z^* is its complex conjugate. The nth order cumulant is given by the coefficient of z^n in a series expansion of the logarithm of G_n. The function G_n is evaluated in a few points in the complex plane around the origin $z = 0$, and by taking the logarithm at each of these points it is possible to interpolate the next derivatives of $\ln(G_n)$ and obtain the cumulants $c_2\{n\}$.

The Lee–Yang zero method. In the cumulant method, non-flow effects are better removed with higher order cumulants. Thus nth order cumulants remove non-flow effects due to $(n - 1)$-particle correlation. In the event $n \to \infty$ we obtain the Lee–Yang zero method [99–101].

Similarly to the cumulant method, one defines a generating function,

$$G_n^\theta(r) = \left\langle \prod_{j=1}^{M} \left[1 + irw_j \cos\left(\phi_j - \theta\right) \right] \right\rangle \tag{7.123}$$

where r is a real positive number and θ is an angle between 0 and π/n. The product involves all the particles in each event, and the average is taken over the events. The behavior of the zeros of the generating function G_n^θ reflects the presence of collective flow in the system. In particular, the first zero of G_n^θ is related to the magnitude of anisotropic flow. In practice, for various values of r and θ (typically 4–5), one calculates

$$g_n^\theta(r) = \prod_{j=1}^{M} \left[1 + irw_j \cos\left(\phi_j - \theta\right) \right] \tag{7.124}$$

and averages over the events to obtain $G_n^\theta(r)$. For each θ, the position r_0^θ of the first positive minimum of the modulus $|G^\theta(r)|$ is determined. An estimate of integrated flow is then given by

$$v_n^\theta\{\infty\} \equiv \frac{j_{01}}{r_0^\theta} \tag{7.125}$$

where $j_{01} \approx 2.40483$ is the first zero of Bessel function j_0. Finally, one averages over θ to estimate flow coefficient v_n.

7.5.6 Event-by-event hydrodynamics

In the preceding section, the hydrodynamical simulations discussed used a smooth initial condition, either the Glauber model or CGC model. In recent years, there has been much interest in event-by-event hydrodynamics. Unlike in smooth hydrodynamics, in event-by-event hydrodynamics, the initial conditions fluctuate event by event. Event-by-event hydrodynamics takes into account the possibility that participant positions can fluctuate from event to event. It was also realized in recent years that the participating nucleons, rather than the reaction plane, determine the symmetry plane of the initial collision zone [112]. In figure 7.12, a schematic representation of participating nucleons in the transverse plane in a Monte-Carlo (MC) event is shown. The geometric overlap region does not coincide with the participating nucleons. It is obvious that the symmetry plane of the participating nucleons is tilted with respect to the reaction plane. In such a situation, the azimuthal angle should be measured with respect to the participant plane rather than the reaction plane.

The participant plane can fluctuate, event by event, and give rise to novel phenomena like triangular flow, the third flow harmonic in the Fourier expansion of aziumthal distribution, which will be absent in smooth initial conditions. Triangular flow is instrumental in explaining the peculiar structures (known as ridges) seen in the two-particle correlation in the $\Delta\phi-\Delta\eta$ plane [113–115]. In figure 7.13 two-particle correlation in the $\Delta\phi-\Delta\eta$ plane, in d+Au and Au+Au collisions are shown. One notices the marked differences in the correlation. Unlike in d+Au collisions, in Au+Au collisions, two particles are correlated over many units of pseudo-rapidity $\Delta\eta$. Correlation in the azimuth, however, is narrow. The ridge structures in two-particle correlations have been observed both in the STAR and the PHENIX experiments. They are also observed in $p\bar{p}$ collisions [117, 118]. The ridge structure has a most compelling explanation, provided that the third flow harmonic, the triangular flow v_3,

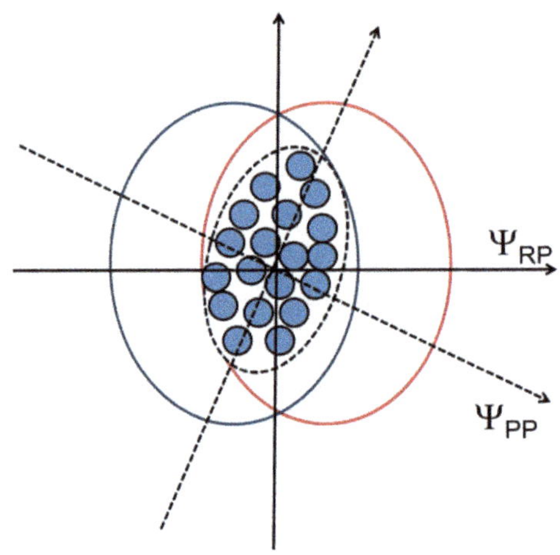

Figure 7.12. Schematic picture of MC Glauber model simulation for participating nucleons in the transverse plane. The overlap region of participating nucleons is tilted with respect to the reaction plane.

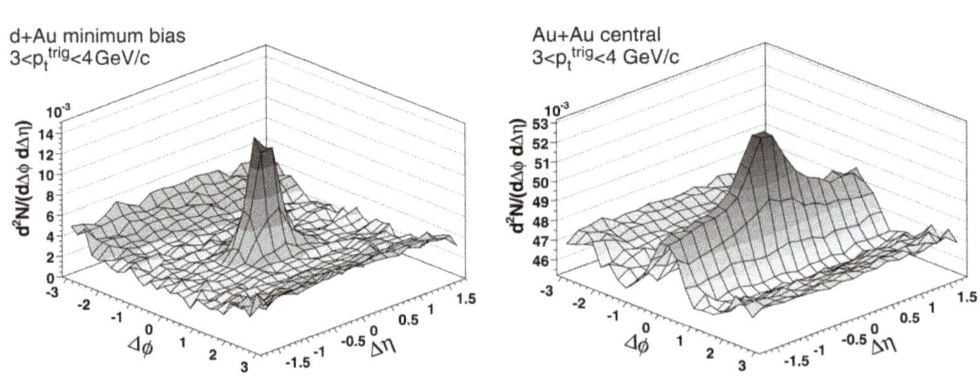

Figure 7.13. Two-particle correlations in $\Delta\phi-\Delta\eta$ in d+Au and Au+Au collisions are shown. The results are from the STAR experiment [116]. Reproduced with the author's permission.

develops in the collisions [119–121]. Specifically, if the initial condition is parameterized with quadrapole and triangular moments, the response of the medium to these anisotropies is reflected in the two-body correlation as a ridge [119, 120].

For example, the azimuthal correlation function can be Fourier decomposed as

$$\frac{\mathrm{d}N^{\text{pairs}}}{\mathrm{d}\Delta\phi} = \frac{N^{\text{pairs}}}{2\pi}\left(1 + \sum_n 2V_{n\Delta}\cos(n\Delta\phi)\right) \tag{7.126}$$

where the first component $V_{1\Delta}$ is understood to be due to momentum conservation and directed flow, the second component $V_{2\Delta}$ is dominated by the contribution from elliptic flow and the third component $V_{3\Delta}$ is dominated by the triangular flow. In figure 7.14, PHOBOS and STAR measurements of long range azimuthal correlation in 200 GeV Au+Au collisions are shown. In the top panel of figure 7.14, the first three Fourier components of the azimuthal correlations are shown. The bottom panels show the residual after these components are taken out. Evidently, experimental data on two-particle correlation are very well described by the three Fourier components. The analysis indicates that the two-particle correlation in the $\Delta\eta - \Delta\phi$ plane is consistent with hydrodynamic models, if triangular flow develops during the evolution, which is possible only when the initial condition fluctuates. Recently, the ALICE collaboration has measured triangular flow in $\sqrt{s}_{\text{NN}} = 2.76$ TeV Pb+Pb collisions [87]. In most central collisions, the elliptic flow (v_2) and triangular flow (v_3) are of similar magnitude. In peripheral collisions however, elliptic flow dominates. More recently, the PHENIX collaboration [122–124] measured triangular flow in $\sqrt{s}_{\text{NN}} = 200$ GeV Au+Au collisions.

In event-by-event hydrodynamics, one generally uses the MC–Glauber model or MC–CGC (the KLN version) to obtain the initial conditions, event by event. Recently, the MC–CGC model was improved by combining the impact parameter dependent saturation model [136, 137] of the high-energy nucleon or nuclear wave function with the classical Yang–Mills description of glasma fields [138].

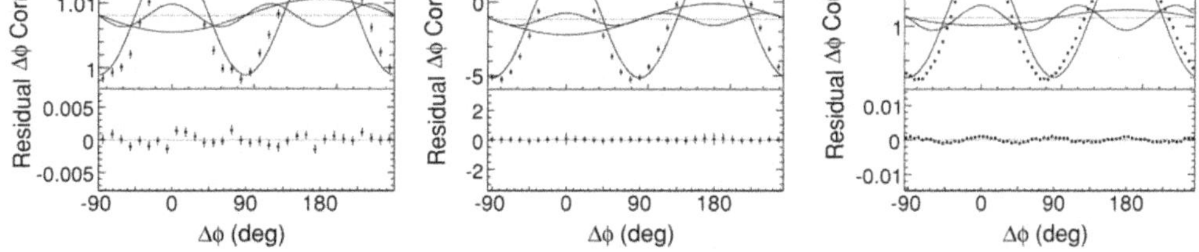

Figure 7.14. Azimuthal correlation functions for mid-central (10–20%) Au+Au collisions at $\sqrt{s}_{\text{NN}} = 200$ GeV obtained from projections of two-dimensional $\Delta\phi - \Delta\eta$ correlation measurements by PHOBOS [19, 25] and STAR [41]. The residual correlation functions after the first three Fourier components are subtracted are shown in the bottom panel. The figure is reproduced with the author's permission from [119].

The model is called the IP-glasma model [125] for the initial condition. In addition to fluctuations of nucleon positions, IP-glasma's description includes quantum fluctuations of color charges on the length-scale determined by the inverse nuclear saturation scale $Q_{\rm sat}$, missed in MC–KLN models [140]. In figure 7.15, the initial energy density distribution in three types of model, MC–Glauber, MC–KLN and IP-glasma is shown. Color charge fluctuations in the length scale $Q_{\rm sat}^{-1}$ introduce additional fluctuations and in the IP-glasma model, the initial density has finer structures than in the MC–KLN model or in the MC–Glauber model.

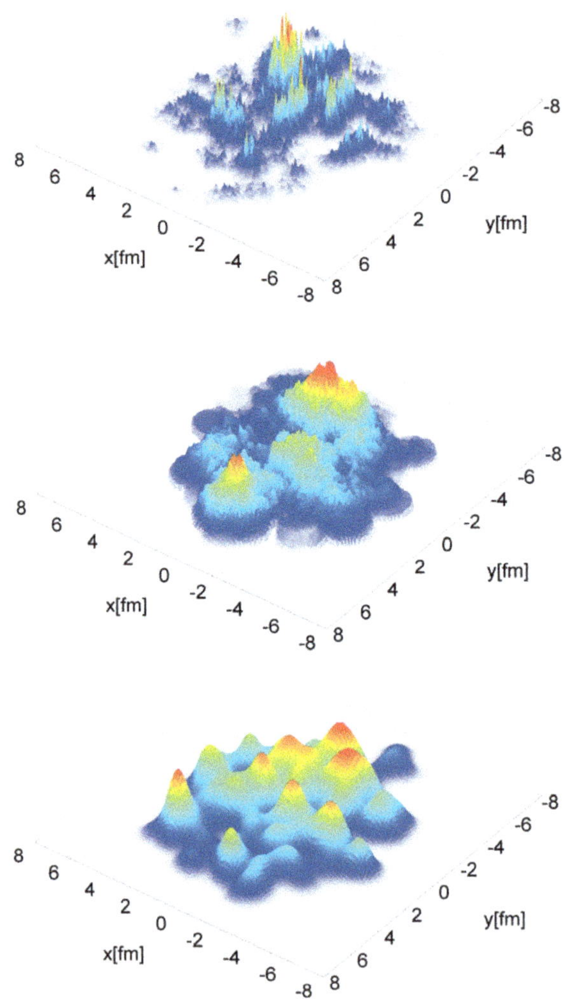

Figure 7.15. Initial energy density (arbitrary units) in the transverse plane in three different heavy ion collision events: (from top to bottom, IP-glasma) the MC–KLN and MC–Glauber models. The figure is reproduced with permission from [125].

Fluid dynamical models require continuous density distribution, however, in event-by-event hydrodynamics, initial conditions are generated using the MC algorithm and in general the distribution is not continuous. For example, in the MC–Glauber model, in an event, participant nucleon positions can be obtained. If a particular MC event has N_{part} participants, the participants' positions in the transverse plane can be labeled as $(x_1, y_1), (x_2, y_2), \ldots(x_{\text{npart}}, y_{\text{npart}})$. The energy density in the transverse plane can be approximated as,

$$\varepsilon(x, y) \propto \sum_{i=1}^{N_{\text{part}}} \delta(x - x_i, y - y_i). \tag{7.127}$$

However, the discrete distribution as in equation (7.127) cannot be evolved. To be used in a hydrodynamic model, the discrete density distribution has to be converted into a smooth energy density distribution. This can be done by smearing the discrete participant positions by some smoothing function,

$$\delta(x - x_i, y - y_i) \to g\left(x - x_i, y - y_i, \zeta_1, \zeta_2 ..\right), \tag{7.128}$$

where ζ_i are parameters of the smoothing function g. For the smoothing function, one generally uses a Gaussian function. Other functions may also be considered. The effect of smoothing was studied in [139]. MC–Glauber model initial participant positions were smoothed with a Gaussian function of various widths and a Woods–Saxon function of various diffuseness. It was shown that the effect of smoothing of participant positions on elliptic and triangular flow is minimal.

One important aspect of event-by-event hydrodynamics is the characterization of the asymmetry of the initial collision zone as well as the azimuthal angle of the participant plane. Each flow harmonics can have their own participant plane. One can generalize the definition of eccentricity to give a simple ansatz to characterize the asymmetry of the initial collision zone [132],

$$\epsilon_n e^{in\psi_n^{PP}} = -\frac{\iint \varepsilon(x, y) r^n e^{in2\phi} dxdy}{\iint \varepsilon(x, y) r^n dxdy} \qquad n = 1, 2, 3, 4, 5 \tag{7.129}$$

where $x = r\cos\phi$ and $y = r\sin\phi$. Equation (7.129) also determines the participant plane angle ψ_n^{PP}. Asymmetry measures ϵ_2 and ϵ_3 are called eccentricity and triangularity. ϵ_4 and ϵ_5 essentially measures the square-ness and five-sidedness of the initial distribution. They may be called rectangularity and pentangularity, respectively. Fourth flow coefficient v_4 is generally referred to as hexadecpolar flow and rectangular flow may be more appropriate. Similarly v_5 may be referred to as the pentangular flow.

7.5.7 Some results of hydrodynamic simulations of heavy ion collisions

In recent years, several (ideal and dissipative) hydrodynamics codes have been developed across the world to simulate relativistic energy heavy ion collisions

[60, 73, 102–109]. Various authors have successfully analyzed experimental data in RHIC and LHC energy collisions. More recently, several simulations in event-by-event hydrodynamics, for Au+Au/Pb+Pb collisions at RHIC/LHC, have been performed [102, 103, 125–135]. For completeness, in the following a few representative results are shown.

In figure 7.16, PHOBOS measurements [110] of charged particle rapidity density in $\sqrt{s}_{\rm NN} = 200\,\rm GeV$ Au+Au collisions in two different collision centralities are shown. Experimental data are nicely reproduced in hydrodynamic simulations. For details of hydrodynamic simulations see [111].

In figure 7.17, fits obtained to the identified particle spectra in the STAR and PHENIX experiments in a hydrodynamic model simulation are shown. Note the quality of fit. Data in the collision centrality 0–5% to 60–90% are well explained. The left panel of figure 7.18 shows the fits obtained to the experimental integrated elliptic flow. While the hydrodynamic predictions fail for integrated flow in peripheral collisions, in central and mid-central collisions, they agree well with

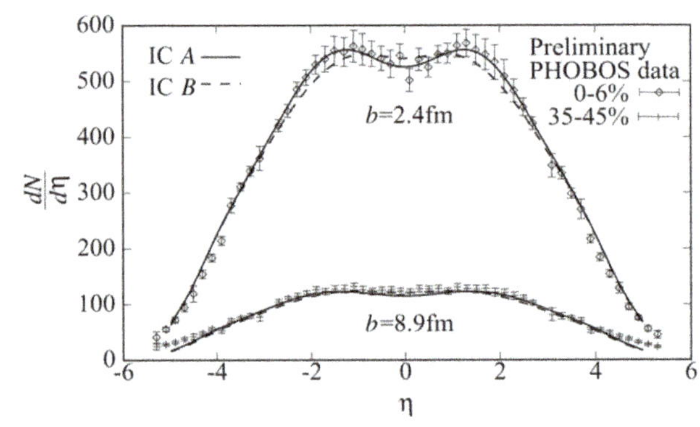

Figure 7.16. Charged particle pseudo-rapidity density in $\sqrt{s}_{\rm NN} = 130$ GeV Au+Au collisions. In a central 0–6% and peripheral 35–45% collision centralities are compared with hydrodynamic simulation with two initial conditions A and B (for details of the initial conditions see [111]). Reproduced with permission from [111].

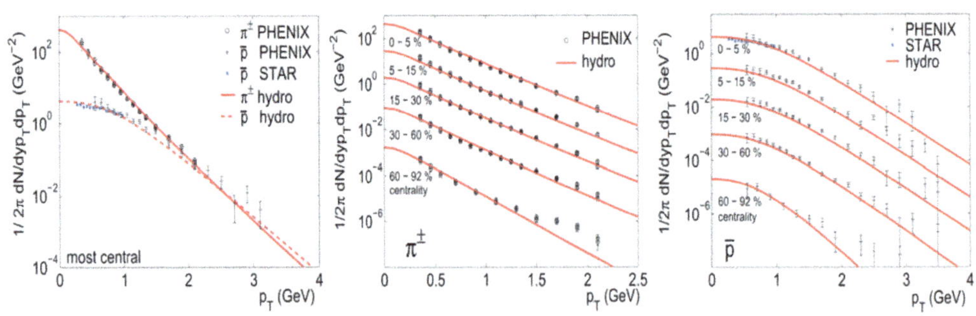

Figure 7.17. Experimental transverse momentum spectra in $\sqrt{s}_{\rm NN} = 130\,\rm GeV$ Au+Au collisions are shown. The left panel shows the pion and antiproton spectra in a central collision. The middle panel shows charged pion spectra semi-central to peripheral collisions and the right panel shows the antiproton spectra. The curves are hydrodynamical calculations. For details see [38]. Reproduced with permission from [38].

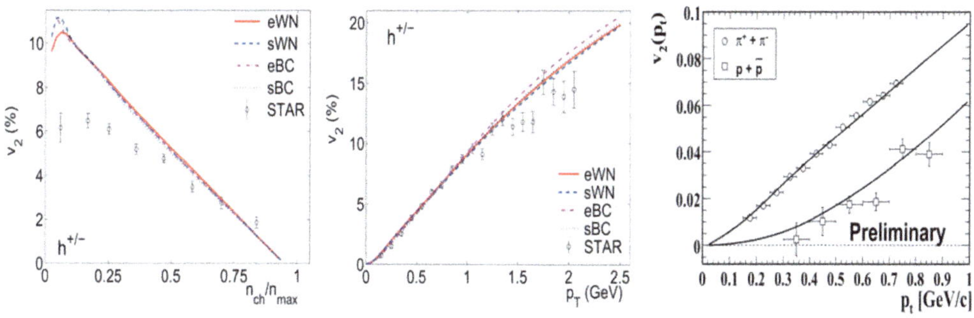

Figure 7.18. Experimental elliptic flow in $\sqrt{s}_{\rm NN} = 130$ GeV Au+Au collisions are shown. The left panel shows $p_{\rm T}$ integrated flow for the charged particles as a function of multiplicity which is a measure of collision centrality. The middle panel shows differential elliptic flow in minimum bias collisions. The right panel shows elliptic flow for charged pions and proton elliptic flow, also in minimum bias collision. For details see [38]. Reproduced with permission from [38].

Table 7.2. Central energy density (ε_i) and temperature (TT_i) at the initial time $\tau_i = 0.6$ fm c^{-1}, for different values of viscosity to entropy ratio $\left(\frac{\eta}{s}\right)$. The bracketed values are estimated central energy density and temperature in $\sqrt{s}_{\rm NN} = 200$ GeV Au+Au collisions [89]. Initial time of the simulations is $\tau_i = 0.6$ fm.

η/s	0	0.08	0.12	0.16
ε_i	89.2 ± 5.0	78.0 ± 4.0	70.5 ± 3.5	61.7 ± 3.0
$\left(\frac{\rm GeV}{\rm fm^3}\right)$	(35.5 ± 5.0)	(29.1 ± 3.6)	(25.6 ± 4.0)	(20.8 ± 2.7)
T_i	486 ± 6	475 ± 5	462 ± 6	447 ± 5
(MeV)	(377 ± 14)	(359 ± 12)	(348 ± 14)	(331 ± 11)

experiments. The middle panel of figure 7.18 shows the differential elliptic flow in minimum bias collisions as a function of $p_{\rm T}$. Here again, agreement with data is reasonable. Finally, in the right panel of figure 7.18, one of the most impressive results of hydrodynamic simulations is shown. Experimental flow of identified particles is mass dependent, with more flow for lighter particles than for heavier particles. It is called mass splitting of flow. At high $p_{\rm T}$ however, the effect of mass splitting is reduced. Experimental mass splitting of flow is correctly reproduced in hydrodynamic simulations. The agreement with experiment is also good. The representative simulation results shown here are for ideal fluid only. As mentioned earlier, there are several simulations with viscous fluid. We will not elaborate on the results of viscous fluid simulations, but will just mention that compared to ideal fluid, a viscous fluid will require lower initial energy density or temperature. This is because entropy is generated during viscous evolution. As an example, in table 7.2, we have noted the central energy density and temperature of the fluid obtained from fits to experimental data in $\sqrt{s}_{\rm NN} = 200$ GeV Au+Au and $\sqrt{s}_{\rm NN} = 2.76$ TeV Pb+Pb collisions. Note that viscous fluid requires lower energy density or temperature. See [89, 91] for details of the simulations.

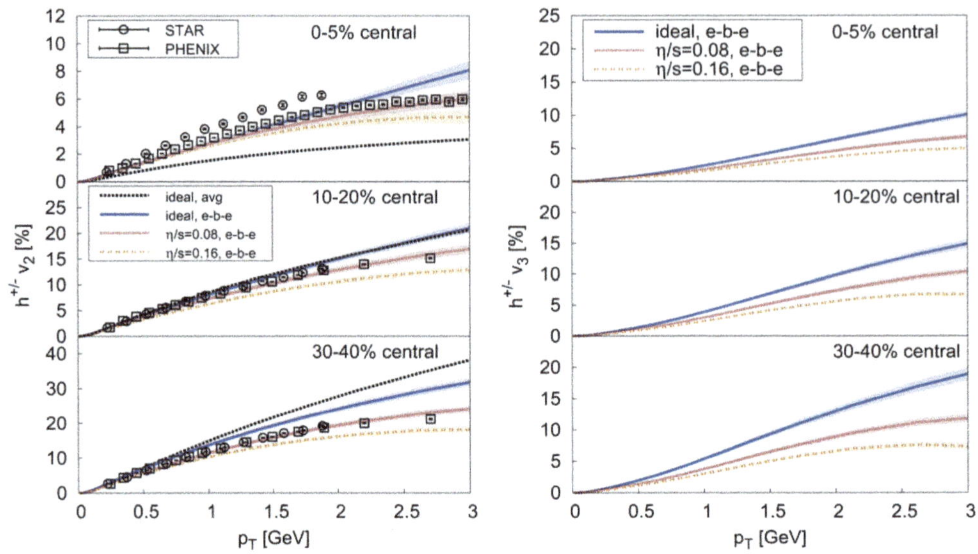

Figure 7.19. Left panel: charged hadron elliptic flow (v_2) for different centralities as a function of transverse momentum for averaged initial conditions (avg) and event-by-event simulations (e-b-e) using different viscosity-to-entropy density ratios compared to STAR and PHENIX data. Right panel: same as in the left panel but for triangular flow. The figures are reproduced with the author's permission from [102].

In recent years, several authors have simulated Au+Au/Pb+Pb collisions at RHIC/LHC, in event-by-event hydrodynamics [102, 103, 125–135]. Some of the simulation results for event-by-event viscous hydrodynamics are briefly discussed here. Viscous effects on elliptic and triangular flow in event-by-event hydrodynamics were studied in [102]. The results are shown in figure 7.19. In [102], a (3 + 1)-dimensional hydrodynamical model was used. As it was in smooth hydrodynamics, the effect of viscosity was to reduce the elliptic flow. The more viscous the fluid, the lower the elliptic flow, except in very central collisions where event-by-event hydrodynamics produces less flow than that in smooth hydrodynamics (the curve labeled as 'avg'). The increase of v_2 in central 0–5% collisions is understood. In event-by-event hydrodynamics, v_2 is determined in each event. Single events have a larger anisotropy with respect to the event plane than the average with respect to the reaction plane and v_2 is increased. In more peripheral collisions, the effect reduces and event by event v_2 becomes smaller than that for the smooth initial condition. Triangular flow is completely fluctuation driven and depends less strongly on collision centrality than the elliptic flow. Triangular flow also reduces with viscosity. It also seems that triangular flow is more sensitive to viscosity than elliptic flow. For example, in 30–40% collisions, at $p_T \approx 1.5$ GeV, elliptic flow reduces by ~15% and 30% when η/s is increased from 0 to 0.08 and 0.12, respectively. The reduction in triangular flow is much larger, respectively, ~30% and 45%. It is to be expected that experimental data on triangular flow can constrain the viscosity over entropy ratio much better than the elliptic flow data.

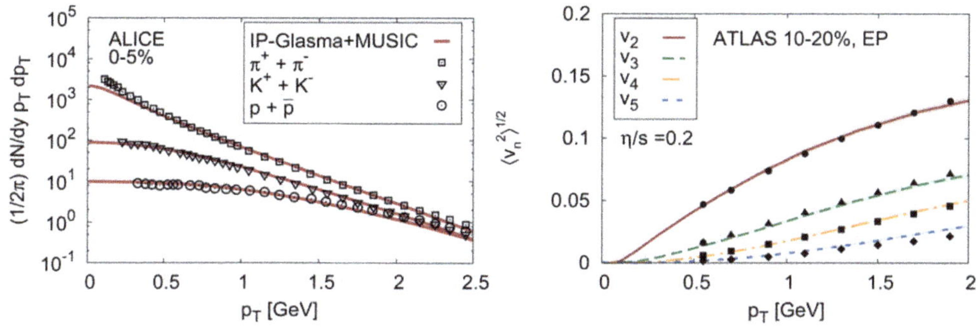

Figure 7.20. Left panel: ALICE measurements for the transverse momentum spectra of pions, kaons and protons in 0–5% Pb+Pb collisions at LHC energy are compared with IP-glasma model predictions with viscosity over entropy ratio $\eta/s = 0.08$. Right panel: ATLAS measurements for the flow coefficients v_n, $n = 2$, 3, 4, 5 are compared with IP-glasma model predictions. The figures are reproduced with the author's permission from [126].

In the left panel of figure 7.20, event-by-event hydrodynamic simulations with an IP-glasma initial condition are compared with ALICE measurements for p_T spectra of pions, kaons and protons [102]. The viscosity over entropy ratio was fixed at $\eta/s = 0.2$. One finds that simulations with IP-glasma initial conditions explain the p_T spectra of charged particles nicely. The IP-glasma initial condition also explains the measured flow coefficients well. IP-glasma model simulation for elliptic flow is shown in the right panel of figure 7.20. They are compared with ATLAS measurements. Charged particle flow coefficients v_2, v_3, v_4 and v_5 appear to agree with the experimental measurements. However, whether or not a similar fit can be obtained with other values of viscosity over entropy ratio was not explored in [102].

7.6 Transport approach based models

Currently the models based on the transport approach can be subdivided into two classes: those which follow the time evolution of the one-body phase space distribution, and those which are based on N-body molecular dynamics or cascade schemes.

7.6.1 VUU-type models

In Vlasov–Uehling–Uhlenbec (VUU) or Boltzmann–Uehling–Uhlenbec (BUU) type models [141–144], one generally solves for the Boltzmann transport equation with external force. In the non-relativistic limit, the equation is

$$\left[\frac{\partial}{\partial t} + \mathbf{v} \cdot \nabla_{\mathbf{r}} - \left(\nabla_{\mathbf{r}} U\right)\nabla_{\mathbf{p}}\right] f(\mathbf{r}, \mathbf{p}, t) = C[f] \tag{7.130}$$

where $f(\mathbf{r}, \mathbf{p}, t)$ is the classical distribution function or rather its quantum analog, the Wigner phase space distribution. $C[f]$ is the collision term, including only binary collisions. The particles move in the mean field potential $U(r)$ which can be identified with the real part of the Brueckner G-matrix[7]. The imaginary part of the G-matrix can be identified with the collision term. In low-energy collisions, a popular choice for the potential is the Skyrme parameterization,

$$U = \alpha\left(\frac{\rho}{\rho_0}\right) + \beta\left(\frac{\rho}{\rho_0}\right)^{\gamma} \tag{7.132}$$

where density is measured in units of the saturation density ρ_0 of cold nuclear matter.

The l.h.s. of equation (7.130) can be regarded as the transport equation (Vlasov equation) for a distribution of classical particles whose time evolution is governed by Hamilton's equations of motion,

$$\dot{\mathbf{p}}_i = -\frac{\partial H}{\partial \mathbf{r}_i} \tag{7.133a}$$

$$\dot{\mathbf{r}}_i = \frac{\partial H}{\partial \mathbf{p}_i}. \tag{7.133b}$$

The collision term is accounted for as follows: the particles move due to their own, self-consistently generated mean field. At the point of closest approach of two particles if the distance,

$$d \leqslant d_0 = \sqrt{\frac{\sigma_{\text{tot}}}{\pi}} \tag{7.134}$$

a $1 + 2 \rightarrow 1' + 2'$ collision is performed. In equation (7.134), σ_{tot} is the total cross section. The final state is chosen randomly according to the relative weights of the different permissible reaction channels. The momenta of the outgoing particles are 'generated' randomly according to the known angular distribution and in agreement with the energy–momentum conservation laws.

One generally solves the transport equation by use of the 'test' particle method. If the projectile and target consist, respectively, of A_{P} and A_{T} particles, the continuous one-body distribution function f at $t = 0$ is represented by an ensemble of $n(A_{\text{P}} + A_{\text{T}})$ point-like particles, i.e. an ensemble of n parallel collisions with $A_{\text{P}} + A_{\text{T}}$ physical particles each. More explicitly, the test particle method corresponds to the replacement of the expectation value of a single particle observable,

$$\langle O(t)\rangle = \int \mathrm{d}^3p\,\mathrm{d}^3r f(\mathbf{r}, \mathbf{p}, t) O(\mathbf{r}, \mathbf{p}) \tag{7.135}$$

[7] The repulsive core of the bare nucleon–nucleon interaction forbids any perturbative approach to nuclear many-body calculation. The problem is circumvented by introducing a reaction or G-matrix. For a Hamiltonian $H = H_0 + V$, with $H_0\psi = \omega\psi$ defining a projection operator Q which prevents collisions into occupied states, the G-matrix is

$$G = V + \frac{Q}{\omega - QH_0Q} G. \tag{7.131}$$

by a MC integration,

$$\langle O(t)\rangle = \frac{1}{n\cdot\left(A_{\mathrm{P}}+A_{\mathrm{T}}\right)}\sum_{i=1}^{n\cdot\left(A_{\mathrm{P}}+A_{\mathrm{T}}\right)} O\left(\mathbf{r}_i(t),\, \mathbf{p}_i(t)\right) \tag{7.136}$$

where $\mathbf{r}_i(t)$ and $\mathbf{p}_i(t)$ are points in phase space which are distributed according to $f(\mathbf{r}, \mathbf{p}, t)$,

$$f(\mathbf{r}, \mathbf{p}, t) = \frac{1}{n\cdot\left(A_{\mathrm{P}}+A_{\mathrm{T}}\right)}\sum_{i=1}^{n\cdot\left(A_{\mathrm{P}}+A_{\mathrm{T}}\right)} \delta\left(\mathbf{r}-\mathbf{r}_i(t)\right)\cdot\delta\left(\mathbf{p}-\mathbf{p}_i(t)\right). \tag{7.137}$$

A large number of test particles is required to reduce numerical noise. In practice the number n lies in the range between 15 and 500 and one employs a grid to obtain a smooth phase space density distribution.

The solution of equation (7.130) requires specification of proper boundary conditions. In heavy ion reactions, the test particles are randomly distributed in a coordinate space sphere of the radius,

$$R = r_0 A^{1/3} \qquad r_0 = 1.12\ \mathrm{fm}, \tag{7.138}$$

ensuring the correct density $\rho_0 \approx 0.17\ \mathrm{fm}^{-3}$ for the nucleus. The initial momenta of the test particles are randomly chosen between 0 and the local Thomas-Fermi momentum:

$$P_{TF} = \left(3\pi^2\rho\right)^{1/3}. \tag{7.139}$$

Once the collision system is initialized with test particles, the latter are then boosted onto every other with proper relative momentum. All the particles are propagated according to the equation of motion and are checked for collision. If the collision criterion is satisfied a collision is performed and the final state is determined from the 'known' angular distribution.

The BUU type model essentially solves for a one-body distribution function and is suitable for one-body observables, e.g. collective flow, stopping and particle spectra. The test particles move on their own in the mean field potential U. It is calculated from the entire distribution, hence the n parallel events are not truly independent and event-by-event correlations cannot be analyzed within this one-body transport model. Clusters or fragments in heavy ion collisions can only be constructed in a statistical way, e.g. with the help of a coalescence model. The reason is that in transport theory the bound states of many nucleons cannot be described directly. The deficiencies of the BUU type of model are corrected in quantum molecular dynamic (QMD) models.

7.6.2 UrQMD: a quantum molecular dynamic model

The QMD model is an N-body theory which simulates heavy ion reactions on an event-by-event basis [145–147]. Below, a brief description of a particular model, ultra-relativistic QMD (UrQMD) is given. For details of UrQMD please see [148, 149]. The UrQMD model is a microscopic transport theory based on the covariant propagation of all hadrons on classical trajectories in combination with stochastic binary scatterings, color string formation and resonance decay. It represents a MC solution of a large set of coupled partial integro-differential equations for the time evolution of various phase space densities $f_i(x, p)$ of particle species $i = N, \Delta, \Lambda$, etc, that non-relativistically assumes the Boltzmann form:

$$\frac{\mathrm{d}f_i(x, p)}{\mathrm{d}t} \equiv \frac{\partial p}{\partial t}\frac{\partial f_i(x, p)}{\partial p} + \frac{\partial x}{\partial t}\frac{\partial f_i(x, p)}{\partial x} + \frac{\partial f_i(x, p)}{\partial t} = Stf_i(x, p) \qquad (7.140)$$

where x and p are the position and momentum of the particle, respectively. $Stf_i(x, p)$ denotes the collision term of these particle species, which are connected to any other particle species $f_k(x, p)$.

In relativistic dynamics, each particle is characterized by four position and four momentum coordinates. Unlike in non-relativistic dynamics in $6N + 1$ dimensions, in relativistic dynamics, the system propagates in $8N$-dimensional phase space, with $6N$ degrees of freedom representing the classical configuration- and momentum space. The remaining $2N$ degrees of freedom contain the eigen time and energy of each particle. In order to reduce the $8N$-dimensional phase space to the commonly used $6N + 1$ dimensions, $(2N - 1)$ Lorentz-covariant constraints have to be introduced. This leads to several complications and we will only state that in the UrQMD model, the constrained dynamics is not properly implemented and there is a systematic frame dependence on the observables.

Unlike the test particle method for BUU-type models, in UrQMD, as in other QMD models, each nucleon is represented by a coherent state of the form

$$\phi_i(\mathbf{x}, \mathbf{q}_i, \mathbf{p}_i, t) = \left(\frac{2}{\pi L}\right)^{3/4}\left[-\frac{2}{L}(\mathbf{x} - \mathbf{q}_i(t))^2 + \frac{\mathrm{i}}{\hbar}\mathbf{p}_i(t)\mathbf{x}\right] \qquad (7.141)$$

which is characterized by six time dependent parameters, $\mathbf{q}_i$ and $\mathbf{p}_i$, respectively. The parameter L, which is related to the extension of the wave packet in phase space, is fixed. Neglecting the anti-symmetrization, the total N-body wave function is assumed to be the direct product of coherent states,

$$\Phi = \prod_i \phi_i(\mathbf{x}, \mathbf{q}_i, \mathbf{p}_i, t). \qquad (7.142)$$

The equations of motion of the many-body system are calculated by means of a generalized variational principle. For a Hamiltonian containing a kinetic term and mutual interactions V_{ij},

$$H = \sum_i T_i + \frac{1}{2}\sum_{ij} V_{ij} \qquad (7.143)$$

and with the coherent state wave function, the Lagrangian can be obtained as [150]

$$\mathcal{L} = \left\langle \Phi \left[i\hbar \frac{\partial}{\partial t} - H \right] \Phi \right\rangle \tag{7.144}$$

$$= \sum_i \left[-\mathbf{q}_i \mathbf{p}_i - T_i - \frac{1}{2} \sum_{j \neq i} \left\langle V_{ij} \right\rangle - \frac{3}{2\pi L} \right] \tag{7.145}$$

with $\langle V_{ij} \rangle = \int \mathrm{d}x_1 \mathrm{d}x_2 \phi_i^* \phi_j^* V_{ij}(x_1, x_2) \phi_i \phi_j$. Variation of the action $S = \int \mathcal{L} \mathrm{d}t$ yields the Euler–Lagrange equation of motion, for each parameter,

$$\dot{\mathbf{p}}_i = -\nabla_{\mathbf{q}_i} \sum_{j \neq i} V_{ij} = -\frac{\partial \langle H \rangle}{\partial \mathbf{q}_i} \tag{7.146a}$$

$$\dot{\mathbf{q}}_i = \frac{\mathbf{p}_i}{m} + \nabla_{\mathbf{p}_i} \sum_j \left\langle V_{ij} \right\rangle = \frac{\partial \langle H \rangle}{\partial \mathbf{p}_i}. \tag{7.146b}$$

As mentioned before, the interaction can be identified with the real part of the Brueckner *G*-matrix. In UrQMD, it is approximated by a non-relativistic density dependent Skyrme potential, supplemented with Yukawa and Coulomb potentials. A Pauli potential[8] may be included optionally.

The UrQMD model consists of three major parts, namely (i) the initialization of projectile and target, (ii) the propagation of nucleons, resonances and newly produced particles due to their mutual potential interactions and (iii) the hard collisions according to the energy dependent cross section for the various channels together with the Pauli blocking.

(i) *Initialization.* At the initial time, the projectile and target nucleus are modeled according to the Fermi-gas ansatz. The wave function of the nucleus is defined as the product wave function of the single nucleon Gaussians. Accounting for the finite width of the Gaussian, in configuration space the centroids of the Gaussians ($\mathbf{q}_i$) are randomly distributed within a sphere with radius R_A,

$$R_A = r_0 \left[\frac{1}{2} \left\{ A + \left(A^{1/3} - 1 \right)^3 \right\} \right]^{1/3} \qquad r_0 = \left(\frac{3}{4\pi\rho_0} \right)^{1/3} \tag{7.147}$$

with the constraint: (i) $\sum_i \mathbf{q}_i = 0$, i.e. in configuration space, nucleons are centered around $\mathbf{0}$.

Each of the nucleons is also given random momentum $\mathbf{p}_i$ between 0 and the local Thomas–Fermi momentum:

$$P_{TF} = \left(3\pi^2 \rho \right)^{1/3} \tag{7.148}$$

[8] A Pauli potential is a repulsive potential which mimics the fermionic properties of a nucleon and allows for the product ansatz (equation (7.142)) for the wave function of the nucleus.

where ρ is the local nucleon density. For the coherent state wave function equation (7.141),

$$\rho(\mathbf{q}_i) = \left(\frac{2}{\pi L}\right)^{\frac{3}{2}}\left[-\frac{4}{L}\left(\mathbf{x} - \mathbf{q}_i(t)\right)^2\right]. \tag{7.149}$$

The momenta are also constrained to $\sum_i \mathbf{v}_i = 0$, i.e. initially, the nucleus is at rest. The potential parameters are chosen such that the nuclei have correct binding energy.

(ii) *Propagation.* To perform a collision, one chooses an impact parameter from a quadratic measure,

$$\mathrm{d}W \propto b\mathrm{d}b \tag{7.150}$$

and places the nuclei appropriately on the collision axis such that the distance between surfaces of the projectile and the target is equal to 3 fm. The target and projectile nucleons are longitudinally boosted by $\frac{\sqrt{s}_{\mathrm{NN}}}{2}$ such that they move in opposite directions and collide. After that the time propagation starts. The Hamiltonian of the total system is computed and particles are propagated according to the equations of motion (7.146).

(iii) *Collision.* In UrQMD, only binary collisions are considered. A collision between two hadrons will occur if the transverse distance fulfils the condition,

$$d_{\mathrm{trans}} < d_0 = \sqrt{\frac{\sigma_{\mathrm{tot}}}{\pi}}, \tag{7.151}$$

σ_{tot} being the total cross section. σ_{tot} depends on the cm energy $\sqrt{s}$ and on the species and quantum numbers of the incoming particles. The final state of a binary collision is randomly chosen according to the relative weights for the different reaction channels. Initial collisions will be nucleonic only, but in the ultra-relativistic regime, very soon the phase space will be populated by all sorts of hadron species, e.g. $\mathrm{NN} \to \mathrm{N}\Delta$, $\Delta \to \mathrm{N}\pi$, $\mathrm{N}\pi\pi$, $\mathrm{K}\Sigma$, etc. The UrQMD model includes a large number of hadrons, ~50 baryon species and ~45 mesons and their anti-particles and all possible reactions, (i) nucleon–nucleon, (ii) meson–meson, (iii) meson–baryon and (iv) baryon–baryon, are considered. UrQMD uses experimental cross sections where available. Where experimental information is not available, the additive quark model[9] and detailed balance method are used to extrapolate to unknown

[9] In the additive quark model, the cross section depends only on the quark content of the colliding hadrons. The model assumes that in any reaction, the forward scattering amplitude is the sum of all possible two-body quark–quark and quark–antiquark scattering amplitudes. For example in a πP collision the scattering amplitude is

$$\langle \pi P|\pi P\rangle = \langle (u\bar{d})(uud)|(u\bar{d})(uud)\rangle = 2\langle uu|uu\rangle + \langle ud|ud\rangle + 2\langle \bar{d}u|\bar{d}u\rangle + \langle \bar{d}d\rangle. \tag{7.152}$$

cross sections. UrQMD also allows for subsequent re-scatterings. After each binary collision or decay, the outgoing particles are checked for further collisions within the respective time step.

At low energy $p_{\text{lab}} < 5$ GeV, particle production is dominated by resonance decays. At high energy $p_{\text{lab}} > 5$ GeV, the UrQMD model considers string excitation and fragmentation processes. In UrQMD, string excitation and fragmentation is implemented in the spirit of the Lund model [151]. The Lund model is a widely used model in high-energy nuclear collisions. Below, a brief description of the model is given.

Lund model. Based on quark confinement, QCD interaction at large distances can be modeled by a string-like potential, $V(r) = \kappa r$. The string-like interaction is supported by lattice QCD simulations. When the separation between two quarks exceeds a critical value, it may be energetically more favorable to break the string by producing $q\bar{q}$ pairs. The process is analogous to the Schwinger mechanism [152] of e^+e^- pair production in strong electric fields. The Lund model is based on the idea of string breaking or fragmentation. A hadron is considered as a string-like object. In hadron–hadron interaction, the strings are excited and subsequently decay, producing new hadrons. A schematic picture of the process is shown in figure 7.21.

To understand the model, let us consider the classical motion of quark and antiquark in a stable meson. In the limit of massless quarks, the Hamiltonian H is given by,

$$H = p_1 + p_2 + \kappa|x_1 - x_2| \tag{7.153}$$

where p_1, p_2 are the momenta of the particles and x_1, x_2 their positions. The particles always move with the speed of light and the equation of motion is

$$\frac{\mathrm{d}p}{\mathrm{d}t} = \pm\kappa \tag{7.154a}$$

where the sign depends on whether the other interacting particle is to the right or to the left. The motion of the $q\bar{q}$ pair is shown in figure 7.22. In the cm frame, the pair will oscillate. The oscillatory motion is called 'yo-yo' motion. If excited, a stable hadron will execute yo-yo motion. For a $q\bar{q}$ system with mass M the maximum relative distance is $L = M/\kappa$. The shaded area spanned by the motion of the quarks in half the time period can be calculated as M^2/κ^2.

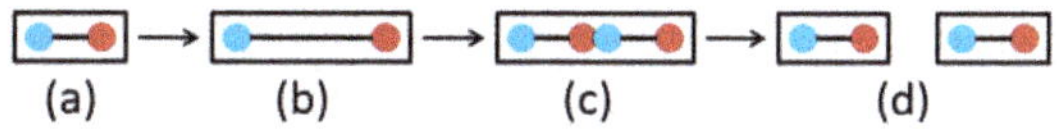

Figure 7.21. Schematic picture of particle production in a string excitation. (a) A ground state meson, a quark–antiquark pair connected by a string; (b) the excited meson, the string is stretched; (c) the stretched string breaks by producing a quark–antiquark pair; (d) two quark–antiquark pairs combine to form two mesons.

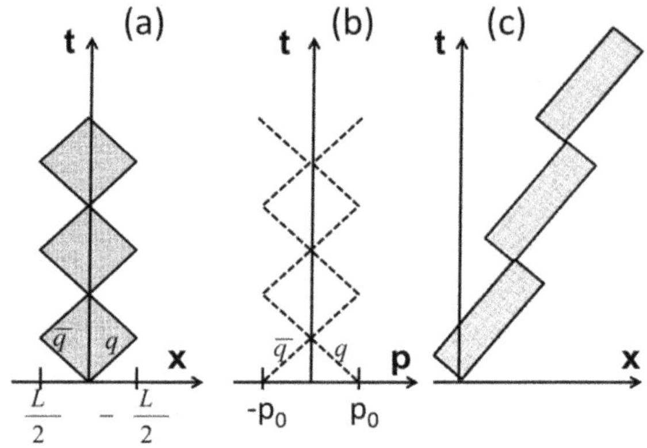

Figure 7.22. (a) Motion of q and $\bar{q}$ in the cm frame. The shaded areas are spanned by the string. (b) Variation of momentum of the pair with time. (c) Motion of the pair in a Lorentz boosted frame.

The constant force is Lorentz invariant: if in the original frame $x = \pm t$ and $E = \pm p$ (+ and − signs for particles moving to the right and left, respectively), then a Lorentz boost β gives

$$t' = \gamma(t - \beta x) = \gamma(1 \mp \beta)t \tag{7.155a}$$

$$p' = \gamma(p - \beta E) = \gamma(1 \mp \beta)p \tag{7.155b}$$

and

$$\frac{dp'}{dt'} = \frac{dp}{dt}. \tag{7.155c}$$

Motion of the $q\bar{q}$ pair in a Lorentz boosted frame is shown in figure 7.21(c). In the boosted frame as well, in half the time period the quarks sweep the same area M^2/κ^2.

Let us now consider hadron production from string fragmentation. Initially, quark and antiquark are moving in opposite directions. As they move apart, energy increases and at a certain point it is energetically favorable to break the string by creating a quark–antiquark pair. Let at (x_1, t_1) the system break into two parts by producing a quark pair $q_1\bar{q}_1$. At a later time (x_2, t_2) it may break further by producing another pair $q_2\bar{q}_2$. The process may continue further producing pairs $q_i\bar{q}_i$ at space–time points (x_i, t_i), (x_{i+1}, t_{i+1}), etc. In figure 7.23, a schematic diagram is shown for fragmentation at (x_1, t_1) and (x_2, t_2) only. String fragmentation in the Lund model is generally described in the light-cone coordinates

$$x^{\pm} = t \pm x. \tag{7.156}$$

In the light-cone coordinates, as depicted in figure 7.23, massless quarks move along the forward light cone x^+ and antiquarks in the backward light-cone x^-. The quarks (antiquarks) can be considered as carriers of forward (backward) light-cone momenta. Space–time descriptions and energy–momentum descriptions are closely related,

$$p^{\pm} = E \pm p = \kappa x^{\pm}. \tag{7.157}$$

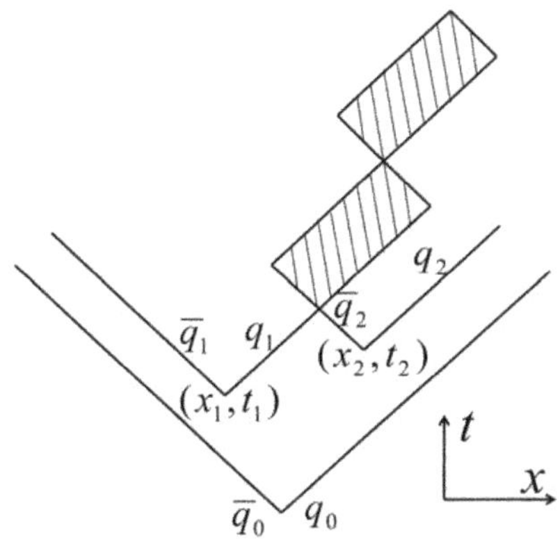

Figure 7.23. Schematic diagram of a string fragmentation. Initial $q_0\overline{q}_0$ move with large energy in opposite directions. After a certain time, quark pairs, $q_1\overline{q}_1$, $q_2\overline{q}_2$, are produced in space–time point (x_1, t_1), (x_2, t_2). Hadrons are formed by combining nearby quarks, $\overline{q}_1 q_2$.

In the Lund model, nearby quark–antiquark $(q_i\overline{q}_{i+1})$ pairs combine to form a hadron. The point of the first crossing of partons is defined as the hadron formation time. The momentum of the hadron formed by combining the quark–antiquark is related to the space–time coordinates of its constituents. For example, the ith meson formed by combining $q_{i-1}\overline{q}_i$ momentum is

$$E_i = \kappa(x_{i-1} - x_i) \tag{7.158a}$$

$$p_i = \kappa(x_i - x_{i-1}) \tag{7.158b}$$

which in light-cone variables are

$$p_i^+ = \kappa(x_{i-1}^+ - x_i^+) \tag{7.159a}$$

$$p_i^- = -\kappa(x_{i-1}^+ - x_i^+). \tag{7.159b}$$

In the Lund model, string breakings are treated stochastically. It is assumed that at each breaking,

$$\text{string}_0 \rightarrow \text{string}_1 + \text{hadron} \tag{7.160}$$

there is a universal method of partitioning of energy between string$_0$ and string$_1$. Let

$$z^\pm = \frac{p^\pm_{\text{hadron}}}{p^\pm_{\text{string}_0}} \quad \left(0 \leqslant z^\pm \leqslant 1\right) \tag{7.161}$$

be the longitudinal momentum fraction transferred to the newly formed hadron. In the Lund model, a fragmentation function $f(z)$ is introduced which gives the probability distribution for hadrons with the longitudinal momentum fraction z from the fragmenting string.

UrQMD implementation of the Lund model. In UrQMD hadron–hadron inelastic collisions with string fragmentation are treated as follows. From the probability distribution,

$$f(\mathbf{p}_T) \propto e^{-p_T^2/\sigma^2} \qquad \sigma = 1.6 \text{ GeV} \tag{7.162}$$

one of the hadrons is given a momentum $\mathbf{p}_T$ (the other gets the same, but in the opposite direction). The momentum transfer excites the strings (between quark and antiquark in the case of mesons and between quark and diquark in the case of baryons). The excitation is assumed to have the continuous mass distribution $f(M) \propto 1/M^2$, limited by the total collision energy, $\sqrt{s} \leqslant M_1 + M_2$. The rest of the $\sqrt{s}$ is equally distributed between the longitudinal momenta of two produced strings: $\mathbf{p}_{1_\parallel} = -\mathbf{p}_{2_\parallel}$. The energy of the strings is calculated as

$$E_i^2 = p_{i_\parallel}^2 + p_T^2 + M_i^2 \qquad i = 1, 2 \tag{7.163a}$$

$$E_1 + E_2 = \sqrt{s}. \tag{7.163b}$$

The excited strings are then allowed to fragment. A quark–antiquark pair is created and placed between the constituent quark–antiquark (quark–diquark) pair. Heavy flavors are suppressed. Suppression factors are,

$$u : d : s : \text{diquark} = 1 : 1 : 0.35 : 0.1. \tag{7.164}$$

UrQMD uses the Field–Feynman [153] fragmentation procedure where strings can fragment from both ends. A hadron is formed randomly on one of the end points of the string. Examples of fragmentation of meson and baryon strings are given below,

$$\text{meson string: } p^+ \underbrace{(q\bar{q}q\bar{q})}_{\text{string}} = z^+p^+ \underbrace{(q\bar{q})}_{\text{meson}} + (p^+ - z^+p^+)\underbrace{(q\bar{q})}_{\text{string}} \tag{7.165a}$$

$$\text{baryon string: } p^- \underbrace{(qqq\bar{q}q)}_{\text{string}} = z^-p^- \underbrace{(qqq)}_{\text{baryon}} + (p^- - z^-p^-)\underbrace{(\bar{q}q)}_{\text{string}}. \tag{7.165b}$$

The quark content of the hadron determines its charge and species. In the string frame, the produced hadron has zero transverse momentum. From the distribution equation (7.162), a transverse momentum is assigned to it stochastically. The fragmentation function is then used to determine the fraction of longitudinal momentum transfer from the string to the hadron. The remaining string is then checked for further fragmentation.

Bibliography

[1] Braun-Munzinger P, Stachel J, Wessels J P and Xu N 1995 *Phys. Lett.* B **344** 43

[2] Cleymans J, Oeschler H and Redlich K 1999 *Phys. Rev.* C **59** 1663

[3] Gazdzicki M and Gorenstein M I 1999 *Acta Phys. Polon.* B **30** 2705 (arXiv: hep-ph/9803462)

[4] Braun-Munzinger P, Heppe I and Stachel J 1999 *Phys. Lett.* B **465** 15

[5] Becattini F, Cleymans J, Keranen A, Suhonen E and Redlich K 2001 *Phys. Rev.* C **64** 024901

[6] Braun-Munzinger P, Magestro D, Redlich K and Stachel J 2001 *Phys. Lett.* B **518** 41

[7] Braun-Munzinger P, Redlich K and Stachel J 2004 Particle production in heavy ion collisions *Quark–Gluon Plasma* vol 3, ed R C Hwa and X-N Wang (Singapore: World Scientific) pp 491–599

[8] Becattini F, Gazdzicki M, Keranen A, Manninen J and Stock R 2004 *Phys. Rev.* C **69** 024905

[9] Cleymans J, Kampfer B, Kaneta M, Wheaton S and Xu N 2005 *Phys. Rev.* C **71** 054901

[10] Becattini F, Manninen J and Gazdzicki M 2006 *Phys. Rev.* C **73** 044905

[11] Cleymans J, Oeschler H, Redlich K and Wheaton S 2006 *Phys. Rev.* C **73** 034905

[12] Cleymans J, Oeschler H, Redlich K and Wheaton S 2006 *J. Phys. G: Nucl. Part. Phys.* **32** S165

[13] Andronic A, Braun-Munzinger P and Stachel J 2006 *Nucl. Phys.* A **772** 167

[14] Andronic A, Braun-Munzinger P and Stachel J 2009 *Acta Phys. Polon.* B **40** 1005 (arXiv:0901.2909 [nucl-th])

[15] Stachel J, Andronic A, Braun-Munzinger P and Redlich K 2013 arXiv: 1311.4662

[16] Tiwari S K and Singh C P 2013 *Adv. High Energy Phys.* **2013** 805413

[17] Becattini F and Cleymans J 2007 *J. Phys. G: Nucl. Part. Phys.* **34** S959

[18] Becattini F, Cleymans J and Strumpfer J 2007 *Proc. Sci. 4th Int. Workshop CPOD (Darmstadt, Germany, July 2007)* PoS (CPOD07)021

[19] Biedron B and Broniowski W 2007 *Phys. Rev.* C **75** 054905

[20] Broniowski W and Biedron B 2008 *J. Phys. G: Nucl. Part. Phys.* **35** 044018

[21] Cleymans J, Strumpfer J and Turko L 2008 *Phys. Rev.* C **78** 017901

[22] Schnedermann E, Sollfrank J and Heinz U W 1993 *Phys. Rev.* C **48** 2462

[23] Xu N and Kaneta M 2002 *Nucl. Phys.* A **698** 306

[24] Retiere F and Lisa M A 2004 *Phys. Rev.* C **70** 044907

[25] Tomasik B 2003 arXiv: nucl-th/0304079

[26] Tang Z, Xu Y, Ruan L, van Buren G, Wang F and Xu Z 2009 *Phys. Rev.* C **79** 051901

[27] Cooper F and Frye G 1974 *Phys. Rev.* D **10** 186

[28] Gyulassy M and McLerran L 2005 *Nucl. Phys.* A **750** 30

[29] McLerran L 2008 arXiv: 0812.4989 [hep-ph]

[30] Iancu E, Leonidov A and McLerran L 2002 arXiv: hep-ph/0202270

[31] McLerran L 2011 *Prog. Theor. Phys. Suppl.* **187** 17 (arXiv:1011.3204 [hep-ph])

[32] McLerran L D 2002 The colour glass condensate and small-x physics *Lecture Notes in Physics* vol 583, ed W Plessas and L Mathelitsch (Berlin: Springer) pp 291–334

[33] Kharzeev D, Levin E and Nardi M 2005 *Nucl. Phys.* A **747** 609

[34] Kolb P F and Heinz U 2004 Hydrodynamic description of ultrarelativistic heavy ion collisions *Quark–Gluon Plasma* vol 3, ed R C Hwa and X-N Wang (Singapore: World Scientific) pp 634–714

[35] Policastro G, Son D T and Starinets A O 2001 *Phys. Rev. Lett.* **87** 081601
[36] Policastro G, Son D T and Starinets A O 2002 *JHEP* **0209** 043
[37] Kovtun P, Son D T and Starinets A O 2003 *JHEP* **0310** 064
[38] Heinz U W and Kolb P F 2002 *Nucl. Phys.* A **702** 269
[39] Heinz U 2005 *J. Phys. G: Nucl. Part. Phys.* **31** S717
[40] Heinz U W and Kolb P F 2002 arXiv: hep-ph/0204061
[41] Danielewicz P and Gyulassy M 1985 *Phys. Rev.* D **31** 53
[42] Eckart C 1940 *Phys. Rev.* **58** 919
[43] Landau L D and Lifshitz E M 1963 *Fluid Mechanics* (Oxford: Pergamon) section 127
[44] Hiscock W A and Lindblom L 1985 *Phys. Rev.* D **31** 725
[45] Hiscock W A and Lindblom L 1983 *Ann. Phys.* **151** 466
[46] Israel W 1976 *Ann. Phys.* **100** 310
[47] Israel W and Stewart J M 1979 *Ann. Phys.* **118** 349
[48] Muronga A 2004 *Phys. Rev.* C **69** 034903
[49] Betz B, Denicol G S, Koide T, Molnar E, Niemi H and Rischke D H 2011 *EPJ Web Conf.* **13** 07005
[50] Denicol G S, Niemi H, Molnar E and Rischke D H 2012 *Phys. Rev.* D **85** 114047
[51] Jaiswal A, Bhalerao R S and Pal S 2013 *Phys. Lett.* B **720** 347
[52] Jaiswal A, Bhalerao R S and Pal S 2013 *Phys. Rev.* C **87** 021901
[53] Karsch F, Kharzeev D and Tuchin K 2008 *Phys. Lett.* B **663** 217
[54] Kharzeev D and Tuchin K 2008 *JHEP* **0809** 093
[55] Fries R J, Müller B and Schäfer A 2008 *Phys. Rev.* C **78** 034913
[56] Monnai A and Hirano T 2009 *Phys. Rev.* C **80** 054906
[57] Dusling K and Schäfer T 2012 *Phys. Rev.* C **85** 044909
[58] Song H and Heinz U 2009 *Nucl. Phys.* A **830** 467C
[59] Song H and Heinz U 2010 *Phys. Rev.* C **81** 024905
[60] Roy V and Chaudhuri A K 2012 *Phys. Rev.* C **85** 024909
Roy V and Chaudhuri A K 2012 *Phys. Rev.* C **85** 049902 (erratum)
[61] Boris J P and Book D L 1973 *J. Comput. Phys.* **11** 38
[62] Rischke D H, Bernard S and Maruhn J A 1995 *Nucl. Phys.* A **595** 346
[63] Bjorken J D 1983 *Phys. Rev.* D **27** 140
[64] Venugopalan R and Prakash M 1992 *Nucl. Phys.* A **546** 718
[65] Aoki Y, Endrodi G, Fodor Z, Katz S D and Szabo K K 2006 *Nature* **443** 675
[66] Fodor Z 2010 *J. Phys: Conf. Ser.* **230** 012013
[67] Aoki Y, Borsanyi S, Durr S, Fodor Z, Katz S D, Krieg S and Szabo K K 2009 *JHEP* **0906** 088
[68] Borsanyi S *et al* 2010 *JHEP* **1011** 077
[69] Huovinen P and Petreczky P 2010 *Nucl. Phys.* A **837** 26
[70] Shen C, Heinz U, Huovinen P and Song H 2010 *Phys. Rev.* C **82** 054904
[71] Roy V and Chaudhuri A K 2011 *Phys. Lett.* B **703** 313
[72] Hirano T and Nara Y 2004 *Nucl. Phys.* A **743** 305
[73] Luzum M and Romatschke P 2008 *Phys. Rev.* C **78** 034915
[74] Roy V, Chaudhuri A K and Mohanty B 2012 arXiv: 1204.2347
[75] Kharzeev D, Levin E and Nardi M 2004 *Nucl. Phys.* A **730** 448
Kharzeev D, Levin E and Nardi M 2004 *Nucl. Phys.* A **743** 329 (erratum)
[76] Jalilian-Marian J, Kovner A, Leonidov A and Weigert H 1997 *Nucl. Phys.* B **504** 415
[77] Jalilian-Marian J, Kovner A, Leonidov A and Weigert H 1998 *Phys. Rev.* D **59** 014014

[78] Iancu E, Leonidov A and McLerran L D 2001 *Phys. Lett.* B **510** 133
[79] Iancu E, Leonidov A and McLerran L D 2001 *Nucl. Phys.* A **692** 583
[80] Adil A, Drescher H-J, Dumitru A, Hayashigaki A and Nara Y 2006 *Phys. Rev.* C **74** 044905
[81] Poskanzer A M and Voloshin S A 1998 *Phys. Rev.* C **58** 1671
[82] Voloshin S A, Poskanzer A M and Snellings R 2008 arXiv: 0809.2949
[83] Ollitrault J-Y 1992 *Phys. Rev.* D **46** 229
[84] Adcox K *et al* (PHENIX Collaboration) 2005 *Nucl. Phys.* A **757**
[85] Adams J *et al* (STAR Collaboration) 2005 *Nucl. Phys.* A **757**
[86] Aamodt K *et al* (The ALICE Collaboration) 20 (arXiv:1011.3914 [nucl-ex])
[87] Aamodt K *et al* (ALICE Collaboration) 2011 *Phys. Rev. Lett.* **107** 032301
[88] Song H and Heinz U W 2009 *J. Phys. G: Nucl. Part. Phys.* **36** 064033
[89] Chaudhuri A K 2009 *Phys. Lett.* B **681** 418
[90] Chaudhuri A K 2010 *J. Phys. G: Nucl. Part. Phys.* **37** 075011
[91] Roy V and Chaudhuri A K 2011 *Phys. Lett.* B **703** 313
[92] Schenke B, Jeon S and Gale C 2011 *Phys. Lett.* B **702** 59–63
[93] Bozek P 2011 *Phys. Lett.* B **699** 283–6
[94] Song H, Bass S A and Heinz U 2011 *Phys. Rev.* C **83** 054912
Song H, Bass S A and Heinz U 2013 *Phys. Rev.* C **87** 019902 (erratum)
[95] Danielewicz P and Odyniec G 1985 *Phys. Lett.* B **157** 146
[96] Borghini N, Dinh P M and Ollitrault J Y 2001 *Phys. Rev.* C **63** 054906
[97] Borghini N, Dinh P M and Ollitrault J Y 2001 Flow analysis from multiparticle azimuthal correlations *Phys. Rev.* C **64** 054901
[98] Borghini N, Dinh P M and Ollitrault J Y 2001 arXiv: 0110016 nucl-ex
[99] Bhalerao R S, Borghini N and Ollitrault J Y 2004 *Phys. Lett.* B **580** 157
[100] Bhalerao R S, Borghini N and Ollitrault J Y 2003 *Nucl. Phys.* A **727** 373
[101] Borghini N, Bhalerao R S and Ollitrault J Y 2004 *J. Phys. G: Nucl. Part. Phys.* **30** S1213
[102] Schenke B, Jeon S and Gale C 2011 *Phys. Rev. Lett.* **106** 042301
[103] Schenke B, Jeon S and Gale C 2012 *Phys. Rev* C **85** 024901
[104] Dusling K and Teaney D 2008 *Phys. Rev.* C **77** 034905
[105] Song H, Bass S A, Heinz U, Hirano T and Shen C 2011 *Phys. Rev. Lett.* **106** 192301
Song H, Bass S A, Heinz U, Hirano T and Shen C 2012 *Phys. Rev. Lett.* **109** 139904 (erratum)
[106] Bozek P 2012 *Phys. Rev.* C **85** 034901
[107] Denicol G S, Kodama T and Koide T 2010 *J. Phys. G Nucl. Part. Phys.* **37** 094040
[108] Romatschke P and Romatschke U 2007 *Phys. Rev. Lett.* **99** 172301
[109] Chaudhuri A K 2008 arXiv: 0801.3180
[110] Back B B *et al* (PHOBOS Collaboration) 2001 *Phys. Rev. Lett.* **87** 102303
[111] Hirano T 2002 *Phys. Rev.* C **65** 011901
[112] Manly S *et al* (PHOBOS Collaboration) 2006 *Nucl. Phys.* A **774** 523
[113] Adams J *et al* (STAR Collaboration) 2005 *Phys. Rev. Lett.* **95** 152301
[114] Putschke J 2007 *J. Phys. G: Nucl. Part. Phys.* **34** S679
[115] Abelev B I *et al* (STAR Collaboration) 2009 *Phys. Rev. Lett.* **102** 052302
[116] Abelev B I *et al* (STAR Collaboration) 2009 *Phys. Rev.* C **80** 064912
[117] Khachatryan V *et al* (CMS Collaboration) 2010 *JHEP* **1009** 091
[118] Velicanu D (CMS Collaboration) 2011 arXiv: 1107.2196

[119] Alver B and Roland G 2010 *Phys. Rev.* C **81** 054905
[120] Alver B H, Gombeaud C, Luzum M and Ollitrault J-Y 2010 *Phys. Rev.* C **82** 034913
[121] Sorensen P 2010 *J. Phys. G: Nucl. Part. Phys.* **37** 094011
[122] Adare A *et al* (PHENIX Collaboration) 2010 *Phys. Rev. Lett.* **105** 062301
[123] Adare A *et al* (PHENIX Collaboration) 2011 *Phys. Rev. Lett.* **107** 252301
[124] Lacey R (PHENIX Collaboration) 2011 *J. Phys. G: Nucl. Part. Phys.* **38** 124048
[125] Schenke B, Tribedy P and Venugopalan R 2012 *Phys. Rev. Lett.* **108** 252301
[126] Gale C, Jeon S, Schenke B, Tribedy P and Venugopalan R 2012 arXiv: 1209.6330
[127] Petersen H, Qin G-Y, Bass S A and Muller B 2010 *Phys. Rev.* C **82** 041901
[128] Holopainen H, Niemi H and Eskola K J 2011 *Phys. Rev.* C **83** 034901
[129] Niemi H, Denicol G S, Holopainen H and Huovinen P 2012 arXiv: 1212.1008
[130] Werner K, Karpenko I, Pierog T, Bleicher M and Mikhailov K 2010 *Phys. Rev.* C **82** 044904
[131] Aguiar C E, Hama Y, Kodama T and Osada T 2002 *Nucl. Phys.* A **698** 639
[132] Qiu Z and Heinz U W 2011 *Phys. Rev.* C **84** 024911
[133] Bozek P and Broniowski W 2012 *Phys. Rev.* C **85** 044910
[134] Gardim F G, Grassi F, Luzum M and Ollitrault J-Y 2012 *Phys. Rev.* C **85** 024908
[135] Gardim F G, Grassi F, Luzum M and Ollitrault J-Y 2012 arXiv: 1210.8422
[136] Bartels J, Golec-Biernat K J and Kowalski H 2002 *Phys. Rev.* D **66** 014001
[137] Kowalski H and Teaney D 2003 *Phys. Rev.* D **68** 114005
[138] Kovner A, McLerran L D and Weigert H 1995 *Phys. Rev.* D **52** 6231 (arXiv:hep-ph/9502289)
[139] Rihan Haque M, Roy V and Chaudhuri A K 2012 *Phys. Rev.* C **86** 037901
[140] Dumitru A and Nara Y 2012 *Phys. Rev.* C **85** 034907
[141] Bertsch G F, Kruse H and Gupta S D 1984 *Phys. Rev.* C **29** 673
Bertsch G F, Kruse H and Gupta S D 1986 *Phys. Rev.* C **33** 1107 (erratum)
[142] Aichelin J and Bertsch G 1985 *Phys. Rev.* C **31** 1730
[143] Kruse H, Jacak B V and Stoecker H 1985 *Phys. Rev. Lett.* **54** 289
[144] Bertsch G F and Das Gupta S 1988 *Phys. Rep.* **160** 189
[145] Aichelin J 1991 *Phys. Rep.* **202** 233
[146] Hartnack C, Puri R K, Aichelin J, Konopka J, Bass S A, Stoecker H and Greiner W 1998 *Eur. Phys. J.* A **1** 151
[147] Aichelin J and Stoecker H 1986 *Phys. Lett.* B **176** 14
[148] Bass S A *et al* 1998 *Prog. Part. Nucl. Phys.* **41** 255
[149] Bleicher M *et al* 1999 *J. Phys. G: Nucl. Part. Phys.* **25** 1859
[150] Kerman A K and Koonin S E 1976 *Ann. Phys.* **100** 332
[151] Andersson B, Gustafson G and Soderberg B 1983 *Z. Phys.* C **20** 317
[152] Schwinger J S 1951 *Phys. Rev.* **82** 664
[153] Field R D and Feynman R P 1978 *Nucl. Phys.* B **136** 1

IOP Publishing

A Short Course on Relativistic Heavy Ion Collisions

A K Chaudhuri

Chapter 8

Two-particle correlation

8.1 Introduction

In relativistic heavy ion collisions, the final state contains a large number of particles and one- and two-particle invariant distributions are routinely measured in experiments. A one-particle or one-body distribution given by

$$E\frac{\mathrm{d}N}{\mathrm{d}^3p} \tag{8.1}$$

essentially measures the probability of emission of a particle with momentum $\mathbf{p}$. Likewise, two-particle or two-body distribution,

$$E_1E_2\frac{\mathrm{d}N}{\mathrm{d}^3p_1\,\mathrm{d}^3p_2}, \tag{8.2}$$

measures the emission probability that a particle with momentum $\mathbf{p}_2$ is emitted simultaneously with a particle with momentum $\mathbf{p}_1$. One can see that if the particle emission processes are independent of each other, two-particle invariant distribution will be the product of two single-particle distributions,

$$E_1E_2\frac{\mathrm{d}N}{\mathrm{d}^3p_1\,\mathrm{d}^3p_2} = \left(E_1\frac{\mathrm{d}N}{\mathrm{d}^3p_1}\right)\left(E_2\frac{\mathrm{d}N}{\mathrm{d}^3p_2}\right). \tag{8.3}$$

Any deviation from the above relation will indicate that the emission probability of the second particle is affected by the emission probability of the first particle, i.e. the particle emission process is not independent but is correlated. Now there are some trivial reasons for such correlations, e.g. energy–momentum conservation. Thus once a particle is detected with momentum $\mathbf{p}_1$, energy–momentum conservation imposes certain restrictions on the second particle momentum. Similarly, if the

doi:10.1088/978-0-750-31060-4ch8

particles are from decay of a resonance, they will exhibit a certain correlation. However, apart from those 'trivial' correlations, the quantum nature of the particles introduces certain correlations. For example, if the particles are bosons, then given that one particle is detected with momentum $\mathbf{p}_1$, the probability of detecting the second particle at a nearby momentum $\mathbf{p}_2$ is enhanced. In contrast, for fermions, the probability of detecting the second particle at a nearby momentum $\mathbf{p}_2$ is reduced. The study of such correlations is an integral part of heavy ion reactions as one can obtain some information about the physical size of the particle emitting source.

The study of Bose–Einstein or Fermi–Dirac correlations is generally called intensity interferometry or femtoscopy when applied to sub-atomic systems. Acknowledging the two scientists, Robert Hanbury Brown and Richard Twiss, who invented the methodology [1, 2], it is also called Hanbury Brown–Twiss (HBT) interferometry.

In the 1950s, Hanbury Brown was measuring the spatial extent of astronomical objects. The prevailing method was to use a Michelson interferometer, where the amplitude interference of light rays from astronomical objects causes a fringe (interference) pattern. Study of fringe structures can give an idea of the angular size of the astronomical objects. In amplitude interference, a fringe pattern is caused due to the difference in path lengths of the light rays. Since for distant objects atmospheric disturbances affect the light paths, accurate measurements of angular size are not possible. Hanbury Brown and Twiss showed that one could make accurate measurements of the angular sizes of the astronomical objects from the correlations of signal intensities in independent detectors. The two physicists developed the theory for intensity interferometry, built the required apparatus, made measurements, analyzed and published the results. Initially, their results were faced with criticism. A large community of physicists declined to accept that two photons can interfere. The reason is Dirac's apparent simple statement in his classic book, *The Principles of Quantum Mechanics*: 'Interference between two different photons can never occur'.

It took quite some time to realize that the statement is applicable only for coherent photons. Astronomical sources produce incoherent photons and HBT correlation, which is essentially due to the quantum nature of the photons, can be observed.

Around the same time, Goldhaber, Goldhaber, Lee and Pais [3], in $p_{\mathrm{lab}} = 1.05\ \mathrm{GeV}/c\, p\bar{p}$ collisions, observed that the angular distribution of like-charge pions and unlike-charge pions differ. The observed difference could not be accommodated in the then prevailing theory, Fermi's statistical model. They could reproduce the experimental data by a detailed multi-pion phase space calculation with symmetrized wave function for like-charge pions. They concluded that the effect is due to the Bose–Einstein nature of pions. Goldhaber, Goldhaber, Lee and Pais were unaware of the HBT experiment. Unknowingly, they had discovered the counterpart of the HBT effect in high-energy collisions.

While in heavy ion collisions we are interested in intensity interferometry, for completeness, in the following, amplitude interferometry will be briefly discussed. It will be followed by a brief discussion on intensity interferometry. There are a few good reviews on HBT interferometry [4–12]. Interested readers can look into these for more detailed discussions.

8.2 Amplitude interferometry

In optics, we have studied interference phenomena. The famous Young's double slit experiment is shown in figure 8.1. When slit 1 is illuminated with a monochromatic light source, an interference pattern or fringe structure is seen on the screen behind the double slit. The interference is called amplitude interference. When two waves from slits S_1 and S_2 superimpose at point P (see figure 8.1), the intensity at point P is proportional to the time averaged square of the sum of the amplitudes of the waves from S_1 and S_2,

$$I \propto \langle |A_1 + A_2|^2 \rangle. \tag{8.4}$$

For simplicity, we represent the amplitudes by sine functions,

$$A_1 = A_0 \sin(\omega t) \tag{8.5a}$$

$$A_2 = A_0 \sin(\omega t + \phi), \tag{8.5b}$$

where we have added a phase ϕ to the wave from slit S_2 acknowledging that it has to traverse an extra path δ to reach P. The extra path or path difference between the two waves can be calculated from geometrical considerations. For $d \ll L$, i.e. when the slit separation d is much smaller than the screen distance L

$$\delta = d \sin\theta. \tag{8.6}$$

We also understand that when the path difference is equal to the multiple of the wavelength λ, the wave will acquire an extra phase 2π. Thus,

$$\frac{\delta}{\lambda} = \frac{\phi}{2\pi}. \tag{8.7}$$

For the sine waves, the sum of the amplitudes at P can be written as

$$A = A_1 + A_2 = 2A_0 \sin\left(\omega t + \frac{\phi}{2}\right) \cos\left(\frac{\phi}{2}\right), \tag{8.8}$$

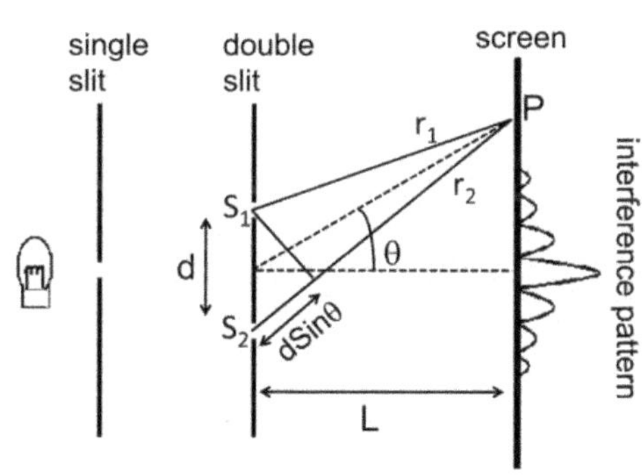

Figure 8.1. Young's double slit experiment. A monochromatic light source in front of the single slit produces two monochromatic, coherent point sources at the double slit at S_1 and S_2. Waves from the two coherent sources result in an interference pattern on the screen.

and the intensity at P is

$$I \propto 4A_0^2 \left\langle \sin^2\left(\omega t + \frac{\phi}{2}\right)\right\rangle \cos^2\left(\frac{\phi}{2}\right)$$

$$= 2A_0^2 \cos^2\left(\frac{\phi}{2}\right) = 2A_0^2 \cos^2\left(\frac{\pi d \sin\theta}{\lambda}\right). \quad (8.9)$$

As expected, equation (8.9) shows an interference pattern. One can also conclude that constructive interference occurs when

$$d \sin\theta_n = n\lambda \qquad n = 1, 2, 3, \ldots \quad (8.10)$$

where θ_n is the angle of the nth order maximum. The equation is generally used to measure the wavelength of a monochromatic light source. However, given a monochromatic light source of known wavelength, equation (8.10) can also be used to obtain unknown slit separation ('source size') by measuring θ_n.

Michelson stellar interferometry is based on the principle of amplitude interferometry. In figure 8.2, a schematic diagram of a Michelson interferometer is shown. Light from a stellar object is received by two small mirrors, M_1 and M_2. The separation between the two mirrors can be adjusted as desired. Two fixed mirrors M_3 and M_4 reflect the light from M_1 and M_2 into a telescope in which the two beams are brought to a focus at O. In effect, the image of the star at O consists of two superimposed images seen through the two mirrors M_1 and M_2. When the separation of the two mirrors M_1, M_2 is suitably adjusted, the image of the star is seen to be crossed by alternate light and dark bands which are called interference fringes. One can define a Michelson's fringe visibility parameter [13],

$$\mathcal{V} = \frac{P_{\max} - P_{\min}}{P_{\max} + P_{\min}} \quad (8.11)$$

where $P_{\max}$ and $P_{\min}$ are, respectively, maximum and minimum power (or equivalently intensity) with respect to the mean power (or intensity). Fringe visibility essentially

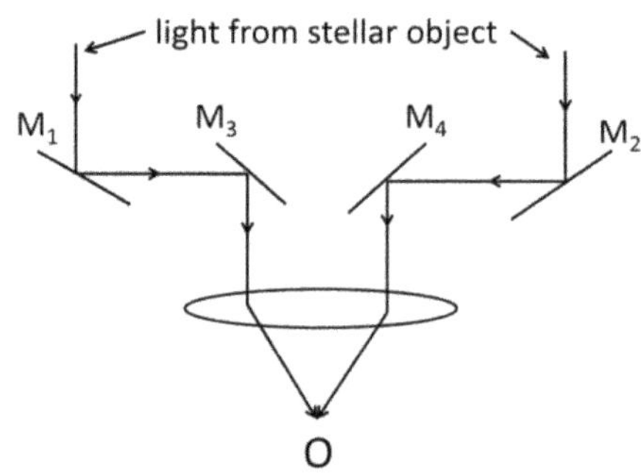

Figure 8.2. Schematic diagram of a Michelson interferometer for stellar objects.

measures the coherence length[1] of the source and is a function of the distance between the two mirrors M_1 and M_2 and the angular diameter of the star. By varying the separation between the mirrors, one can obtain the angular size of the star.

Just as in Young's double slit experiment, in Michelson's interferometer the fringe or interference pattern is due to the phase difference between the two light rays (the distance between the mirrors is adjusted to introduce a delay between the light rays). For astronomical objects, the relative phase of two paths is not just the physical path difference but also contains uncontrollable contributions from atmospheric distortions. Uncertainty in the measurements cannot be controlled.

8.3 Intensity interferometry

Hanbury Brown and Twiss overcame the difficulties associated with amplitude interferometry and developed intensity interferometry for astronomical objects. In intensity interferometry, one measures two photons in *coincidence* as a function of their relative momentum, and measures the correlation function, defined as

$$C(k_1, k_2) = \frac{\langle n_{12} \rangle}{\langle n_1 \rangle \langle n_2 \rangle} \tag{8.12}$$

where $\langle n_{12} \rangle$ is the probability of detecting two coincident photons of momentum k_1 and k_2 in detectors 1 and 2. $\langle n_1 \rangle$ and $\langle n_2 \rangle$ are, respectively, the probability of detecting a photon of momentum k_1 in detector 1 and momentum k_2 in detector 2. From the correlation function, one can obtain information about the (angular) size of the distant object.

The physical basis of HBT interferometry can be understood from the following simple example. As depicted in figure 8.3, light waves emitting from point sources A and B can both be received by detector D_1 as well as by detector D_2. The amplitude at detector D_1 can be written as

$$A_1 = \frac{1}{L}\left(\alpha e^{ikr_{1A}+i\phi_A} + \beta e^{ikr_{1B}+i\phi_B}\right), \tag{8.13}$$

where r_{1A} and r_{2B} are the distance to detector D_1 from source A and source B, respectively. Total intensity at detector D_1 is obtained by squaring the amplitude,

$$I_1 = \frac{1}{L^2}\left(|\alpha|^2 + |\beta|^2 + \alpha^*\beta e^{ik(r_{1A}-r_{1B})+i(\phi_B-\phi_A)} + \alpha\beta^* e^{-ik(r_{1A}-r_{1B})-i(\phi_B-\phi_A)}\right). \tag{8.14}$$

Similarly, we can write for intensity I_2 at detector D_2,

$$I_2 = \frac{1}{L^2}\left(|\alpha|^2 + |\beta|^2 + \alpha^*\beta e^{ik(r_{2A}-r_{2B})+i(\phi_B-\phi_A)} + \alpha\beta^* e^{-ik(r_{2A}-r_{2B})-i(\phi_B-\phi_A)}\right). \tag{8.15}$$

[1] Coherence length is the propagation distance from a coherent source to a point where an electromagnetic wave maintains a specified degree of coherence. The significance is that interference will be strong within a coherence length of the source, but not beyond it.

The last two terms in equations (8.14) and (8.15) contain information about the source distance to the detector. If the phases are random, i.e. a chaotic source, then when averaged over the detection time, those terms will vanish and

$$\langle I_1 \rangle = \langle I_2 \rangle = \frac{1}{L^2}\left(|\alpha|^2 + |\beta|^2\right), \tag{8.16}$$

and one obtains

$$\langle I \rangle = \langle I_1 \rangle \langle I_2 \rangle = \frac{1}{L^4}\left(|\alpha|^2 + |\beta|^2\right)^2, \tag{8.17}$$

which is independent of the separation between the two sources.

The same is not true for the time averaged *coincidence* rate. The time averaged coincidence rate is obtained by multiplying the two intensities and then averaging,

$$\begin{aligned}\langle I_1 I_2 \rangle &= \frac{1}{L^4}\Big(|\alpha|^4 + |\beta|^4 + 2|\alpha|^2|\beta^2| + |\alpha|^2|\beta|^2 e^{ik[(r_{1A}-r_{1B})-(r_{2A}-r_{2B})]} \\ &\quad + |\alpha|^2|\beta|^2 e^{-ik[(r_{1A}-r_{1B})-(r_{2A}-r_{2B})]}\Big) \\ &= \langle I_a \rangle \langle I_2 \rangle + \frac{2}{L^4}|\alpha|^2|\beta|^2 \cos k(r_{1A} - r_{1B} - r_{2A} + r_{2B}).\end{aligned} \tag{8.18}$$

Two-photon intensity correlation is then defined as

$$\begin{aligned}C(R, d) &= \frac{\langle I_1 I_2 \rangle}{\langle I_1 \rangle \langle I_2 \rangle} \\ &= 1 + \frac{2|\alpha|^2|\beta|^2}{\left(|\alpha|^2 + |\beta|^2\right)^2} \cos k(r_{1A} - r_{1B} - r_{2A} + r_{2B}).\end{aligned} \tag{8.19}$$

For large separation between the detectors and sources, the argument of the cosine function can be written as

$$k(r_{1A} - r_{1B} - r_{2A} + r_{2B}) \to k(\mathbf{r}_A - \mathbf{r}_B) \cdot \left(\hat{\mathbf{r}}_2 - \hat{\mathbf{r}}_1\right) = \mathbf{R} \cdot \left(\mathbf{k}_2 - \mathbf{k}_1\right) \tag{8.20}$$

where $\mathbf{k}_1(\mathbf{k}_2) = k\hat{\mathbf{r}}_{1,2}$ is the wave vector of the light seen in detector 1(2). For large separation between the detectors and sources, $\alpha \approx \beta$ and the two particle correlation function can be written as

$$C(R, d) = 1 + \frac{1}{2}\cos\left\{\mathbf{R} \cdot \left(\mathbf{k}_2 - \mathbf{k}_1\right)\right\}. \tag{8.21}$$

The cosine function in the correlation function oscillates and two photons largely separated in momentum or space can apparently be strongly correlated. However, this unrealistic result is due to the assumption of 'only' two point sources. In real life there will be a continuum of sources. Defining a source distribution $\rho(\mathbf{r})$, the correlation function can be written as

$$C\left(\mathbf{k}_1, \mathbf{k}_2\right) = 1 + \frac{1}{2}\frac{\int d^3r\rho(\mathbf{r}) \cos\left\{\mathbf{r} \cdot \left(\mathbf{k}_2 - \mathbf{k}_1\right)\right\}}{\int d^3r\rho(\mathbf{r})}. \tag{8.22}$$

For photons largely separated in momentum or space, the correlation function dies down. Equation (8.22) also indicates that the two-photon correlation function essentially measures the Fourier transform of the source distribution.

8.3.1 A quantum mechanical reason for HBT interferometry

In quantum mechanics, two-photon correlation, as was observed in HBT interferometry, is due to the bosonic nature of the photons, the light quanta. Quantum mechanics requires that the wave function of a bosonic many-body system is symmetric under exchange of one particle. Quantum mechanical reasoning for HBT correlation can be ascertained if figure 8.3 is analyzed properly. Figure 8.3 essentially depicts four different possibilities (see figure 8.4):

(i) light source at A emits two photons, one detected by D_1, the other by D_2;
(ii) light source at B emits two photons, one detected by D_1, the other by D_2;
(iii) light source A emits a photon which is detected by D_1 and light source B emits a photon which is detected by D_2;

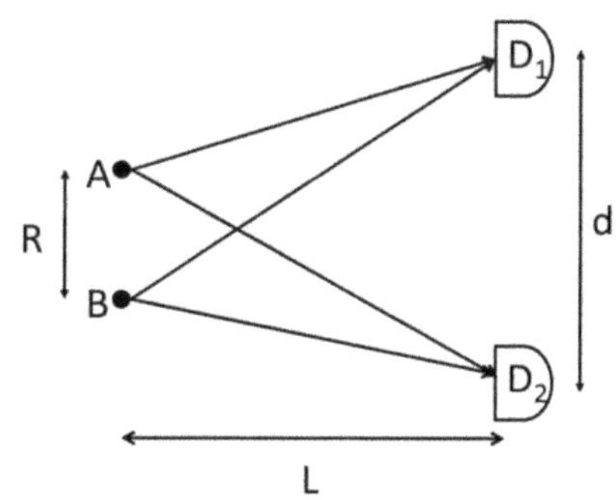

Figure 8.3. Schematic diagram of Bose–Einstein correlation. Light emitted from point sources A and B is detected by detectors D_1 and D_2.

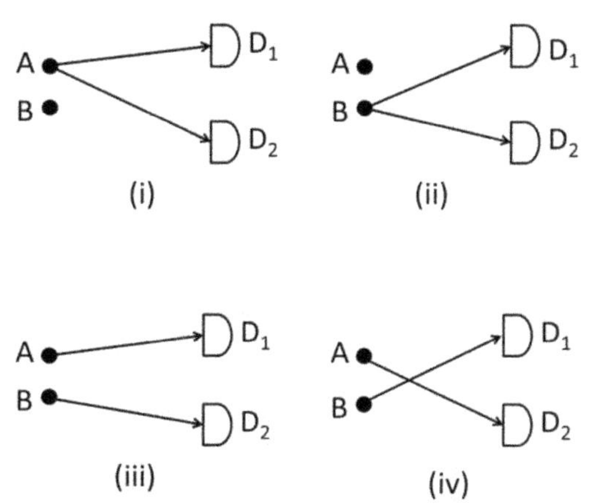

Figure 8.4. Analysis of figure 8.3. Four independent photon production and detection processes are included in figure 8.3. Only processes (iii) and (iv) are responsible for HBT correlation.

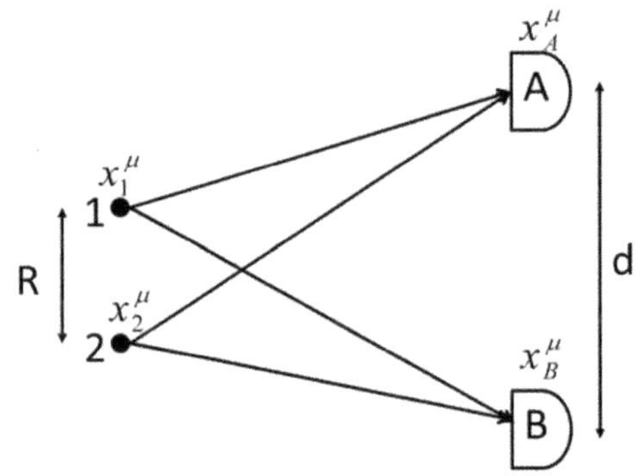

Figure 8.5. Schematic diagram of two-particle correlation.

(iv) light source A emits a photon which is detected by D_2 and light source B emits a photon which is detected by D_1.

The first two processes correspond to photon emission from either source 1 or source 2. They are distinguishable and do not produce interference phenomena. Interference occurs only from processes (iii) and (iv) and one can see that the process (iv) is just the exchange process for (iii). Quantum mechanically, the HBT effect is a consequence of the one-boson exchange effect, or symmetrization of the two-boson wave function. It is well-known that for a pair of bose particles in orthogonal states ϕ_α and ϕ_β, the two-particle state is symmetric under exchange of one particle and can be written as

$$\phi(1, 2) = \frac{1}{2}\left[\phi_\alpha(1)\phi_\beta(2) + \phi_\alpha(2)\phi_\beta(1)\right]. \tag{8.23}$$

On the other hand, for a fermionic system, the wave function is antisymmetric under exchange of one particle. The two-particle wave function can be written as

$$\phi(1, 2) = \frac{1}{2}\left[\phi_\alpha(1)\phi_\beta(2) - \phi_\alpha(2)\phi_\beta(1)\right]. \tag{8.24}$$

8.3.2 Two-particle correlation function

In stellar astronomy, source sizes are static. But in heavy ion collisions, the particle emitting source evolves in space–time. The two-photon correlation function equation (8.22) can be easily generalized to treat such sources. As before, consider two point sources 1 and 2, respectively at positions x_1^μ and x_2^μ, separated by a distance R, emitting particles (see figure 8.5). Approximating the wave functions by a plane wave, the amplitude for the process that a particle of momentum k^μ emitted from either of the sources is detected at the detector A at x_A^μ is

$$A_1(k) = \frac{1}{\sqrt{2}}\left[\mathrm{e}^{-\mathrm{i}k_\mu\left(x_A^\mu - x_1^\mu\right)}\mathrm{e}^{\mathrm{i}\phi_1} \pm \mathrm{e}^{-\mathrm{i}k_\mu\left(x_A^\mu - x_2^\mu\right)}\mathrm{e}^{\mathrm{i}\phi_2}\right], \tag{8.25}$$

where the (±) sign is for bosons and fermions, respectively. Squaring it we obtain

$$|A_1(k)|^2 = \frac{1}{2}\left[2 \pm \left(e^{ik_\mu(x_1^\mu - x_2^\mu)}e^{i(\phi_1-\phi_2)} + e^{-ik_\mu(x_1^\mu - x_2^\mu)}e^{-i(\phi_1-\phi_2)}\right)\right]. \tag{8.26}$$

The probability of detecting one particle at the detector at x_A^μ is obtained by averaging $|A_1(k)|^2$ over many events,

$$P(k) = \langle |A_1(k)|^2 \rangle = \frac{1}{2}\left[2 \pm \left(e^{ik_\mu(x_1^\mu - x_2^\mu)}\left\langle e^{i(\phi_1-\phi_2)}\right\rangle + e^{-ik_\mu(x_1^\mu - x_2^\mu)}\left\langle e^{-i(\phi_1-\phi_2)}\right\rangle\right)\right]. \tag{8.27}$$

For incoherent or chaotic sources,

$$\left\langle e^{\pm i(\phi_1-\phi_2)}\right\rangle = 0 \tag{8.28}$$

and we obtain

$$P(k) = \langle |A_1(k)|^2 \rangle = 1. \tag{8.29}$$

The quantum mechanical amplitude for the process that two detectors at x_A^μ and x_B^μ simultaneously detect two particles with momenta k_1^μ and k_2^μ can be written as

$$A_2(k_1, k_2) = \frac{1}{\sqrt{2}}\left[e^{-ik_{1\mu}(x_A^\mu - x_1^\mu)}e^{i\phi_1}e^{-ik_{2\mu}(x_B^\mu - x_2^\mu)}e^{i\phi_2} \pm e^{-ik_{1\mu}(x_A^\mu - x_2^\mu)}e^{i\phi_2}e^{-ik_{2\mu}(x_B^\mu - x_1^\mu)}e^{i\phi_1}\right]. \tag{8.30}$$

Squaring the amplitude, we can obtain for the probability of detecting two particles with momenta k_1 and k_2,

$$\begin{aligned} P(k_1, k_2) = \langle |A(k_1, k_2)|^2 \rangle &= \frac{1}{2}\left[2 \pm \left(e^{i(k_{1\mu}-k_{2\mu})(x_1^\mu - x_2^\mu)} + e^{-i(k_{1\mu}-k_{2\mu})(x_1^\mu - x_2^\mu)}\right)\right. \\ &= 1 \pm \cos\left(q_\mu R^\mu\right) \end{aligned} \tag{8.31}$$

with $q_\mu = k_{1\mu} - k_{2\mu}$ and $R^\mu = x_1^\mu - x_2^\mu$. The correlation function depends on the relative momenta of the two particles and on the spatial separation of the two sources. Two-particle correlation can be obtained as

$$C(k_1, k_2) = \frac{\langle n_{12}\rangle}{\langle n_2\rangle\langle n_2\rangle} = \frac{P(k_1, k_2)}{P(k_1)P(k_2)} = 1 \pm \cos\left(q_\mu R^\mu\right). \tag{8.32}$$

The cosine function oscillates and two particles, largely separated in momentum or space, can be strongly correlated. This unrealistic situation arises due to the assumption of 'only' two point sources. In real life there will be a continuum of sources. Defining a source distribution $\rho(x)$ normalized to unity, the probability,

$$\begin{aligned} P(k_1, k_2) &= \frac{1}{2}\left[2 \pm \left(\int d^4x_1\rho(x_1)e^{iq_\mu x_1^\mu}\int d^4x_2\rho(x_2)e^{-iq_\mu x_2^\mu}\right.\right. \\ &\quad \left.\left. + \int d^4x_1\rho(x_1)e^{-iq_\mu x_1^\mu}\int d^4x_2\rho(x_2)e^{iq_\mu x_2^\mu}\right)\right] \\ &= 1 \pm |\tilde{\rho}(q)|^2 \end{aligned} \tag{8.33}$$

where

$$\tilde{\rho}(q) = \int d^4 x e^{iq_\mu x^\mu} \rho(x), \tag{8.34}$$

is the Fourier transform of the source distribution $\rho(x)$.

The two-body correlation function,

$$C(k_1, k_2) = \frac{P(k_1, k_2)}{P(k_1)P(k_2)} = 1 \pm |\tilde{\rho}(q)|^2, \tag{8.35}$$

then measures the Fourier transform of the source distribution. If the source distribution is approximated by a Gaussian function,

$$\rho(x) = e^{-x_\mu x^\mu / (2R)^2}, \tag{8.36}$$

the Fourier transform is also a Gaussian,

$$\tilde{\rho}(q) = e^{-q^2 R^2/2}, \tag{8.37}$$

and the correlation function can be written as

$$C(k_1, k_2) = C(K, q) = 1 \pm e^{-q^2 R^2/2}. \tag{8.38}$$

The typical correlation function for bosons and fermions, as a function of momentum difference q, is shown in figure 8.6. Given that a boson of certain momentum is emitted, the probability of emitting a second boson with adjacent momentum is enhanced. In contrast, for fermions, the probability is

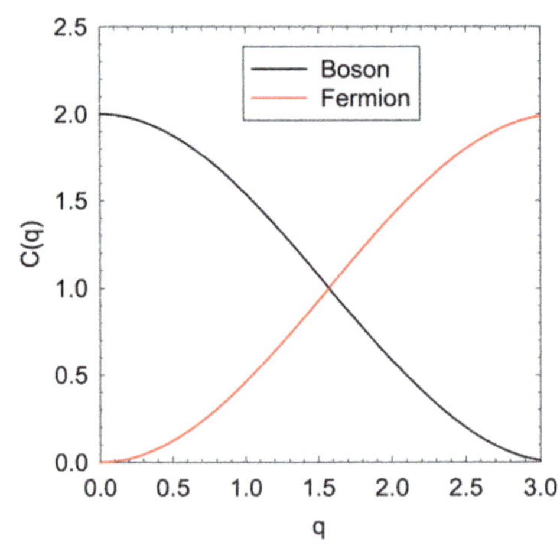

Figure 8.6. Two-particle correlation for an extended source. The black line is for bosons and the red line is for fermions.

diminished. One also notes that for a Gaussian source, the size of the source can be deduced from the correlation function. This is the important realization that coincidence measurements of two particles can give information about the source size. There are several ways one can define an effective source size. One standard procedure is to use the inverse of the full width at half maximum (FWHM), which for equation (8.38) is

$$R_{\mathrm{eff}} = \frac{\sqrt{2\ln 2}}{q}. \tag{8.39}$$

8.3.3 Two-particle correlation in heavy ion collisions

In heavy ion collisions, the source emits many particles. The two-particle state is a subset of the many-particle state and is a mixed state. Quantum mechanically mixed states are described by density matrices. For example, the single-particle density matrix for pions of a given charge at equal time is given by

$$\left\langle \pi_{\mathbf{p}}^{\dagger}(x_1, t)\pi_{\mathbf{p}}(x_1', t)\right\rangle = \int \mathrm{d}x_2\mathrm{d}x_3\ldots\mathrm{d}x_N\,\Psi^*(x_1, x_2,\ldots x_N, t)\Psi(x_1', x_2,\ldots x_N, t) \tag{8.40}$$

where $\pi_{\mathbf{p}}^{\dagger}$ and $\pi_{\mathbf{p}}$ are creation and annihilation operators. The single-particle density matrix $\langle \pi^{\dagger}(x_1, t)\pi(x_1', t)\rangle$ is the amplitude for the process where, at time t, a pion with momentum $\mathbf{p}$ is annihilated at space–time x'_1 and created at x_1. Similarly, the two-particle density matrix,

$$\begin{aligned}&\left\langle \pi_{\mathbf{p}_2}^{\dagger}(x_2, t)\pi_{\mathbf{p}_1}^{\dagger}(x_1, t)\pi_{\mathbf{p}_2}(x_1', t)\pi_{\mathbf{p}_1}(x_1', t)\right\rangle\\ &\quad= \int \mathrm{d}x_3\ldots\mathrm{d}x_N\,\Psi^*(x_1, x_2, x_3, \ldots x_N, t)\Psi(x_1', x_2', x_3, \ldots x_N, t),\end{aligned} \tag{8.41}$$

is the amplitude for the process where, at time t, a pion is annihilated at space–time x_1', a second pion is annihilated at x_2' followed by creation of a pion at x_2 and then at x_1. With the density matrix, the one- and two-particle invariant distribution can be written as

$$E\frac{\mathrm{d}N}{\mathrm{d}^3p} = E\left\langle \pi_{\mathbf{p}}^{\dagger}\pi_{\mathbf{p}}\right\rangle \tag{8.42}$$

$$\begin{aligned}E_1E_2\frac{\mathrm{d}N}{\mathrm{d}^3p_1\mathrm{d}^3p_2} &= E_1E_2\left\langle \pi_{\mathbf{p}_2}^{\dagger}\pi_{\mathbf{p}_1}^{\dagger}\pi_{\mathbf{p}_2}\pi_{\mathbf{p}_1}\right\rangle\\ &= E_1E_2\left[\left\langle \pi_{\mathbf{p}_1}^{\dagger}\pi_{\mathbf{p}_1}\right\rangle\left\langle \pi_{\mathbf{p}_2}^{\dagger}\pi_{\mathbf{p}_2}\right\rangle + \left\langle \pi_{\mathbf{p}_1}^{\dagger}\pi_{\mathbf{p}_2}\right\rangle\left\langle \pi_{\mathbf{p}_2}^{\dagger}\pi_{\mathbf{p}_1}\right\rangle\right].\end{aligned} \tag{8.43}$$

Defining a covariant quantity,

$$\tilde{S}(\mathbf{p}_1, \mathbf{p}_2) = \sqrt{E_1E_2}\left\langle \pi_{\mathbf{p}_1}^{\dagger}\pi_{\mathbf{p}_2}\right\rangle, \tag{8.44}$$

the two-particle correlation function can be written as,

$$C(p_1, p_2) = 1 + \frac{|\tilde{S}(\mathbf{p}_1, \mathbf{p}_2)|^2}{\tilde{S}(\mathbf{p}_1, \mathbf{p}_1)\tilde{S}(\mathbf{p}_2, \mathbf{p}_2)}. \tag{8.45}$$

It is possible to show that $\tilde{S}$ is a Fourier transform of a Wigner function,

$$\tilde{S}(\mathbf{p}_1, \mathbf{p}_2) = \tilde{S}(\mathbf{q}, \mathbf{K}) = \int \mathrm{d}^4 x \mathrm{e}^{-iq\cdot x} S(x, K), \tag{8.46}$$

where the off-shell four-vector $K = \frac{1}{2}(p_1 + p_2)$ is the average of two on-shell ($p_i^0 = E_i$) four-momenta and $q = p_1 - p_2$ is the off-shell difference of two momenta so that the scalar product vanishes, $K^\mu q_\mu = 0$. Only for a special case is $\mathbf{p}_1 = \mathbf{p}_2 = \mathbf{p}$, K on-shell, $K = p$ and

$$\tilde{S}(\mathbf{p}, \mathbf{p}) = \tilde{S}(\mathbf{0}, \mathbf{p}) = \int \mathrm{d}^4 x S(x, p). \tag{8.47}$$

The two-particle correlation function can now be evaluated as

$$C(p_1, p_2) = C(q, K) = 1 + \frac{\left|\int \mathrm{d}^4 x \mathrm{e}^{-iq\cdot x} S(x, K)\right|^2}{\int \mathrm{d}^4 x S\left(x, K + \frac{q}{2}\right) \int \mathrm{d}^4 x S\left(x, K - \frac{q}{2}\right)}. \tag{8.48}$$

For small momentum transfers, dependence of the emission function on q is weak and one can make the smoothness approximation

$$S\left(x, K + \frac{q}{2}\right) \approx S\left(x, K - \frac{q}{2}\right) \approx S(x, K) \tag{8.49}$$

and write the correlation function as

$$C(q, K) = 1 + \frac{\left|\int \mathrm{d}^4 x \mathrm{e}^{-iq\cdot x} S(x, K)\right|^2}{\left|\int \mathrm{d}^4 x S(x, K)\right|^2} = 1 + |\langle \mathrm{e}^{-iq.x}\rangle(K)|^2 \tag{8.50}$$

where the angular bracket $\langle \ldots \rangle$ indicates space–time averaging over the emission function.

One generally models the emission function $S(x, K)$. Noting that the Wigner function is the quantum analog of the classical distribution function, one simple model is to use the classical phase space distribution function. In hydrodynamical models, the phase space distribution function is the local equilibrium Bose–Einstein distribution, localized on a three-dimensional freeze-out hyper surface $\Sigma(x)$.

$$f(x, p) = \frac{g}{(2\pi)^3} \frac{1}{\exp\left\{\beta(x)\left[p^\mu u_\mu(x) - \mu(x)\right]\right\} - 1}. \tag{8.51}$$

The emission function can be modeled as

$$S(x, K) = \int K^\mu \mathrm{d}\sigma_\mu(x') f(x', K) \delta^{(4)}(x - x'). \tag{8.52}$$

The denominator and numerator of equation (8.50) can be written as

$$\int \mathrm{d}^4 x S(x, K) = \int K^\mu \mathrm{d}\sigma_\mu(x) f(x, K) \tag{8.53}$$

and

$$\left|\int \mathrm{d}^4 x \mathrm{e}^{-\mathrm{i}q\cdot x} S(x, K)\right|^2 = \left(\int K^\mu \mathrm{d}\sigma_\mu(x) \cos(q\cdot x) f(x, K)\right)^2 + \left(\int K^\mu \mathrm{d}\sigma_\mu(x) \sin(q\cdot x) f(x, K)\right)^2. \tag{8.54}$$

8.4 Gaussian parameterization of the correlation function

The two-particle correlation function is usually parameterized by a Gaussian function. Using a generic Gaussian function,

$$\mathrm{e}^{-q_\mu q_\nu \langle \tilde{x}^\mu \tilde{x}^\nu \rangle}, \tag{8.55}$$

where the variances are written as

$$\langle \tilde{x}^\mu \tilde{x}^\nu \rangle = \langle (x - \bar{x})^\mu \rangle \langle (x - \bar{x})^\nu \rangle, \tag{8.56}$$

the two-particle correlation function in equation (8.50) can also be written as

$$\begin{aligned} C(q, K) &= 1 \pm |\langle \mathrm{e}^{-\mathrm{i}q\cdot x} \rangle (K)|^2 \\ &= 1 \pm \mathrm{e}^{-q_\mu q_\nu \langle \tilde{x}^\mu \tilde{x}^\nu \rangle (K)}, \end{aligned} \tag{8.57}$$

where the angular bracket $\langle \ldots \rangle$ indicates space–time averaging over the emission function. In practical applications, one generally uses the on-mass-shell condition. For a two-pion correlation function,

$$q^0 = E_1 - E_2 = \sqrt{m_\pi^2 + p_1^2} - \sqrt{m_\pi^2 + p_2^2} \tag{8.58}$$

$$K^0 = \frac{E_1 + E_2}{2} = \frac{1}{2}\left(\sqrt{m_\pi^2 + p_1^2} + \sqrt{m_\pi^2 + p_2^2}\right), \tag{8.59}$$

the on-mass-shell condition yields the relation

$$q \cdot K = \frac{1}{2}(p_1 - p_2) \cdot (p_1 + p_2) = \frac{1}{2}\left(p_1^2 - p_2^2\right) = \frac{1}{2}\left(m_\pi^2 - m_\pi^2\right) = 0. \tag{8.60}$$

The condition can be used to eliminate one component of q, namely,

$$q^0 = \frac{\mathbf{K} \cdot \mathbf{q}}{K^0} = \boldsymbol{\beta} \cdot \mathbf{q}, \tag{8.61}$$

where $\boldsymbol{\beta}$ is the velocity of the pair.

With the on-mass-shell condition,

$$\begin{aligned} q_\mu q_\nu \langle \tilde{x}^\mu \tilde{x}^\nu \rangle(\mathbf{K}) &= \sum_{i,j=1}^{3} q_i q_j \big[\beta^i \beta^j \langle \tilde{t}\tilde{t} \rangle + \langle \tilde{x}^i \tilde{x}^j \rangle - 2\beta^j \langle \tilde{t}\tilde{x}^i \rangle \big](\mathbf{K}) \\ &= \sum_{i,j=1}^{3} q_i q_j \big[\beta^i \beta^j \langle \tilde{t}\tilde{t} \rangle + \langle \tilde{x}^i \tilde{x}^j \rangle - \beta^j \langle \tilde{t}\tilde{x}^i \rangle - \beta^i \langle \tilde{t}\tilde{x}^j \rangle \big](\mathbf{K}) \\ &= \sum_{i,j=1}^{3} q_i q_j \langle (\tilde{x}^i - \beta^i \tilde{t})(\tilde{x}^j - \beta^j \tilde{t}) \rangle(\mathbf{K}). \end{aligned} \tag{8.62}$$

Defining Gaussian width parameters R_{ij} as

$$R_{ij}^2(\mathbf{K}) = \Big\langle (\tilde{x}_i - \beta_i \tilde{t})(\tilde{x}_j - \beta_j \tilde{t}) \Big\rangle(\mathbf{K}), \tag{8.63}$$

the two-particle correlation function can be written as

$$C(q, K) = 1 \pm \mathrm{e}^{-q^i q^j R_{ij}^2(\mathbf{K})} \qquad i, j = 1, 2, 3. \tag{8.64}$$

8.4.1 Bertsch–Pratt parameterization

Gaussian source (8.64) is characterized by six independent variances, R_{ij}, $i, j = 1, 2, 3$. In Cartesian or Bertsch–Pratt parameterization [14–16], the Z-axis is along the beam direction defining the longitudinal direction. In the transverse plane, one of the axes (the x-axis in figure 8.6) is chosen parallel to the pair momentum $\mathbf{K}$ defining the out direction. The remaining axis (the y-axis in figure 8.7) then defines the side direction. In this coordinate system one can write

$$\mathbf{K} = (K_T, 0, K_L) \tag{8.65}$$

$$\boldsymbol{\beta} = \frac{\mathbf{K}}{K^0} = (\beta_T, 0, \beta_L). \tag{8.66}$$

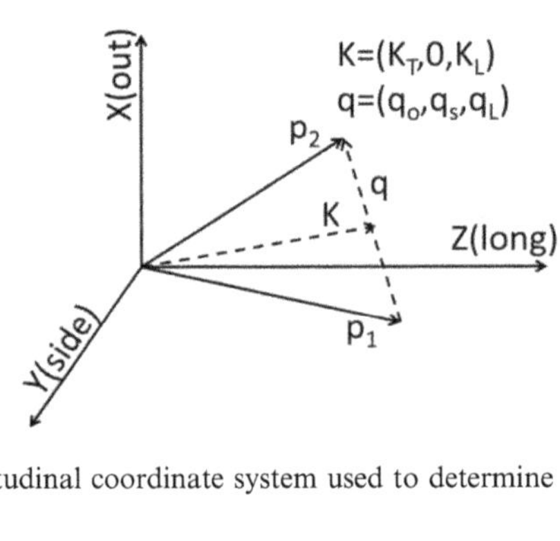

Figure 8.7. The out–side longitudinal coordinate system used to determine HBT radii R_{side}, R_{out} and R_{long}.

The six components of the variances are

$$R_{11}^2(\mathbf{K}) = R_{\text{out}}^2(\mathbf{K}) = \left\langle\left(\tilde{x} - \beta_T\tilde{t}\right)^2\right\rangle(\mathbf{K}), \tag{8.67a}$$

$$R_{22}^2(\mathbf{K}) = R_{\text{side}}^2(\mathbf{K}) = \langle\tilde{y}^2\rangle(\mathbf{K}), \tag{8.67b}$$

$$R_{33}^2(\mathbf{K}) = R_{\text{long}}^2(\mathbf{K}) = \left\langle(\tilde{z} - \beta_L\tilde{t})^2\right\rangle(\mathbf{K}), \tag{8.67c}$$

$$R_{12}^2(\mathbf{K}) = R_{\text{out-side}}^2(\mathbf{K}) = \left\langle\left(\tilde{x} - \beta_T\tilde{t}\right)\tilde{y}\right\rangle(\mathbf{K}), \tag{8.67d}$$

$$R_{13}^2(\mathbf{K}) = R_{\text{out-long}}^2(\mathbf{K}) = \left\langle\left(\tilde{x} - \beta_T\tilde{t}\right)\left(\tilde{z} - \beta_L\tilde{t}\right)\right\rangle(\mathbf{K}), \tag{8.67e}$$

$$R_{23}^2(\mathbf{K}) = R_{\text{side-long}}^2(\mathbf{K}) = \left\langle\tilde{y}\left(\tilde{z} - \beta_L\tilde{t}\right)\right\rangle(\mathbf{K}). \tag{8.67f}$$

It is evident that HBT parameters R_{ij} mix the spatial and temporal information on the source in a non-trivial way. They depend on the pair momentum **K** and their interpretation will depend on the frame of reference for **K**. In general, identification of HBT widths with physical source size is not correct. The size parameters from HBT correlation can be interpreted as the *region of homogeneity*, the region from which particle pairs with momentum **K** are most likely to be emitted. The space–time variances $\langle\tilde{x}^\mu\tilde{x}^\nu\rangle$ coincide with total source extensions only in the special case that the emission function shows no position–momentum correlations and factorizes $S(x,K)=f(x)g(K)$.

If the system has additional symmetries, they will impose further restrictions on the variances. For example, for azimuthally symmetric source functions, there will be reflection symmetry with respect to the y-axis and $\langle\tilde{y}\rangle=0$. Consequently, two variances, R_{12} and R_{23}, which are linear in $\tilde{y}$ vanish identically (note that R_{22}, which is a quadratic function of $\tilde{y}$, does not vanish). The correlation function can then be parameterized in terms of four Gaussians,

$$C(q, K) = 1 + \exp\left[-R_{\text{out}}^2 q_{\text{out}}^2 - R_{\text{side}}^2 q_{\text{side}}^2 - R_{\text{long}}^2 q_{\text{long}}^2 - R_{\text{out-long}}^2 q_{\text{out}} q_{\text{long}}\right]. \tag{8.68}$$

If the system possesses the additional symmetry of boost-invariance, then $R_{13} = R_{\text{out-long}}$ also vanishes (the linear term in $\tilde{z}$ in R_{13} vanishes, additionally, one can choose $\beta_L = 0$). The correlation function can then be parameterized in terms of three Gaussians,

$$C(q, K) = 1 + \exp\left[-R_{\text{out}}^2 q_{\text{out}}^2 - R_{\text{side}}^2 q_{\text{side}}^2 - R_{\text{long}}^2 q_{\text{long}}^2\right]. \tag{8.69}$$

This is the most common form for the parameterized correlation function. It is common to associate R_{side} with the transverse size of the system (or more accurately, to the transverse homogeneity length). The width R_{out} measures a combination of transverse size and duration of emission. Similarly, R_{long} measures a combination of longitudinal size and duration of emission. The difference

$$R_{\text{out}}^2 - R_{\text{side}}^2 = \beta_T^2 \langle \tilde{t} \rangle - 2\beta_T \langle \tilde{t} x \rangle + \left(\langle \tilde{x}^2 \rangle - \langle \tilde{y}^2 \rangle \right) \tag{8.70}$$

gives a measure of the emission duration $\Delta t = \sqrt{\langle t^2 \rangle - \langle t \rangle^2}$ if the last two terms in the above equation are comparatively smaller than the first term. However, it is difficult to ensure smallness of the last two terms and any extraction of emission duration from the difference $R_{\text{out}}^2 - R_{\text{side}}^2$ is model dependent. We do note that the emission time Δt should not be confused with the physical emission time, the time between nuclear collision and freeze-out.

8.4.2 Yano–Koonin–Podgoreskiı́ parameterization

An alternative Gaussian parameterization is Yano–Koonin–Podgoreskiı́ (YKP) parameterization [17]. It uses the on-mass-shell condition equation (8.60) with $q_{\text{trans}} = \sqrt{q_{\text{side}}^2 + q_{\text{out}}^2}$, q_{long} and q^0 to write

$$C(\mathbf{q}, \mathbf{K}) = 1 + \exp\left[-R_{\text{trans}}^2 q_{\text{trans}}^2 - R_{\text{long}}^2 \left(q_{\text{long}}^2 - \left(q^0\right)^2 \right) - \left(R_{\text{out}}^2 + R_{\text{long}}^2 \right) (q \cdot U)^2 \right]. \tag{8.71}$$

The parameterization has three radius parameters, R_{trans}, R_{long} and R_{out}, and a velocity parameter U, which is assumed to be longitudinal only,

$$U(\mathbf{K}) = \gamma(\mathbf{K})(1, 0, 0, v(\mathbf{K})), \gamma(\mathbf{K}) = \frac{1}{\sqrt{1 - v(\mathbf{K})^2}}. \tag{8.72}$$

YKP parameterization has the advantage that the radius parameters R_{trans}, R_{long} and R_{out} do not depend on the longitudinal velocity of the measurement frame, while the fourth fit parameter $v(\mathbf{K})$ is simply boosted by that velocity. The frame in which $v(\mathbf{K}) = 0$ is called the Yano–Koonin (YK) frame where R_{trans}, R_{long} and R_{out} can be interpreted as the transverse, longitudinal and temporal homogeneity lengths of the source.

8.4.3 The incoherence factor

The correlation function is written for completely chaotic or incoherent sources. The possibility that the emission of particles is neither perfectly chaotic nor completely coherent can be taken into account by adding the parameter λ to the correlation function, which, in general, depends on k_T.

$$C(q, K) = 1 \pm \lambda e^{-q^i q^j R_{ij}^2(\mathrm{K})} \qquad i, j = 1, 2, 3. \tag{8.73}$$

For a fully chaotic source the parameter $\lambda = 1$ and $\lambda < 1$ for a partially chaotic source. λ is called the incoherence parameter. In spite of its name, it also includes several other effects as well as incoherence. For example, in experiments, the momenta of the particles are binned. Due to finite bin width, for projection of the correlation function in, say, the transverse momentum direction, the smallest longitudinal momentum is no longer zero as required. Effects like particle mis-identification, resonance decay contributions, final state Coulomb interactions, etc, also contribute to the incoherence parameter.

Of particular interest is the Coulomb interaction. Repulsive Coulomb interaction between like-charge pairs causes a reduction in the number of real pairs at low q. The experimental correlation function at low q is then reduced. For correct estimation of the source size, the correlation function needs to be corrected for the final state Coulomb interaction. See [7, 8, 15] for details of Coulomb correction. Traditionally, the Bose–Einstein correlation function with Coulomb correction is expressed as

$$C(q, K) = K_{\mathrm{Coul}}(q_{inv})\left[1 + \lambda e^{-q^i q^j R_{ij}^2(\mathbf{K})}\right] \tag{8.74}$$

where $K_{\mathrm{Coul}}(q_{inv})$ is the square of the Coulomb wave function, integrated over the whole source. The standard procedure assumes that all the pairs are primary pairs and Coulomb correction is applied to all of them. The procedure generally over-corrects the Coulomb reduction. One then dilutes the Coulomb correction requiring that only a fraction f of the pairs are primary pairs requiring Coulomb correction. The correlation function, with 'diluted' Coulomb correction is

$$C(q, K) = K'_{\mathrm{Coul}}(q_{inv})\left[1 + \lambda e^{-q^i q^j R_{ij}^2(\mathbf{K})}\right] \tag{8.75}$$

with

$$K'_{\mathrm{Coul}}(q_{inv}) = (1 - f) + f K_{\mathrm{Coul}}(q_{inv}). \tag{8.76}$$

A reasonable assumption is to take $f = \lambda$ assuming that λ is the fraction of primary pairs requiring the correction. In a further improvement [18, 19], called the Bowler–Sinyukov procedure, only the Bose–Einstein correlated pairs are corrected for the Coulomb interaction,

$$C(q, K) = (1 - \lambda) + K_{\mathrm{Coul}}(q_{inv})\lambda e^{-q^i q^j R_{ij}^2(\mathbf{K})}. \tag{8.77}$$

8.5 Experimental results

In relativistic heavy ion collisions, the two-pion correlation function has been measured at the Alternating Gradient Synchrotron (AGS) [20], SPS [22], RHIC [21, 22] and more recently in LHC energy collisions [23]. In the following, a few experimental results on two-pion correlation measurements at RHIC energy will be discussed briefly. In relativistic energy heavy ion collisions, one generally records the data event-wise. An event corresponds to one collision. Experimentally, two-particle correlation functions are constructed as the ratio of actual pairs and mixed or reference pairs. When two particles are selected from the same event, they are called actual pairs. The reference or mixed pairs are selected randomly from different events. The correlation function is then constructed as

$$C(\Delta q, \Delta K) = \frac{D_A(\Delta q, \Delta K)}{D_R(\Delta q, \Delta K)} \tag{8.78a}$$

$$D_A(\Delta q, \Delta K) = \frac{\text{number of actual pairs in bin } \Delta q, \Delta K}{\text{number of actual pairs in sample}} \tag{8.78b}$$

$$D_R(\Delta q, \Delta K) = \frac{\text{number of reference pairs in bin } \Delta q, \Delta K}{\text{number of actual pairs in sample}}. \tag{8.78c}$$

In figure 8.8, STAR measurements [23] for the two-pion correlation function in a 0–5% Au+Au collision at RHIC are shown. The shape of the correlation function is generic, i.e. a qualitatively similar function is obtained at different collision centrality and energy. The Bose–Einstein enhancement at low p_T is manifested. The figure also manifests the effect of Coulomb interaction. The repulsive Coulomb interaction among the like-charged pairs causes a reduction in the number of real pairs at low q and the experimental correlation function is reduced. When Coulomb interactions among all the pairs are included (see [23] for details of the Coulomb correction), as manifest in the data, the correlation at low q is increased.

The main aim of correlation measurements is to determine the HBT radii. The experimental correlation functions are fitted with a parameterized form of correlation function to determine various HBT radii. The fitted values of λ, R_{out}, R_{side} and

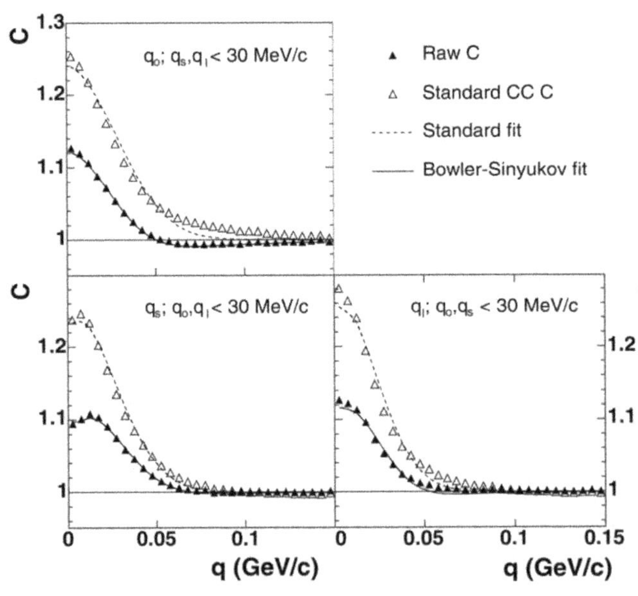

Figure 8.8. STAR measurements of the two-pion correlation function in $\sqrt{s}_{\text{NN}} = 200$ GeV, 0–5% Au+Au collisions. Projections of the three-dimensional $\pi^-\pi^-$ correlation function on 'out', 'side' and 'long' directions, for $K_T = [150,250]$ MeV. The filled symbols are data without Coulomb correction. The open symbols include Coulomb interaction among all pairs. Reproduced with permission from [23].

R_{long}, for $\pi^+\pi^+$ and $\pi^-\pi^-$ correlations, in 0–5% Au+Au collisions are shown in figure 8.9. For details of the fitting procedure see [23]. The HBT parameters for $\pi^+\pi^+$ and $\pi^-\pi^-$ are in excellent agreement with each other. The momentum dependence of the HBT parameters indicates that while the incoherence parameter λ increases with m_T, the HBT radii R_{out}, R_{side} and R_{long} decrease with m_T. The results can be qualitatively understood. The effects of long-lived resonances are included in the incoherence parameter. The greater the resonance contribution, the greater λ. At larger m_T, resonance contributions are reduced, and λ is increased. Any collective expansion, longitudinal or radial, reduces the region of homogeneity and HBT radii decrease. The decrease in HBT radii with m_T can be associated with the development of collective flow. Collective flow increases with m_T.

In HBT measurements, the ratio R_{out}/R_{side} is of particular interest. Under certain approximations, the ratio is a measure of emission duration (see equation (8.70)). The ratio is expected to be greater than unity if the emission duration is large. For example, if there is a first order quark–gluon plasma (QGP) phase transition with a finite transition width, system expansion will be delayed which will manifest as $R_{out}/R_{side} > 1$ [25, 27]. In figure 8.9 m_T dependence of the ratio R_{out}/R_{side} is also shown.

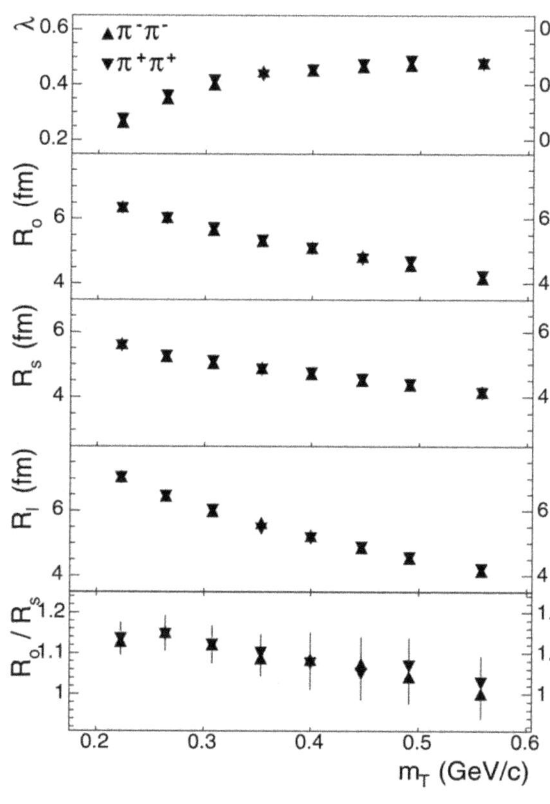

Figure 8.9. STAR measurements for HBT parameters for $\pi^+\pi^+$ and $\pi^-\pi^-$ correlation functions in 0–5% Au+Au collisions at RHIC ($\sqrt{s}_{NN} = 200$ GeV), as a function of the transverse mass of the pion pairs. Reproduced with permission from [23].

The ratio $R_{out}/R_{side} \sim 1$. Approximately constant R_{out}/R_{side} is inconsistent with time-delayed expansion.

HBT radii measure the homogeneity region rather than the source size. In figure 8.10 the centrality dependence of the HBT radius parameters in $\sqrt{s}_{NN} = 200$ GeV Au+Au collisions are shown. Centrality dependence is shown as a function of the cube root of the Glauber model participant number N_{part}. HBT radii parameters R_{out}, R_{side} and R_{long} approximately increase linearly with $N_{part}^{1/3}$. The more central the collisions, the greater the homogeneity region for correlated pion emission. The ratio R_{out}/R_{side} does not show appreciable dependence on the collision centrality and is approximately constant. From geometrical considerations, the pion emitting source is expected to be larger in central collisions than in peripheral collisions. Figure 8.10 then indicates that the physical source size and homogeneity region of correlated pion emission are related, though the exact relation is obscure.

The product of the three HBT radii $R_{out}R_{side}R_{long}$ can be considered as the 'homogeneity' volume. In figure 8.11, the product is shown as a function of multiplicity. The collision energies are noted in the figure. It increases approximately

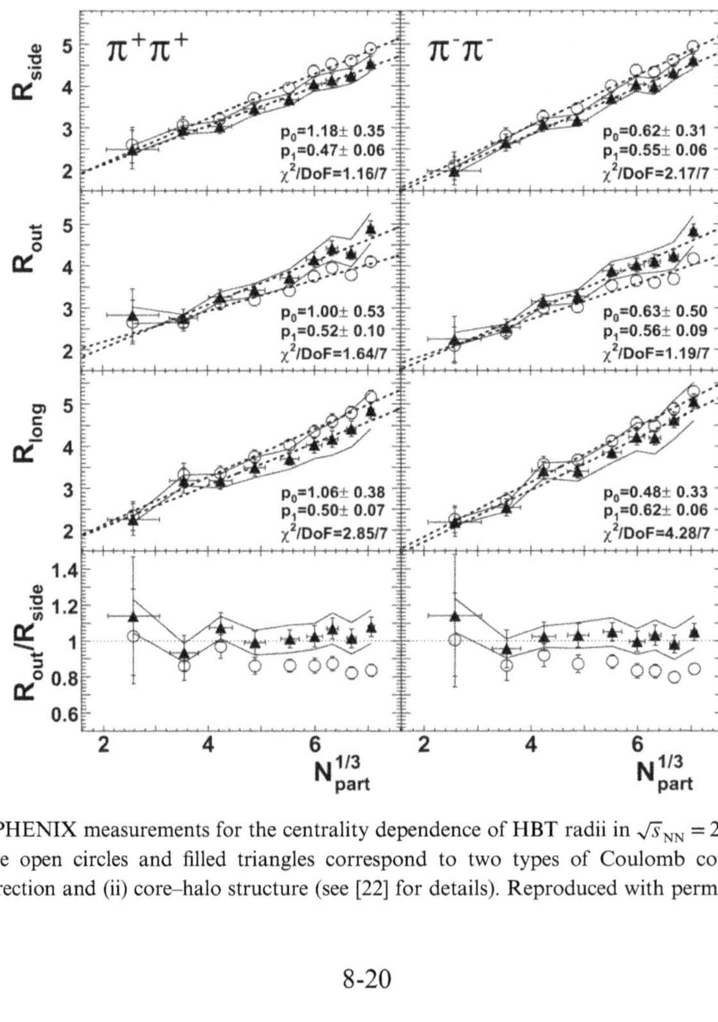

Figure 8.10. PHENIX measurements for the centrality dependence of HBT radii in $\sqrt{s}_{NN} = 200$ GeV Au+Au collisions. The open circles and filled triangles correspond to two types of Coulomb correction, (i) full Coulomb correction and (ii) core–halo structure (see [22] for details). Reproduced with permission from [22].

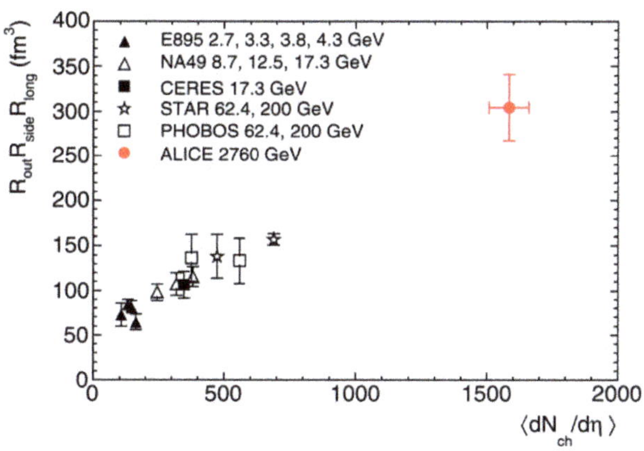

Figure 8.11. Multiplicity dependence of the 'homogeneity volume' $R_{out}R_{side}R_{long}$ in AGS, SPS, RHIC and LHC energy collisions. The homogeneity volume increases approximately linearly with multiplicity. Reproduced with permission from [26].

linearly with multiplicity. From RHIC to LHC, the homogeneity volume is increased approximately by a factor of 2. Considering that from RHIC to LHC, center-of-mass energy is increased by a factor of 14, the increase in 'homogeneity' volume is not large.

Initially, it was envisaged that two-pion correlation measurements would provide a direct measure for the size of the medium produced in heavy ion collisions. However, it was soon realized that rather than the physical size, HBT radii determine the size of the homogeneity region (the region from where correlated particles are emitted). Even though HBT radii are not directly related to the system size, nevertheless they are important as the momentum space information extracted from the single particle spectra can be supplemented with the coordinate space information. HBT measurements also provide for a stringent test for dynamical models for heavy ion collisions.

In heavy ion collisions, the dynamical models in use are: (i) transport models based on Boltzmann equations, (ii) hydrodynamical models and (iii) a combination of Boltzmann and hydrodynamic models. One of the widely used transport models is the ultra-relativistic quantum molecular dynamics model (UrQMD) [28, 29]. The UrQMD model employs only hadronic and string degrees of freedom. Any possibility of QGP formation is neglected. The model partially explains the experimental HBT data in $\sqrt{s}_{NN} = 200$ GeV Au+Au collisions [30]. The calculated transverse momentum, centrality and system size dependence of the Pratt–HBT radii R_{long} and R_{side} are reasonably well reproduced. The model, however, over-predicts R_{out} in central collisions. The corresponding quantity $\sqrt{R_{out}^2 - R_{side}^2}$ of the pion emission source is also larger than experimental estimates.

Hydrodynamical models which are quite successful in explaining many features of ultra-relativistic heavy ion collisions, however, fail to reproduce HBT measurements [31–33]. The HBT radii are largely over-predicted. The fact that hydrodynamic models over-predict HBT radii came to be known as the 'HBT puzzle'. Indeed, it appeared that models with phase transition or soft equations of state give worse descriptions of the HBT data than models without phase transition or with stiff equations of state. However, it is now understood that HBT correlations are particularly sensitive to the freeze-out condition. In most of the early hydrodynamic models, freeze-out was treated somewhat unrealistically. Also, early hydrodynamic calculations neglected the effect of viscosity. With realistic freeze-out, viscosity and HBT data can be reasonably reproduced in present-day hydrodynamic model calculations [34].

In hybrid models, hydrodynamic evolution is followed by a hadronic cascade [35–37]. The model has the advantage of more realistic treatment of freeze-out conditions than in a purely hydrodynamic model. Model calculations where the ideal hydrodynamic evolution of QGP fluid is supplemented by a hadronic cascade however do not fare well with respect to HBT measurements. HBT radii are over-predicted by a factor of ~2.

Bibliography

[1] Hanbury Brown R and Twiss R Q 1956 *Nature* **178** 1046
[2] Hanbury Brown R and Twiss R Q 1956 *Nature* **177** 27
[3] Goldhaber G, Goldhaber S, Lee W-Y and Pais A 1960 *Phys. Rev.* **120** 300
[4] Weiner R M 2000 *Introduction to Bose–Einstein Correlations and Subatomic Interferometry* (New York: Wiley)
[5] Baym G 1998 *Acta Phys. Polon.* B **29** 1839 (arXiv:nucl-th/9804026)
[6] Boal D H, Gelbke C K and Jennings B K 1990 *Rev. Mod. Phys.* **62** 553
[7] Wiedemann U A and Heinz U W 1999 *Phys. Rep.* **319** 145
[8] Heinz U W and Jacak B V 1999 *Annu. Rev. Nucl. Part. Sci.* **49** 529
[9] Gyulassy M, Kauffmann S K and Wilson L W 1979 *Phys. Rev.* C **20** 2267
[10] Csorgo T 2002 *Heavy Ion Phys.* **15** 1
[11] Padula S S 2005 *Braz. J. Phys.* **35** 70
[12] Lisa M A, Pratt S, Soltz R and Wiedemann U 2005 *Annu. Rev. Nucl. Part. Sci.* **55** 357
[13] Mandel L and Wolf E 1965 *Rev. Mod. Phys.* **37** 231
[14] Pratt S 1984 *Phys. Rev. Lett.* **53** 1219
Pratt S 1986 *Phys. Rev.* D **33** 72
[15] Bertsch G F, Gong M and Tohyama M 1988 *Phys. Rev.* C **37** 1896
[16] Chapman S, Scotto P and Heinz U 1995 *Phys. Rev. Lett.* **74** 4400
[17] Yano F B and Koonin S E 1978 *Phys. Lett.* B **78** 556
Podgoretskii M I 1983 *Sov. J. Nucl. Phys.* **37** 272
[18] Bowler M G 1991 *Phys. Lett.* B **270** 69
[19] Sinyukov Yu M *et al* 1998 *Phys. Lett.* B **432** 249
[20] Ahle L *et al* (E802 Collaboration) 2002 *Phys. Rev.* C **66** 054906
[21] Adamova D *et al* (CERES Collaboration) 2003 *Nucl. Phys.* A **714** 124
[22] Adler S S *et al* (PHENIX Collaboration) 2004 *Phys. Rev. Lett.* **93** 152302
[23] Adams J *et al* (STAR Collaboration) 2005 *Phys. Rev.* C **71** 044906
[24] Aamodt K *et al* (ALICE Collaboration) 2011 *Phys. Lett.* B **696** 328
[25] Rischke D H and Gyulassy M 1996 *Nucl. Phys.* A **608** 479
[26] Aamodt K *et al* (ALICE Collaboration) 2011 *Phys. Lett.* B **696** 328
[27] Heinz U W and Kolb P F 2002 arXiv: hep-ph/0204061

[28] Bass S A *et al* 1998 *Prog. Part. Nucl. Phys.* **41** 225–370
[29] Bleicher M *et al* 1999 *J. Phys. G: Nucl. Part. Phys.* **25** 1859–96
[30] Li Q, Bleicher M and Stoecker H 2006 *Phys. Rev.* C **73** 064908
[31] Hirano T and Tsuda K 2002 *Phys. Rev.* C **66** 054905
[32] Hirano T and Tsuda K 2003 *Nucl. Phys.* A **715** 821
[33] Kolb P F and Heinz U W 2004 *Hydrodynamic description of ultrarelativistic heavy ion collisions Quark–Gluon Plasma* vol 3, ed R C Hwa and X-N Wang (Singapore: World Scientific) pp 634–714
[34] Bozek P 2012 *Phys. Rev.* C **85** 034901 (arXiv:1110.6742 [nucl-th]).
[35] Soff S, Bass S A and Dumitru A 2001 *Phys. Rev. Lett.* **86** 3981 (arXiv:nucl-th/0012085)
[36] Bass S A and Dumitru A 2000 *Phys. Rev.* C **61** 064909 (arXiv:nucl-th/0001033)
[37] Teaney D, Lauret J and Shuryak E V 2002 *Nucl. Phys.* A **698** 479 (arXiv:nucl-th/0104041)
[38] Teaney D, Lauret J and Shuryak E V (arXiv:nucl-th/0110037)

Chapter 9

Signals of quark–gluon plasma

9.1 Introduction

In ultra-relativistic heavy ion collisions, a deconfined state of quarks and gluons (quark–gluon plasma; QGP) is expected to be produced. However, detection of QGP is not simple. One may ask what the observables that will signal formation of QGP are. Unlike in other phase transitions, in confinement–deconfinement phase transition (or cross-over), the constituents of the high temperature phase (QGP) do not exist freely. They are confined within the hadrons. QGP is a transient state; it expands, cools, hadronizes and cools further until interactions between the hadrons become too weak to continue the evolution. Any information about the QGP phase, if produced in high-energy nuclear collisions, has to be obtained from the observed hadrons only. Hadronization is a non-perturbative process. Currently it is not properly understood. Whether or not the hadronization process erases any memory of the constituent quarks is uncertain. If the hadronization process erases the memory, one cannot comment on the initial QGP phase from the observed hadrons. The current search for QGP at RHIC/LHC is on the premise that the hadronization process does not erase the memory and from the observed hadrons, one can comment on the possible existence of QGP. Signals of QGP are a very important subject in the study of QGP. In the following, some important signals of QGP are discussed.

9.2 J/ψ suppression

Quarkonium is the bound state of heavy (c or b) quarks and anti-quarks. One generally does not consider the bound state of top and anti-top quarks. Top quarks are very heavy, mass ($m_t \approx 175$ GeV) and they can decay by electroweak interaction, with a short lifetime of 0.5×10^{-24} sec. The top quark will decay before a bound state $t\bar{t}$ is formed. Unlike that of a light hadron (e.g. pion, kaon or nucleon) quarkonium mass arises mostly from the current quark masses. Charmonium is the bound state of $c\bar{c}$ quarks. In figure 9.1, some of the low lying charmonium states

doi:10.1088/978-0-750-31060-4ch9

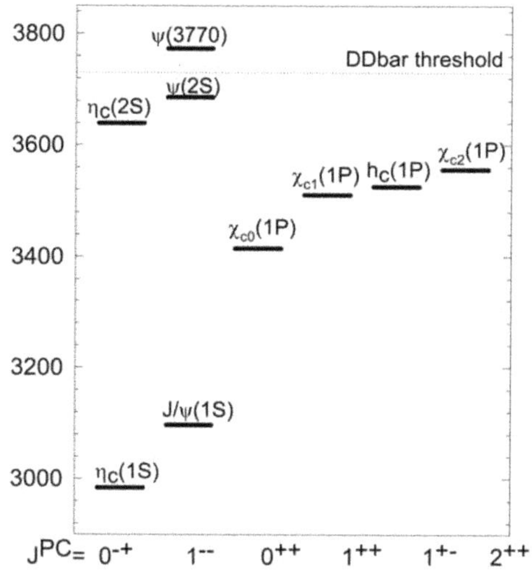

Figure 9.1. Low lying states of charmonium. The $D\overline{D}$ threshold is also shown.

are shown. It has hydrogen-atom-like spectra. Note that the charmonium ground state is η_c. Similar spectra are obtained for bottonium, the bound state of $b\bar{b}$ quarks.

The production and decay of heavy quarkonia is the ideal testing ground for quantum chromodynamics (QCD). It has also been suggested [1] that they are ideal probes for confinement–deconfinement transition. Due to their heavy mass, they are produced in initial collisions (thermal production is suppressed by $e^{-m_Q/T}$), in a short time scale $\tau \sim 1/m_Q$. In the following, the idea behind the suggestion will be briefly described. For many facets of quarkonia physics, the reader can consult [2, 3].

9.2.1 Production mechanism

Even though the J/ψ meson was discovered some 40 years ago, there is large uncertainty about theoretical modeling of the production process. Quarkonium production is assumed to be a two step process (shown schematically in figure 9.2 for J/ψ). In the first step a heavy quark pair $Q\overline{Q}$ is produced in a hard partonic collision. This step is accurately calculable in perturbative QCD (pQCD) in an expansion in powers of the strong coupling constant α_s. In the second step, the $Q\overline{Q}$ pair evolves into a quarkonium state (H). The second step is essentially non-perturbative and needs to be modeled. The two step process implicitly assumes that the short distance or perturbative effects can be separated from the long distance or non-perturbative effects.

An inclusive cross section for production of a quarkonium state H in collisions of two hadrons h_A and h_B, in leading order in α_s, can be written as

$$\sigma_{h_A h_B \to HX} = K \sum_{i,j} \int d\hat{s} \int dx_1 dx_2 f_i^{h_A}(x_1, \mu) f_j^{h_B}(x_2, \mu) \times \hat{\sigma}_{ij} \delta(\hat{s} - x_1 x_2 s) F_{Q\overline{Q} \to H}(\hat{s}) \tag{9.1}$$

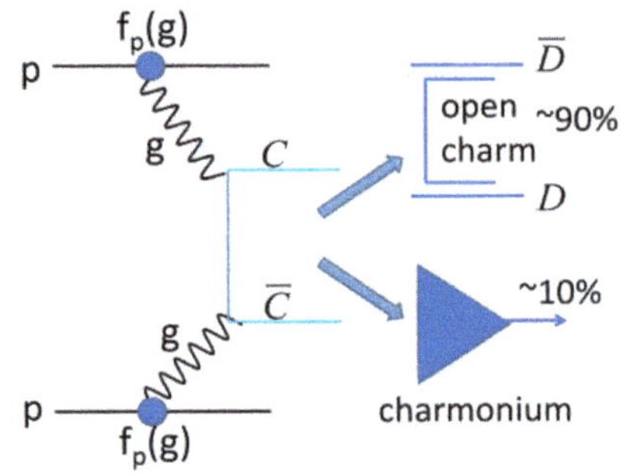

Figure 9.2. Schematic diagram of J/ψ production in elementary collisions. Note that most of the $c\bar{c}$ quarks go to the open charm channel.

where $ij = q\bar{q}$ or gg. s is the square of the partonic center-of-mass energy, and $\hat{\sigma}_{ij}$ is the sub-process $ij \to Q\overline{Q}$ cross section. The factor K takes into account the effect of higher order processes. The factor $F_{Q\overline{Q}\to H}(\hat{s})$ is the transition probability that the $Q\overline{Q}$ pairs transform into a quarkonium state H. It has to be modeled. Different treatments of this non-perturbative transition probability $F_{Q\overline{Q}\to H}(\hat{s})$ have led to various theoretical models, e.g. the color singlet model (CSM), the color evaporation model (CEM), the non-relativistic QCD (NRQCD) factorization approach and the fragmentation function approach, brief discussions of which are given below.

Color singlet model. In the CSM [4–8] the $Q\overline{Q}$ pair that evolves into a quarkonium state H is in a color-singlet state and has the same spin and angular momentum quantum numbers as those of H. Experiments indicate that J/ψ production is dominated by gluon fusion. However, fusing two gluons with the spin and angular momentum of J/ψ has a problem. The quantum number of J/ψ is $J^{PC} = 1^{--}$, i.e. the total J/ψ spin is 1, and is odd parity. The charge conjugation is also odd. Now consider the decay, $J/\psi \to gg$. Now J/ψ is C-odd but gg is C-even, the decay is forbidden so is $gg \to J/\psi$. The three-gluon fusion will not have a sufficient rate. It appears that there is no direct mechanism to fuse two gluons to produce a J/ψ. One possibility is to fuse two gluons to produce the higher state χ_c. It is shown in figure 9.1 that χ_c are C-even. In the CSM, two gluons fuse to produce a χ_c which then decays to J/ψ. As one can observe from figure 9.1, the decay of χ_c to ψ is kinematically forbidden $m_\psi > m_{\chi c}$. All the ψ have to be produced from decay of B-mesons. Experimental data do not support the CSM.

Color evaporation model. Two gluons can fuse in a color octet state,

$$8 \otimes 8 = 1 \oplus 8 \otimes 8 \otimes 10 \otimes \overline{10}. \tag{9.2}$$

In the CEM [9–13] initial partonic collisions produce $Q\overline{Q}$ pairs in the octet state which then radiate a gluon to become colorless,

$$gg \to {}^3P_2^8 \to g + {}^3S_1(H). \tag{9.3}$$

The radiated gluon is soft and the color shedding process is non-perturbative. In CEM, it is assumed that every produced $Q\overline{Q}$ pair, with invariant mass less than the threshold for producing a pair of open-flavor heavy mesons, evolves into a quarkonium state H. It is further assumed that the transition probability for the $Q\overline{Q}$ pair to evolve into a quarkonium state is an energy–momentum and process independent constant which can be fixed by comparison with the measured total cross section.

NRQCD factorization approach. NRQCD is an effective field theory that describes the behavior of the bound states of a heavy quark (Q) and a heavy anti-quark ($\overline{Q}$) when the velocity v of the quarks in the quarkonia $Q\overline{Q}$ rest frame is non-relativistic ($v \ll c$). In NRQCD factorization, the probability of a $Q\overline{Q}$ pair evolving into a quarkonium state is expressed in terms of the matrix elements of four-fermion NRQCD operators [14, 15]. The inclusive cross section for the direct production of a quarkonium state H is written as a sum of the products of these NRQCD matrix elements with the corresponding $Q\overline{Q}$ production cross sections:

$$\sigma(H) = \sum_n \sigma_n(\Lambda)\langle 0|O_n^H(\Lambda)|0\rangle. \tag{9.4}$$

Here Λ is the ultraviolet cut off of the effective theory and the σ_n are cross sections to produce a $Q\overline{Q}$ pair in the color, spin and orbital angular momentum state n. They are convolutions of parton level cross sections at the scale μ with parton distribution and can be organized as an expansion in powers of α_s. The NRQCD long-distance matrix elements are vacuum expectation values of the four-fermion operator,

$$O_n^H(\Lambda) = \chi^\dagger \kappa_n \psi(\Lambda)\left(a_H^\dagger a_H\right)\psi^\dagger \kappa_n' \chi(\Lambda) \tag{9.5}$$

where $a_H^\dagger$ is the creation operator for state H, $\psi^\dagger$ and χ are two-component (Pauli) fields of a heavy quark and a heavy antiquark. κ_n and κ_n' are direct products of color and spin matrices. A key feature of NRQCD factorization is that the production can occur through color octet, as well as color singlet channels. The NRQCD matrix element for the color singlet state $\left\langle O\left({}^3S_1^1\right)\right\rangle$ can be determined from electromagnetic decay rates. The color octet matrix elements must be determined through comparisons of theoretical predictions with measurements. Equation (9.4) contains an infinite number of 'unknown' NRQCD matrix elements. One generally organizes the sum in equation (9.4) as an expansion in powers of v. Thus the NRQCD factorization formula is a double expansion in powers of v and powers of α_s.

The fragmentation function approach. Earlier, we discussed the perturbative QCD (pQCD) factorization for the production of hard hadrons. In the fragmentation function approach [16–18] the quarkonium production cross section is written in terms of convolutions of heavy $Q\bar{Q}$ production cross sections with light-cone fragmentation functions. In leading order,

$$d\sigma_{AB\to H+X}(p_T) = \sum_i d\hat{\sigma}_{AB\to i+X}(p_T/z, \mu) \otimes D_{i\to H}(z, \mu). \tag{9.6}$$

It describes the process in which initial partonic collisions $a + b \to i + X$ produce a parton i which fragments into the quarkonium state H via the fragmentation function $D_{i\to H}$. The symbol $\otimes$ represents convolution with parton distributions of the colliding hadrons. In the next to leading order, processes in which initial collisions produce a quark pair $a + b \to [Q\bar{Q}] + X$ can be included and the quark pair fragmentation function $D_{[Q\bar{Q}]\to H}$ can be used to produce the quarkonium state H.

9.2.2 A mechanism for quarkonia suppression

Let us now discuss Matsui and Satz's idea of using charmonium suppression as a signal of QGP formation. Quarkonium is the bound state of heavy quarks ($m_{\text{charm}} \approx 1.15-1.35$ GeV, $m_{\text{bottom}} \approx 4.0-4.4$ GeV) and the non-relativistic Schrödinger equation,

$$\left\{2m_c - \frac{1}{m_c}\nabla^2 + V(r)\right\}\phi_i = M_i\phi_i(r), \tag{9.7}$$

is expected to describe the bound state properties, with the inter-quark potential $V(r)$. Indeed, charmonium and bottonium spectra are well explained by the phenomenological 'Cornell potential',

$$V(r) = \sigma r - \frac{\alpha_{\text{eff}}}{r}, \tag{9.8}$$

where r is the inter-quark separation, $\sigma \approx 0.192\,\text{GeV}^2$ is the string constant and $\alpha_{\text{eff}} = \frac{4}{3}\alpha_s(r) \approx 0.471$. Charmonium mass, radius and formation time can be obtained from solution of the non-relativistic Schrödinger equation and are given in table 9.1. Now in QGP at high temperature, the 'color' interaction potential will be screened,

$$V(r, T) = \frac{\sigma}{\mu(T)}\left(1 - e^{-\mu(T)r}\right) - \frac{\alpha_{\text{eff}}}{r}e^{-\mu(T)r}. \tag{9.9}$$

Table 9.1. J/ψ mass, radii and formation time from solution of the Schrödinger equation.

	J/ψ	ψ'	χ_c
M (GeV)	3.07	3.698	3.5
R (rm)	0.453	0.875	0.696
τ_F (fm)	0.89	1.5	2.0

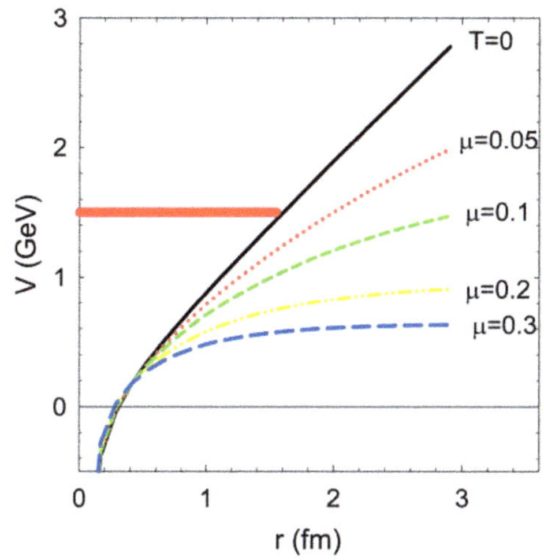

Figure 9.3. Quark potential as a function of screening mass (or temperature).

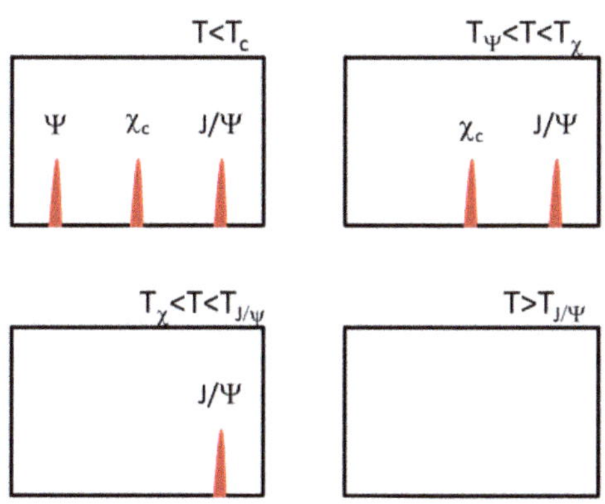

Figure 9.4. Sequential suppression of charmonium states in QGP. As the temperature increases, more and more tightly bound states get dissociated.

$\mu(T)$ is the screening mass, the inverse of the Debye screening radius r_c. For $\mu \to 0$, equation (9.8) is recovered. For $\mu \neq 0$, the screened potential satisfies

$$\operatorname*{Lim}_{r \to 0}[rV(r, T)] \sim -\alpha, \tag{9.10}$$

the 1/r behavior in the short-distance limit. For large r,

$$\operatorname*{Lim}_{r \to \infty}\frac{1}{r}\ln\left[\frac{\sigma}{\mu(T)} - V(r, T)\right] \sim -\mu(T) \tag{9.11}$$

indicating that the range of the binding force decreases exponentially with screening mass. When the screening radius r_c falls below the binding radius of a given charmonium state, c and $\bar{c}$ no longer 'see' each other and a binding for that state is not possible. Screening mass μ is an increasing function of temperature T. From lattice

Table 9.2. Indicative melting or dissociation temperature T_d in unit of critical temperature T_c for various charmonium and bottonium states from lattice calculations.

state	$J/\psi(1S)$	$\chi_c(1P)$	$\psi'(2S)$	$\Upsilon(1S)$	$\chi_b(1P)$	$\Upsilon(2S)$	$\chi_b(2P)$	$\Upsilon(2S)$
T_d/T_c	1.5	1.1	1.1	4.0	1.8	1.60	1.2	1.1

simulations, $\frac{\mu}{T_c} \approx 4\frac{T}{T_c}$. In figure 9.3, inter-quark potential (equation 9.9) as a function of screening mass (or equivalently temperature) is shown. At high μ or T, the potential barrier decreases. A state which is bound in $T=0$ potential may become unbound in finite T potential. An example is shown in figure 9.3. The bound state in $T=0$ potential shown as the red line is no longer a bound state in the $\mu = 0.2$ GeV potential.

In 1986, Matsui and Satz [1] suggested that if QGP is formed in nuclear collisions, J/ψ, the bound state of $c\bar{c}$ will be suppressed, with respect to pp collisions. The idea is simple. In the presence of QGP, J/ψ production will be inhibited due to color screening of the potential. A $c\bar{c}$ pair, which otherwise could transform into a J/ψ, is now unable to do so. It will be melted, eventually forming a pair of open charm mesons. Now only ~60% of experimentally observed J/ψ are directly produced, the rest are from the decay for excited states, ~30% are from decay of χ_c and ~10% from decay of ψ. Higher states are less tightly bound and in a QGP environment, it is expected that the dissociation temperature for the excited states will be less than that for J/ψ and there will be a hierarchy of melting or dissociation temperature (T_d), $T_{d_{J/\psi}} > T_{d_{\chi_c}} > T_{d_{\psi}}$. There are large uncertainties about the melting temperatures. In table 9.2, indicative values as obtained from lattice studies are shown. The hierarchical dissociation or melting temperature of the charmonium states suggests that it can be used as a QGP thermometer. For example, consider a hypothetical experiment where a beam of J/ψ, χ_c and ψ' pass through a QGP medium at temperature T and the outcome is analyzed (see figure 9.4). At low temperatures, all the charmonium states will pass through. As the temperature of the medium is increased, the state ψ' will fail to pass through, and they will be dissociated in the medium. At still higher temperatures the state χ_c, and at yet higher temperature J/ψ, will melt. In checking the temperature at which ψ', χ_c and J/ψ melt, we have a way of specifying the temperature of the QGP medium. Sequential melting of charmonium states will have a characteristic imprint on J/ψ survival probability. If we plot the J/ψ cross section as a function of energy, a step-function-like behavior, as shown in figure 9.5, is expected.

Over the years, several groups have measured the J/ψ yield in heavy ion collisions. For a review of the data prior to RHIC energy collisions, and their interpretations, see [19, 20]. In brief, experimental data do show suppression. However, dissociation in QGP is not the only source of J/ψ suppression. There are various reasons due to which J/ψ production can be suppressed in nuclear collisions. Indeed, in experiments, J/ψ are suppressed in pA collisions as well, where one does not expect QGP formation. J/ψ suppression unrelated to dissociation in QGP is termed the cold nuclear matter (CNM) effect. It is discussed below. Prior to the NA50 measurements

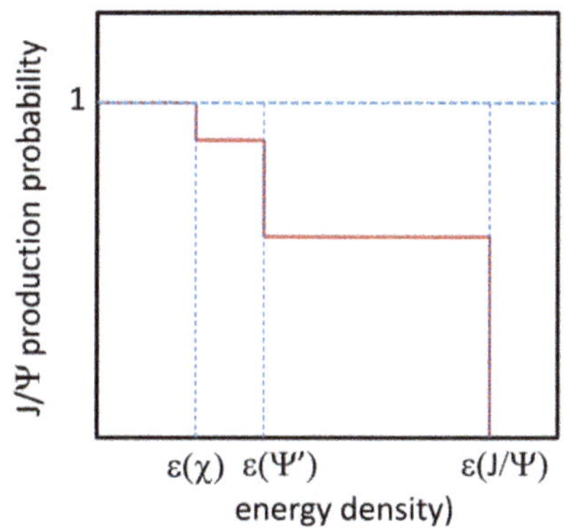

Figure 9.5. Expected J/ψ suppression pattern as a function of energy if charmonium states undergo sequential melting.

[21] of J/ψ suppression in 158 AGeV Pb+Pb collisions, all the experimental data were explained in terms of the CNM effect. The NA50 collaboration first observed anomalous suppression which goes beyond the CNM effect. In recent years, J/ψ suppression has been observed by the PHENIX collaboration in $\sqrt{s}_{\mathrm{NN}} = 200$ GeV Au+Au collisions at RHIC [25, 26] and also by the ALICE collaboration in $\sqrt{s}_{\mathrm{NN}} =$ 2.76 TeV Pb+Pb collisions at LHC [24]. PHENIX made systematic measurements of J/ψ production in nuclear collisions. They have measured J/ψ yield in p+p collisions at RHIC to obtain the basic 'reference' invariant yield [22, 23]. Measurements of J/ψ production in d+Au collisions [23] give a reference for 'CNM effects'. Cold and hot nuclear matter effects are studied in Au+Au and Cu+Cu collisions, where yields are measured as a function of collision centrality [25, 26]. In figure 9.6, the PHENIX measurements for nuclear modification factor,

$$R_{AA} = \frac{1}{N_{\mathrm{coll}}} \frac{\sigma^{AA \to J/\psi X}}{\sigma^{pp \to J/\psi X}}, \tag{9.12}$$

in $\sqrt{s}_{\mathrm{NN}} = 200$ GeV Au+Au collisions are shown. The data show suppression. The CNM effect alone cannot explain the data. If J/ψ are suppressed in the QGP, the data are explained.

There have been arguments that rather than being suppressed, charmonium will be enhanced in very high-energy collisions [27, 28]. The reasoning is simple. Due to large initial energy, a large number of $c\bar{c}$ pairs will be produced in initial hard scatterings. Recombination of $c\bar{c}$ can occur enhancing charmonium production. Apparently, PHENIX data on J/ψ production in Au+Au are not consistent with models which predict J/ψ enhancement. More recently, the ALICE collaboration [24] at LHC has measured J/ψ suppression in $\sqrt{s}_{\mathrm{NN}} = 2.76$ TeV Pb+Pb collisions. They took measurements at forward rapidity. Interestingly, compared to $\sqrt{s}_{\mathrm{NN}} =$ 200 GeV Au+Au collisions, J/ψ are less suppressed in $\sqrt{s}_{\mathrm{NN}} = 2.76$ TeV Pb+Pb collisions. One possible interpretation could be the regeneration effect. Compared to

RHIC, at LHC, due to much higher collision energy, initially heavy quarks are produced in large numbers. In both RHIC and LHC energy collisions, the medium is produced at very high temperatures. If regeneration effects are not included, in central collisions, J/ψ suppression will be similar at both RHIC and at LHC. If the regeneration effect is included, its effect will be more on J/ψ yield at LHC than at RHIC (the initial number of heavy quarks is comparatively smaller at RHIC than at LHC). The enhanced regeneration effect at LHC will be manifested as one J/ψ suppression at LHC than at RHIC.

9.2.3 CNM effect on J/ψ suppression

As discussed, J/ψ suppression in pA or AA collisions, unrelated to QGP formation, is termed the CNM effect. There are several sources for the CNM effect which are broadly classified as (i) the initial state effect and (ii) the final state effect. A major initial state effect is the modification of parton distribution in nuclei. $Q\overline{Q}$ pairs are produced in initial hard scattering which evolve into the quarkonium state. The production cross section of $Q\overline{Q}$ depends on the parton density. Experimental data in deep inelastic collisions indicate that the parton distribution is significantly modified inside a nucleus. These nuclear modifications depend on the fraction of the total hadron momentum carried by the parton, x, on the momentum scale, Q^2, and on the

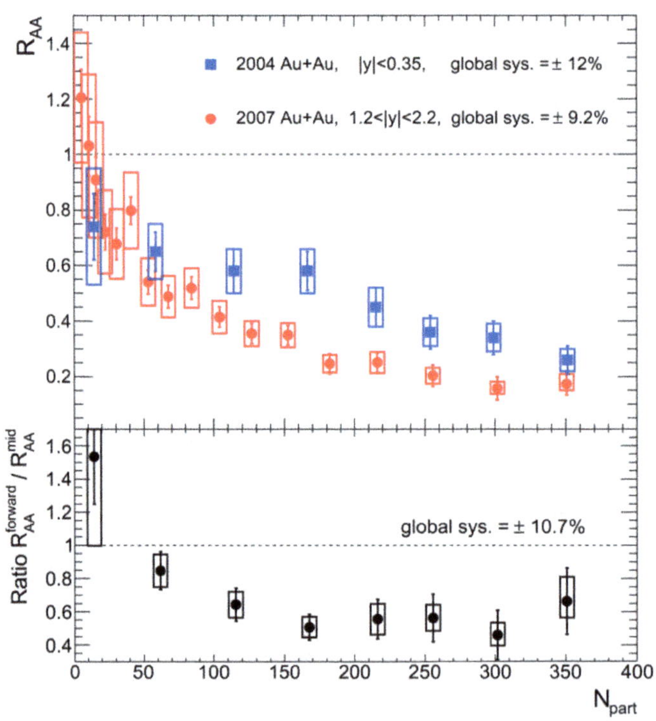

Figure 9.6. Nuclear modification factor R_{AA} for J/ψ in $\sqrt{s}_{\rm NN} = 200$ GeV Au+Au collisions and in $\sqrt{s}_{\rm NN} = 2.76$ TeV Pb+Pb collisions. The figure is reproduced with the author's permission from [26].

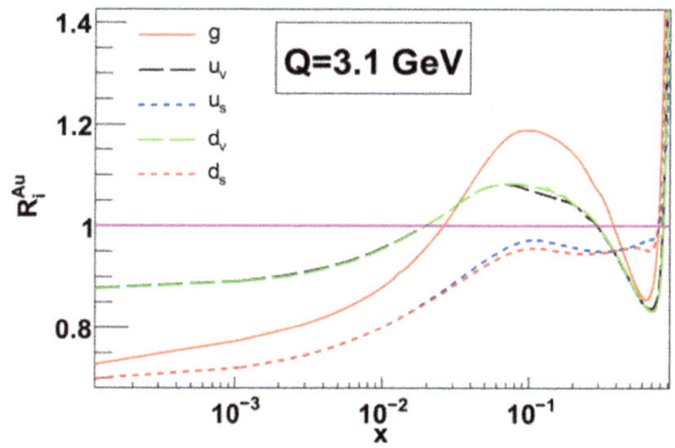

Figure 9.7. Nuclear modification of parton distribution functions. The ratio of parton distribution in *Au* and in protons are shown. If there is no parton modification the ratio will be unity.

mass number of the nucleus, A. While the mechanisms governing these modifications are not yet well understood, several groups have parameterized the modification in terms of the ratio

$$R_i\left(A, x, Q^2\right) = \frac{f_i^A\left(x, Q^2\right)}{f_i^p\left(x, Q^2\right)} \tag{9.13}$$

where $f_i^A\left(x, Q^2\right)$ and $f_i^p\left(x, Q^2\right)$ are parton distributions in a nucleus and nucleon, respectively. One such parameterization is shown in figure 9.7. One observes that at low x, the ratio $R_i < 1$, they are called the shadowing region. There are also anti-shadowing $R_i > 1$. In nuclear collisions, J/ψ yield will be enhanced due to the anti-shadowing effect and suppressed due to the shadowing effect. The overall effect is suppression. The other initial state effect is parton energy loss. A parton traversing the nucleus before the hard scattering can suffer energy loss leading to a reduction in quarkonium yield. Another CNM effect is absorption (i.e. destruction) of the quarkonium state as it passes through the nucleus. Since the latter occurs after the $Q\overline{Q}$ pair has been produced and while it is traversing the nuclear medium, this absorption is typically referred to as a final state effect.

The CNM effect due to nuclear absorption can be calculated in a Glauber type model. For example, in a Glauber model, in a pA collision at impact parameter **b**, the J/ψ production cross section can be written as

$$\sigma_{pA\to J/\psi} = \sigma_{pp\to J/\psi} \times T_A(\mathbf{b}) \tag{9.14}$$

where $T_A(\mathbf{b}) = \int_{-\infty}^{\infty} dz\rho(\mathbf{b}, z)$ is the nuclear thickness function, $\rho(\mathbf{b},z)$ being the nuclear density distribution. Now if J/ψ interact inelastically with the target nucleons with cross section $\sigma_{\text{abs}}^{J/\psi N}$, the thickness function can be modified as,

$$T_A(\mathbf{b}) \to T_A^{\text{eff}}(\mathbf{b}) = \int_{-\infty}^{\infty} dz\rho(\mathbf{b}, z) \exp\left(-\sigma_{abs}^{J/\psi N} \int_{z}^{\infty} dz'\rho(\mathbf{b}, z')\right) \tag{9.15}$$

For finite $J/\psi N$ inelastic cross sections, the thickness function is reduced, effectively reducing the J/ψ production cross section in pA collisions, in comparison to pp collisions. The equation can be easily generalized to nucleus–nucleus (AB) collisions,

$$\sigma_{AA\to J/\psi} = \sigma_{pp\to J/\psi} \times \int d^2s T_A^{\mathrm{eff}}(\mathbf{s}) T_B^{\mathrm{eff}}(\mathbf{b}-\mathbf{s}). \tag{9.16}$$

Experimental data prior to the NA50 experiment [21] on J/ψ production in 158 A GeV Pb+Pb collisions were consistent with the Glauber model of the nuclear absorption cross section $\sigma_{J/\psi N} = 4-5$ mb. NA50 data showed suppression beyond the Glauber model of nuclear absorption. If in addition there exists an anomalous suppression such as suppression in a deconfined medium, the NA50 data are explained.

9.3 Strangeness enhancement

For a long time, strangeness enhancement has been considered a signature of QGP formation [29, 30] in ultra-relativistic collisions. In QGP, strangeness can be produced in reactions,

$$\text{quark annihilation:}\ q + \bar{q} \to s + \bar{s} \tag{9.17a}$$

$$\text{gluon fusion:}\ g + g \to s + \bar{s}. \tag{9.17b}$$

Gluon fusion dominates the strangeness production process. Consider QGP in equilibrium. Strange quark density (with degeneracy 2(spin) × 3(color), can be obtained as (see equation (4.92)),

$$n_s = n_{\bar{s}} = 3 \times 2 \int \frac{d^3p}{(2\pi)^3} e^{-\sqrt{p^2+m_s^2}/T} = \frac{3Tm_s^2}{\pi^2} K_2(m_s/T). \tag{9.18}$$

The density of non-strange anti-quarks $\bar{q}$ ($\bar{q}$ stands for $\bar{u}$ or $\bar{d}$), in the limit of small mass, is

$$n_{\bar{q}} = 3 \times 2 \int \frac{d^3p}{(2\pi)^3} e^{-|p|/T} e^{-\mu_q/T} = e^{-\mu_q/T} \frac{6}{\pi^2} T^3 \tag{9.19}$$

where μ_q is the quark chemical potential, one third of the baryonic chemical potential μ_B. The ratio of strange quarks to non-strange quarks is

$$\frac{n_s}{n_{\bar{q}}} = \frac{1}{2}\left(\frac{m_s}{T}\right)^2 K_2\left(\frac{m_s}{T}\right) e^{\mu_B/3T}. \tag{9.20}$$

For $\mu_B > 0$, the ratio is greater than one. Strange quarks will be produced in more abundance than the non-strange quarks. During hadronization, numerous strange quarks can be bound with available non-strange quarks and strange hadron production will enhance. Qualitatively, $\frac{n_s}{n_q} > 1$ can be understood as follows: the medium produced in the collision is rich in light quarks u and d, inherited from the colliding nuclei. In such an environment the Pauli exclusion principle prohibits further production of $u\bar{u}$, $d\bar{d}$ light quark pairs. $s\bar{s}$ production however is not Pauli blocked. If strange quark mass is not too heavy, $s\bar{s}$ can easily be produced.

The above estimate assumes that the strange quarks are equilibrated. It also neglects that QGP is an evolving medium. Approaches to equilibration can be studied by solving the master equation for the change in density of strange quarks. For the reaction mentioned above, one can write the master equation for strange quark density,

$$\frac{\mathrm{d}\rho_{s,\bar{s}}}{\mathrm{d}t} = \frac{1}{2}\rho_g^2(t)\langle v\sigma\rangle_T^{gg\rightarrow s\bar{s}} + \rho_q(t)\rho_{\bar{q}}(t)\langle v\sigma\rangle_T^{q\bar{q}\rightarrow s\bar{s}} - \rho_s(t)\rho_{\bar{s}}(t)\langle v\sigma\rangle_T^{s\bar{s}\rightarrow q\bar{q},gg} \tag{9.21}$$

where $\langle\sigma v\rangle_T$ is the momentum averaged cross section for the processes indicated,

$$\langle\sigma v\rangle_T = \frac{\int \mathrm{d}^3p_1\mathrm{d}^3p_2 f_1(p_1)f_2(p_2)\sigma v}{\int \mathrm{d}^3p_1\mathrm{d}^3p_2 f_1(p_1)f_2(p_2)}. \tag{9.22}$$

When the loss term in the master equation balances the gain terms, the (chemical) equilibrium of strangeness is achieved. Explicit calculations indicate that strange quarks achieve near complete equilibrium [30, 31]. The onset of transverse expansion however impedes the equilibration process [32]

QGP matter ultimately evolves into a hadron gas. Strangeness can also be produced in hadronic reactions,

$$\begin{array}{ll} \pi + N \rightarrow Y + \overline{K} & \pi + \overline{N} \rightarrow \overline{Y} + K \\ \pi + Y \rightarrow \Xi + \overline{K} & \pi + \overline{Y} \rightarrow \overline{\Xi} + K \\ \pi + \Xi \rightarrow \Omega + \overline{K} & \pi + \overline{\Xi} \rightarrow \overline{\Omega} + K \\ \pi + \pi \rightarrow K + \overline{K} & \end{array} \tag{9.23}$$

where Y represents a hyperon, $\Lambda(uds)$, $\Sigma^+(uus)$, $\Sigma^0(uds)$, $\Sigma^-(dds)$. K represents a kaon, $K^+(u\bar{s})$, $K^0(\mathrm{d}\bar{s})$, $K^-(\bar{u}s)$. The common feature of all the processes listed in equation (9.23) is the reaction $q\bar{q} \rightarrow s\bar{s}$ (see figure 9.8). The produced strangeness can be redistributed among various hadrons in exchange reactions, e.g. $\Lambda + \pi \rightarrow N + K$,

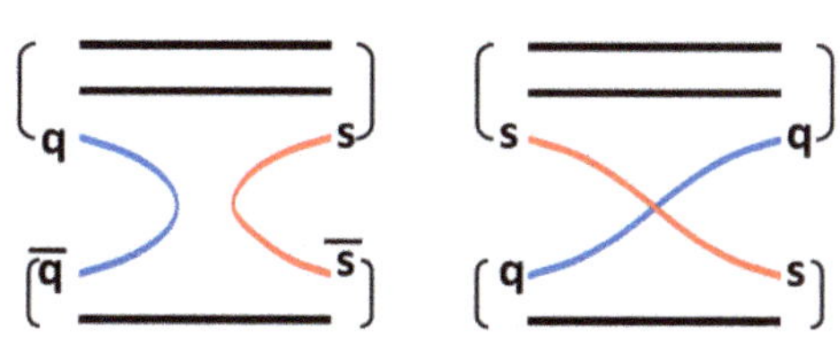

Figure 9.8. Left panel: direct production of strangeness in (non-strange) meson–baryon reactions. Right panel: exchange of strangeness in (strange) meson–baryon reactions.

schematically shown in the right panel of figure 9.8. The strangeness producing reactions are OZI-rule[1] suppressed. Compared to gluon fusion, hadronic reactions are slow and there is also a threshold effect. A master equation can be set up for the study of strangeness equilibration in hadronic reactions [31]. The equilibration process is much slower than in QGP. It is then argued that strangeness phase space is saturated during the QGP phase of the evolution.

To measure the enhancement experimentally, one defines a strangeness enhancement factor (E),

$$E = \frac{\frac{1}{N_{\text{part}}}\frac{\mathrm{d}N^{AA}}{\mathrm{d}y}}{\frac{1}{2}\frac{\mathrm{d}N^{pp}}{\mathrm{d}y}}. \tag{9.24}$$

In figure 9.9, experimental data [35, 36] in $\sqrt{s}_{\text{NN}} = 200$ GeV Au+Au collisions for strangeness enhancement are shown. The data do show enhancement. ϕ meson (which is a hidden strange meson) production is also enhanced [37]. However, it is uncertain whether or not the enhancement is due to increased production in QGP or due to less strangeness production in pp collisions. One useful way of looking for strangeness enhancement is the Wroblewski parameter [33]. It is defined as the ratio of newly created strange quarks to light quarks,

$$\lambda_s = \frac{2\langle s\bar{s}\rangle}{\langle u\bar{u}\rangle + \langle d\bar{d}\rangle}. \tag{9.25}$$

The factor has been studied in model calculations. Statistical hadronization models indicate that from elementary pp collisions to heavy ion collisions, the Wroblewski parameter increases from ~0.2 to ~0.5 [34]. The canonical suppression (less strangeness in pp collisions than in heavy ion collisions) arises from the need to conserve strangeness within a small, local volume, which limits the strangeness production in pp collisions relative to AA collisions [38–40]. In the language of statistical mechanics, while canonical ensemble is applicable in pp collisions, in AA collisions, the grand canonical ensemble is applicable. For example, consider $K^+ K^-$ production in a pionic thermal system of volume V at temperature T. For large T kaons are abundantly produced and density reaches the grand canonical value,

$$n_{K^+} = \left[\frac{1}{2\pi^2} m_{K^+}^2 T K_2\left(\frac{m_{K^+}}{T}\right)\right]. \tag{9.26}$$

[1] The OZI rule was proposed independently by Susumu Okubo, George Zweig and Jugoro Iizuka in the 1960s. It states that hadronic reactions, where quark lines in the Feynman diagram cannot be connected continuously from initial hadrons to final state hadrons, are suppressed.

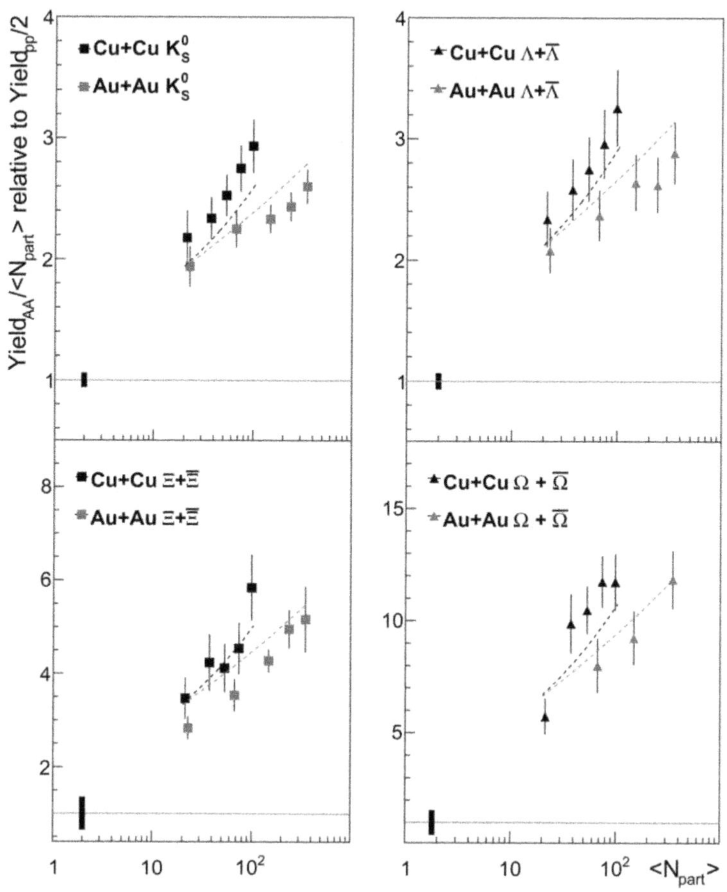

Figure 9.9. STAR measurements for enhancement factor in $\sqrt{s}_{\mathrm{NN}} = 200$ GeV Cu+Cu and Au+Au collisions, for various multi-strange particles. The figure is reproduced with permission from [36].

In the limit of low temperature, K^+, K^- are rarely produced. In order to satisfy the strangeness neutrality condition, the kaon pair must appear in the near vicinity. The density of K^+ then reaches the canonical ensemble value,

$$n_{K^+} = \left[\frac{1}{2\pi^2} m_{K^+}^2 T K_2\!\left(\frac{m_{K^+}}{T}\right)\right] \times \left[V_0 \frac{1}{2\pi^2} m_{K^-}^2 T K_2\!\left(\frac{m_{K^-}}{T}\right)\right]. \tag{9.27}$$

The first term coincides with the grand canonical value. The second term describes the phase space suppression since with each K^+ a K^- has to appear in the vicinity to conserve strangeness. The parameter V_0 is the correlation volume where K^+ and K^- should be created to preserve strangeness conservation. In heavy ion collisions, $V_0 \propto N_{\text{part}}$ and in proton induced collisions, $V_0 \sim V_{\text{proton}}$. The canonical suppression then arises due to the fact that (i) particles are produced in pairs that restrict the available momentum phase space and (ii) they appear in near vicinity in space to fulfil locality of conservation laws.

9.4 Electromagnetic probes

Photons and dileptons are considered to be important probes for QGP diagnostics. They are called electromagnetic probes as they only interact electromagnetically. Unlike the hadrons, which are emitted only from the freeze-out surface, photons and dileptons have a large mean free path and are emitted from the entire volume. Total production is then obtained by convoluting their production rate over the four-volume.

9.4.1 Photons

For QGP diagnostic purposes, one is interested only in 'direct photons', i.e. photons produced as a result of some collision process. However, one of the problems with the photon as a probe is the large background. The transient QGP state ultimately transforms into hadrons. A large number of hadrons decay into photons, e.g. $\pi^0 \to 2\gamma$. Decay photons constitute the background and need to be eliminated to get the direct photon signal. Several methods have been devised experimentally to eliminate the decay photons, e.g. invariant mass analysis, mixed event analysis, etc. However, due to the overwhelmingly large number of background photons, the elimination could not be exact and experimental data on 'direct' photon production could not be obtained very accurately. There is definitely scope for further improvement.

In a nucleus–nucleus collision, there are various sources of direct photons. They are discussed briefly below.

(i) *Prompt photons.* Initial hard collision produce prompt photons. Prompt photon production in a nucleon–nucleon reaction $a + b \to \gamma + X$ can be calculated in leading order pQCD,

$$E\frac{\mathrm{d}\sigma}{\mathrm{d}^3p^\gamma} = K \sum_{ij=q,\bar{q},g} \int \mathrm{d}x_i \mathrm{d}x_j f_{i/a}\left(x_i, Q^2\right) f_{j/b}\left(x_j, Q^2\right) \times \delta\left(\hat{s} + \hat{t} + \hat{u}\right) \frac{\hat{s}}{\pi} \frac{\mathrm{d}\hat{\sigma}}{\mathrm{d}\hat{t}}(ij \to \gamma + X) \tag{9.28}$$

where $f_{i/a,b}(x, Q^2)$ is the parton distribution function and $\frac{\mathrm{d}\hat{\sigma}}{\mathrm{d}\hat{t}}$ is the elementary partonic cross section. The above equation scaled by the binary collision number gives the prompt photon production in nucleus–nucleus collisions.

(ii) *Fragmentation photons.* Initial hard scattered partons may fragment into photons ($q \to q + \gamma$). For example, in partonic collisions, $i + j \to k + l$, the scattered parton k may fragment into photons. The invariant cross section for the process can be written as

$$E\frac{\mathrm{d}\sigma}{\mathrm{d}^3p^\gamma} = K \sum_{ijk=q,\bar{q},g} \int \mathrm{d}x_i \mathrm{d}x_j \frac{\mathrm{d}z}{z^2} f_{i/a}\left(x_i, Q^2\right) f_{j/b}\left(x_j, Q^2\right) \times \frac{\hat{s}}{\pi} \frac{\mathrm{d}\hat{\sigma}}{\mathrm{d}\hat{t}}(ij \to k + l) \delta\left(\hat{s} + \hat{t} + \hat{u}\right) D_{\gamma/k}\left(z, Q_F^2\right). \tag{9.29}$$

In equation (9.29), $D_{\gamma/k}$ is the fragmentation function. As before, the above equation should be scaled by the binary collision number to obtain the fragmentation photons in nucleus–nucleus collisions.

(iii) *Pre-equilibrium photons*. In nucleus–nucleus collisions, an extended QCD medium is produced. Before achieving local thermal equilibrium the medium is in the pre-equilibrium stage. Photons will also be emitted in the pre-equilibrium stage. However, it is difficult to distinguish pre-equilibrium photons from thermal photons. Simulations with the parton cascade model [41] indicate that pre-equilibrium photon production equals the thermal photon at $p_T = 2$ GeV. Low p_T photons are predominantly thermal.

(iv) *Thermal photons*. Photons emitted from the (locally) equilibrated QGP and hadronic matter are called thermal photons. For QGP diagnostic purposes, these photons are most important. In the following, we will briefly discuss thermal photons. For more detailed information, see [42–44].

Thermal photons from the QGP phase. In the QGP phase, the important reactions for direct photons (schematically shown in figure 9.10) are; (i) annihilation: $q + \bar{q} \to g + \gamma$ ($q + \bar{q} \to \gamma$ is forbidden due to energy momentum conservation), (ii) the Compton process: $q + g \to q + \gamma$ and (iii) bremsstrahlung: $q + q \to q + q + \gamma$.

The thermal emission rate of photons with energy E and momentum p is related to the photon self-energy by the relation

$$E\frac{\mathrm{d}R}{\mathrm{d}^3p} = -\frac{2}{(2\pi)^3}\,\mathrm{Im}\Pi^\mu_\mu \frac{1}{e^{E/T} - 1} \tag{9.30}$$

where $\mathrm{Im}\Pi^\mu_\mu$ is the imaginary part of the retarded photon self-energy at temperature T. The relation is valid perturbatively as well as non-perturbatively, in all orders in strong interaction, but in order e^2 in electromagnetic interactions (scattering of emerging photons was neglected). Equation (9.30) is formally equivalent to kinetic theory if the photon self-energy is approximated by loop diagrams of some finite order.

The photon production rate due to annihilation and Compton scattering was first calculated in [45–47]. The procedure will be discussed briefly. In kinetic theory, the

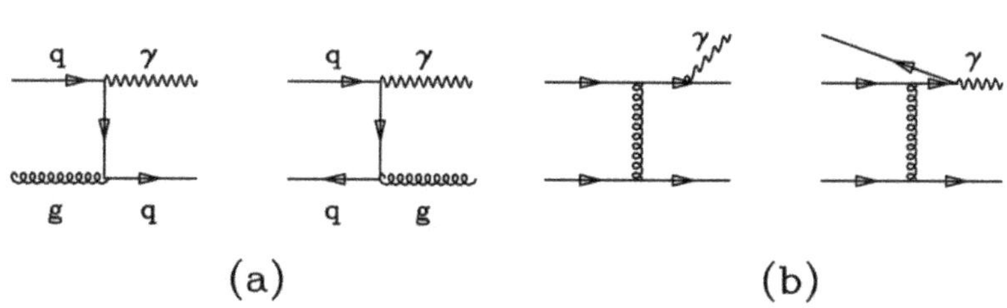

Figure 9.10. Leading order Feynman diagrams contributing to direct photon production in QGP. (a) Compton and annihilation process, (b) bremsstrahlung process.

photon production rate (per unit time per unit volume) from the $1 + 2 \rightarrow 3 + \gamma$ process can be written as

$$\mathcal{R}_i = \mathcal{N} \int \frac{d^3p_1}{2E_1(2\pi)^3} \frac{d^3p_2}{2E_2(2\pi)^3} f_1(E_1) f_2(E_2) (2\pi)^4 \delta^4(p_1^\mu + p_2^\mu - p_3^\mu - p^\mu) \\ \times |\mathcal{M}_i|^2 \frac{d^3p_3}{2E_3(2\pi)^3} \frac{d^3p}{2E(2\pi)^3} \left[1 \pm f_3(E_3)\right] \tag{9.31}$$

where $\mathcal{M}_i$ is the amplitude for one of the basic processes, and $f(E)$ are the Fermi–Dirac or Bose distributions, as appropriate. The $\pm$ in the last term corresponds to Bose enhancement or Pauli blocking. Using Mandelstam's variable, s, t, u, the differential rate can be written,

$$E\frac{d\mathcal{R}_i}{d^3p} = \frac{\mathcal{N}}{(2\pi)^7} \frac{1}{16E} \int ds\,dt |\mathcal{M}_i(s, t)|^2 \int dE_1 dE_2 f_1(E_1) f_2(E_2) \\ \times \left[1 \pm f_3(E_1 + E_2 - E)\right] \theta(E_1 + E_2 - E) \sqrt{aE_1^2 + bE_1 + c} \tag{9.32}$$

where

$$\begin{aligned} a &= -(s + t)^2 \\ b &= 2(s + t)(Es - E_2 t) \\ c &= st(s + t) - (Es + E_2 t)^2. \end{aligned} \tag{9.33}$$

For massless particles, amplitude $\mathcal{M}$ is related to the differential cross section as

$$\frac{d\sigma}{dt} = \frac{|\mathcal{M}|^2}{16\pi s^2}. \tag{9.34}$$

For Compton and annihilation processes, the differential cross sections are

$$\frac{d\sigma^{\text{annihilation}}}{dt} = \frac{8\pi\alpha\alpha_s}{9\,s^2} \frac{u^2 + t^2}{ut} \tag{9.35}$$

$$\frac{d\sigma^{\text{Compton}}}{dt} = \frac{-\pi\alpha\alpha_s}{3\,s^2} \frac{u^2 + s^2}{us}. \tag{9.36}$$

$\mathcal{N} = 20$ for the annihilation process when summing over u and d quarks and $\mathcal{N} = \frac{320}{3}$ for the Compton process. The cross sections have poles at t and/or $u = 0$. The divergences can be screened following the Braaten, Pisarski formalism [48]. Here is the final result obtained in [45], where quantum statistics was neglected,

$$E\frac{dR}{d^3p} = \frac{5}{9} \frac{\alpha\alpha_s}{2\pi^2} T^2 \exp\left(-\frac{E}{T}\right) \ln\left(\frac{2.912}{g^2} \frac{E}{T}\right). \tag{9.37}$$

Initially it was thought that Compton and annihilation processes would dominate the photon production mechanism. In the lowest order, the $2 \rightarrow 2$ rates are $O(g^2e^2)$. The $2 \rightarrow 3$ bremsstrahlung rates $O(g^4e^2)$ should be negligible. However, it was realized later that bremsstrahlung processes can contribute at the same order as

Compton/annihilation processes. Bremsstrahlung processes can have collinear singularity (the emitted photon is collinear with quarks) when they contribute to the same order $O(g^2e^2)$. The importance of the bremsstrahlung process was first considered in [49–51], however, the Landau–Pomeranchuk–Migdal (LPM) effect (when photon emission is suppressed due to multiple collisions) was neglected. Arnold, Moore and Yaffe [52] made a complete calculation in leading order. The calculations are quite involved and beyond the scope of this book. They also provided a simple parameterized form for easy use in hydrodynamics. The results are listed below.

Leading order photon emission rate from QGP:

$$E^{\gamma}\frac{dR}{d^3k} = A(k)\left[\ln\left(T/m_{\infty}\right) + C_{\rm tot}(k/T)\right] \tag{9.38}$$

with

$$C_{\rm tot} = \frac{1}{2}\ln(2k/T) + C_{2\to 2}(k/T) + C_{\rm brem}(k/T) + C_{\rm annih}(k/T). \tag{9.39}$$

The leading log coefficient $A(k)$ is given by

$$A(k) = 2\alpha\left[d_F\sum_i q_i^2\right]\frac{m_{\infty}}{k}n_f(k) \tag{9.40}$$

where $n_f(k) = [\exp(k/T) + 1]^{-1}$ is the Fermi distribution function, and d_F is the dimension of the quark representation ($d_F = N_c = 3$ for QCD). $q_i = 2/3$ for up quarks and $-1/3$ for down quarks. $m_{\infty} = g_s^2\, T^2/3$ is the thermal quark mass in the leading order, g_s being the strong coupling constant ($\alpha_s = g_s^2/4\pi$). For two-flavor QCD,

$$A(k) = \frac{40\pi T^2}{9}\alpha\alpha_s\frac{n_f(k)}{k}. \tag{9.41}$$

The strong coupling $\alpha_s = \frac{g_s^2}{4\pi}$ is defined at a scale of order of the temperature (T). $C_{2\to 2}(k/T)$, $C_{\rm brem}(k/T)$ and $C_{\rm annih}(k/T)$ all involve multidimensional integrations, which can only be solved numerically. Numerical results for QCD plasmas are reproduced quite accurately by the approximate phenomenological fits [52],

$$C_{2\to 2}(x) = 0.041x^{-1} - 0.3615 + 1.01e^{-1.35x}, \tag{9.42}$$

$$C_{\rm brem}(x) + C_{\rm annih}(x) \approx \sqrt{1 + \frac{N_f}{6}}\,\frac{0.548\log(12.28 + 1/x)}{x^{3/2}} + \frac{0.133x}{\sqrt{1 + x/16.27}}. \tag{9.43}$$

Thermal photons from the hadronic phase. Photons are also emitted from the hadronic phase. For example, as shown in figure 9.11, a π^+ meson can annihilate a π^- meson to produce a photon and a ρ^0 meson. Photons are emitted in several reactions in the hadronic phase. J Kapusta, P Lichard and D Seibert [45, 46] evaluated photon

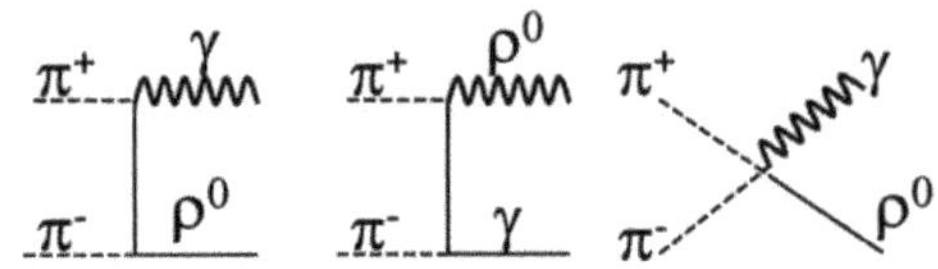

Figure 9.11. Feynman diagrams for photon production in hadron gas from $\pi^+\pi^-$ annihilation.

production rates from hadronic channels: (i) $\pi + \pi \to \rho + \gamma$, (ii) $\pi + \rho \to \pi + \gamma$, (iii) $\pi + \pi \to \eta + \gamma$, (iv) $\pi + \eta \to \pi + \gamma$ and (v) $\omega \to \pi + \gamma$. The role of A_1 resonance in photon production was investigated in [53]. Later, in a more comprehensive analysis [54], hadronic photon production rate in a meson gas consisting of light pseudo-scalar, vector and axial vector mesons (π, K, ρ, K^*, A_1) was obtained. A parameterized form was also provided. Below, we reproduce the parameterized reaction rates.

Photon emission rate from the hadronic phase:

$$E\frac{\mathrm{d}R_{\pi^+\rho\to\pi^+\gamma}}{\mathrm{d}^3k} = F^4(E)T^{2.8} \times \exp\left[\frac{-\left(1.461T^{2.3094} + 0.727\right)}{(2TE)^{0.86}} + \left(0.566T^{1.4094} - 0.9957\right)\frac{E}{T}\right]$$

$$E\frac{\mathrm{d}R_{\pi^+\pi\to\rho^+\gamma}}{\mathrm{d}^3k} = F^4(E)\frac{1}{T^5} \times \exp\left[-\left(9.314T^{-0.584} - 5.328\right)(2TE)^{0.088} + \left(0.3189T^{0.721} - 0.8998\right)\frac{E}{T}\right]$$

$$E\frac{\mathrm{d}R_{\rho\to\pi\pi\gamma}}{\mathrm{d}^3k} = F^4(E)\frac{1}{T^2} \times \exp\left[-\frac{\left(-35.459T^{1.26} + 18.827\right)}{(2TE)^{\left(-1.44T^{0.142}+0.9996\right)}} - 1.21\frac{E}{T}\right]$$

$$E\frac{\mathrm{d}R_{\pi K^*\to K\gamma}}{\mathrm{d}^3k} = F^4(E)T^{3.75} \times \exp\left[-\frac{0.35}{(2TE)^{1.05}} + \left(2.3894T^{0.03435} - 3.222\right)\frac{E}{T}\right]$$

$$E\frac{\mathrm{d}R_{\pi K\to K^*\gamma}}{\mathrm{d}^3k} = F^4(E)\frac{1}{T^3} \times \exp\left[-\left(5.4018T^{-0.6864} - 1.51\right)(2TE)^{0.07} - 0.91\frac{E}{T}\right]$$

$$E\frac{\mathrm{d}R_{\rho K\to K\gamma}}{\mathrm{d}^3k} = F^4(E)T^{3.5} \times \exp\left[-\frac{\left(0.9386T^{1.551} + 0.634\right)}{(2TE)^{1.01}} + \left(0.568T^{0.5397} - 1.164\right)\frac{E}{T}\right]$$

$$E\frac{\mathrm{d}R_{K^*K\to\pi\gamma}}{\mathrm{d}^3k} = F^4(E)T^{3.7} \times \exp\left[\frac{\left(-6.096T^{1.889} + 1.0299\right)}{(2TE)^{\left(-1613T^{12.162}+0.975\right)}} - 0.96\frac{E}{T}\right].$$

In the above equations, the E and T are in GeV and the rates are in units of $\mathrm{GeV}^{-2}\,\mathrm{fm}^{-2}$. The dipole form factor $F(E)$ is

$$F(E) = \left(\frac{2\Lambda^2}{2\Lambda^2 - E}\right)^2 \qquad \Lambda = 1\ \mathrm{GeV}. \tag{9.44}$$

The photon production rate from the QGP phase and the hadronic phase has to be convoluted over the four-volume to obtain total photon production during the evolution of the fireball,

$$E\frac{\mathrm{d}N^\gamma}{\mathrm{d}^3k} = \int \mathrm{d}^4x\left(E\frac{\mathrm{d}R}{\mathrm{d}^3k}\right) = \int \tau\,\mathrm{d}\tau\,\mathrm{d}x\,\mathrm{d}y\,\mathrm{d}\eta\left(E\frac{\mathrm{d}R}{\mathrm{d}^3k}\right). \tag{9.45}$$

There are several simulations for direct photon production in relativistic heavy ion collisions. For demonstration purposes, we will show a few results. In the left panel of figure 9.12, transverse momentum dependence of the invariant cross section in $\sqrt{s}_{\mathrm{NN}} = 200$ GeV p+p collisions and invariant yield in $\sqrt{s}_{\mathrm{NN}} = 200$ GeV Au+Au collisions are shown. The p+p data are from [55]. The Au+Au data are from [56, 57]. The three curves on the p+p data represent NLO pQCD calculations. For $p_T > 2$ GeV, the pQCD calculation is consistent with the p+p data within the theoretical uncertainties. The dashed curves for the Au+Au data are obtained by scaling the photon yield in p+p collisions by the nuclear overlap function T_{AA}. At low $p_T < 2.5$ GeV, experimental Au+Au data are under-predicted. At low p_T, photon production increases faster than the binary NN collision scaled p+p cross section. The solid lines in figure 9.12 are fits with an exponential plus binary collision scaled p+p data.

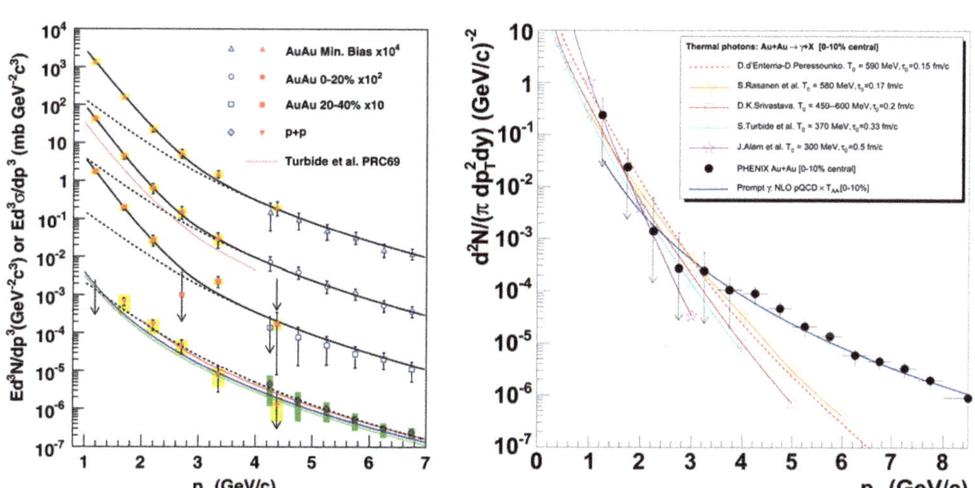

Figure 9.12. Left panel: invariant cross section (p+p) and invariant yield (Au+Au) of direct photons as a function of p_T. See the text for details. The figure is reproduced with the author's permission from [56]. Right panel: thermal photon production in 0–10% Au+Au collisions at $\sqrt{s}_{\mathrm{NN}} = 200$ GeV in different hydrodynamical models [58–62] are compared with experimental data. Reproduced from [62] with permission.

In the right panel of figure 9.12 experimental data on direct photon production in 0–10% Au+Au collisions are compared with different hydrodynamical model calculations. All the hydrodynamical simulations are comparable to the experimental data and with each other within a factor of 2. The results confirm the dominance of thermal radiation in the direct photon spectrum in the low and intermediate p_T range.

9.4.2 Dileptons

Similar to photons, dileptons are also emitted at every stage of heavy ion collisions. In the QGP phase, a $q\bar{q}$ can interact to form a virtual photon, which subsequently decays into a lepton pair or dileptons, $q + \bar{q} \to \gamma^* \to l^+ + l^-$ ($l = e$ or μ). In the hadronic phase, dileptons are produced from interactions of charged hadrons with their anti-particles, e.g. $\pi^+ + \pi^- \to \rho \to l^+ + l^-$, from hadronic decays, e.g. $\pi^0 \to l^+ + l^- + \gamma$, $\rho \to l^+ + l^-$, $\omega \to l^+ + l^-$, $\phi \to l^+ + l^-$, etc. Dileptons are also produced in the Drell–Yan process (in the Drell–Yan process, a valence quark from the projectile hadron interacts with a sea anti-quark from the target hadron to form a virtual photon, which then decays into a lepton pair). Depending on the dilepton invariant mass (M), dilepton emission can be classified into three distinct regimes. In figure 9.13, invariant mass (M) dependence of dilepton production in high-energy nuclear collisions is shown schematically. One can distinguish three distinct regions, (i) $M \leqslant M_\phi (= 1.024\ \text{GeV})$ where the dominating source of dilepton production is vector meson decays, (ii) $M_\phi < M \leqslant M_{J/\psi} (= 3.1\ \text{GeV})$ where the dominant source is the thermal production from QGP and (iii) $M \geqslant M_{J/\psi}$ which is dominated by primordial emission, decay of charmoniums, etc.

In the following, we briefly discuss the dilepton emission rate in QGP and in hadronic resonance gas. As for photons, emission rates are to be convoluted over the space–time volume to obtain production yield in nucleus–nucleus collisions. For more details see [42].

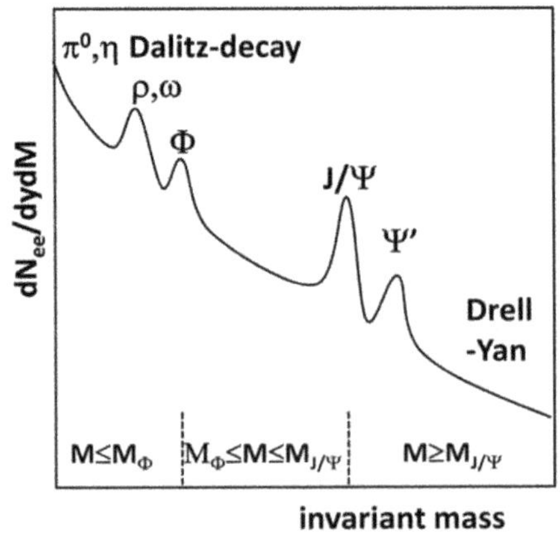

Figure 9.13. A schematic presentation of invariant mass dependence of dilepton production in high-energy nuclear collisions.

The dilepton production rate from a thermal system at temperature T, in a reaction, $a^+ + a^- \rightarrow l^+ + l^-$, where a is either a quark or a pion, was obtained in [63, 64]. From quantum electrodynamics (QED), one can calculate the cross section for $e^+ + e^- \rightarrow \mu^+ + \mu^-$,

$$\bar{\sigma}(M) = \frac{4\pi}{3}\frac{\alpha^2}{M^2}\left[1 + \frac{2m_l^2}{M^2}\right]\left[1 - \frac{4m_l^2}{M^2}\right]^{1/2} \tag{9.46}$$

where M is the invariant mass of the $\mu^+\mu^-$ pair, α is the fine structure constant ($\alpha = 1/137$) and m_l is the mass of μ. For $q\bar{q}$ annihilation, the above is multiplied by the color factor $N_c = 3$, and a factor reflecting the fractional charges of quarks. The modified cross section is

$$\sigma(M) = F_q\bar{\sigma}(M) \tag{9.47}$$

$$F_q = N_c(2s + 1)^2 \sum_f e_f^2, \tag{9.48}$$

where s is the spin of quarks, e_f is the fractional charge and the sum is over the quark flavors. In the hadronic phase, in the vector meson dominance mode, dilepton production goes via the reaction, $\pi^+ + \pi^- \rightarrow \rho \rightarrow l^+ + l^-$. The quantum electrodynamics (QED) cross section is then multiplied by the Breit–Wigner form factor,

$$F_\pi(M) = \frac{m_\rho^4}{\left(m_\rho^2 - M^2\right)^2 + m_\rho^2\Gamma_\rho^2} \qquad M_\rho \sim 770\ \text{MeV} \qquad \Gamma_\rho \sim 150\ \text{MeV}. \tag{9.49}$$

The dilepton cross section from $\pi^+\pi^-$ annihilation then becomes

$$\sigma_\pi(M) = F_\pi(M)\bar{\sigma}(M)\left[1 - \frac{4m_\pi^2}{M^2}\right]^{1/2}. \tag{9.50}$$

Kinetic theory gives the reaction rate (number of reactions per unit time per unit volume),

$$R\left(a^+a^- \rightarrow l^+l^-\right) = \int \frac{d^3p_1}{(2\pi)^3}\frac{d^3p_2}{(2\pi)^3} f(\mathbf{p}_1)f(\mathbf{p}_2) \times \sigma\left(a^+a^- \rightarrow l^+l^-\right)v_{\text{rel}} \tag{9.51}$$

where $f(\mathbf{p})$ is the occupation probability at momentum $\mathbf{p}$ and energy $E = \sqrt{M^2 + m_a^2}$. Relative velocity v_{rel} can be computed as,

$$v_{\text{rel}} = \frac{\left[(p_1 \cdot p_2)^2 - m_a^4\right]^{1/2}}{E_1E_2} \tag{9.52}$$

Approximating $f(\mathbf{p}) = \exp(-E/T)$, and integrating over five of the six variables,

$$R\left(a^+a^- \rightarrow l^+l^-\right) = \frac{T^6}{(2\pi)^4}\int_{z_0}^{\infty} \sigma(z)z^2\left(z^2 - 4z_a^2\right)K_1(z)dz \tag{9.53}$$

with $z = M/T$, $z_a = m_a/T$ and K_1 is the modified Bessel function of the first kind.

Apart from the total number of lepton pairs emitted per unit space–time volume, $R = \frac{dN}{d^4x}$, one is interested in several differential rates. They can be obtained from equation (9.51) by an appropriate change of variables. The rate for producing lepton pairs with invariant mass M is

$$\frac{dN}{d^4x dM^2} = \frac{\sigma(M)}{2(2\pi)^4} M^3\, TK_1(M/T)\left[1 - \frac{4m_a^2}{M^2}\right]. \tag{9.54}$$

The production rate of lepton pairs with invariant mass M, momentum p and energy E ($E = \sqrt{p^2 + M^2}$) can be written as

$$E\frac{dN}{d^4x dM^2 d^3p} = \frac{\sigma(M)}{4(2\pi)^5} M^2 \exp(-E/T)\left[1 - \frac{4m_a^2}{M^2}\right]. \tag{9.55}$$

We will not discuss any hydrodynamic simulation for dilepton production here. Dileptons have been measured in SPS energy ($\sqrt{s}_{NN} = 17.4$ GeV). Hydrodynamical simulations under-predict low invariant mass dilepton yield. In the low invariant mass region, dilepton emission is largely mediated by the ρ meson. Now, properties of a hadron can change in a medium. Due to medium effect, the ρ meson mass may drop, or its width increase. These two effects are considered as a likely cause of underestimation of low mass dilepton yield in hydrodynamic simulations. Currently, experimental data do not distinguish between these two effects. For a more detailed account, please see [42]. We may mention here that dileptons are possibly better suited to probe QGP than direct photons. In contrast to photons which are characterized by the transverse momentum, dileptons are characterized by two parameters, the transverse momentum and invariant mass. For differential diagnosis, an increased degree of freedom may be useful.

9.5 Jet quenching

Jets are clusters of hadrons, all moving approximately in the same direction. In hadron–hadron or e^+e^- collisions, one generally observes two-jet structures with back-to-back momenta. In the left panel of figure 9.14, an experimental reconstruction of the final state particle trajectories is shown. The two-jet structure is clearly evident in p+p collisions. In nucleus–nucleus collisions, the jet structure is not obvious due to the large number of particles (see the right panel of figure 9.14). However, jets are there and one can use a 'jet algorithm' to find them.

Jet structure in hadron–hadron collisions can be understood qualitatively in pQCD. The jet properties depend, in general, on two scales: the energy of the jet and its virtuality, specified by the largest possible transverse momentum of one of its sub-jets. The basic two-body reaction $1 + 2 \rightarrow 3 + 4$ is in the partonic (quark or gluon) level. The scattered partons are highly virtual (squared four-momentum transfer is large) and reduce their virtuality by radiating gluons or by splitting into quark–antiquark pairs. Such parton branching is governed by Dokshitzer–Gribov–Lipatov–Altarelli–Parisi equation. Finally the partons fragment into hadrons. The

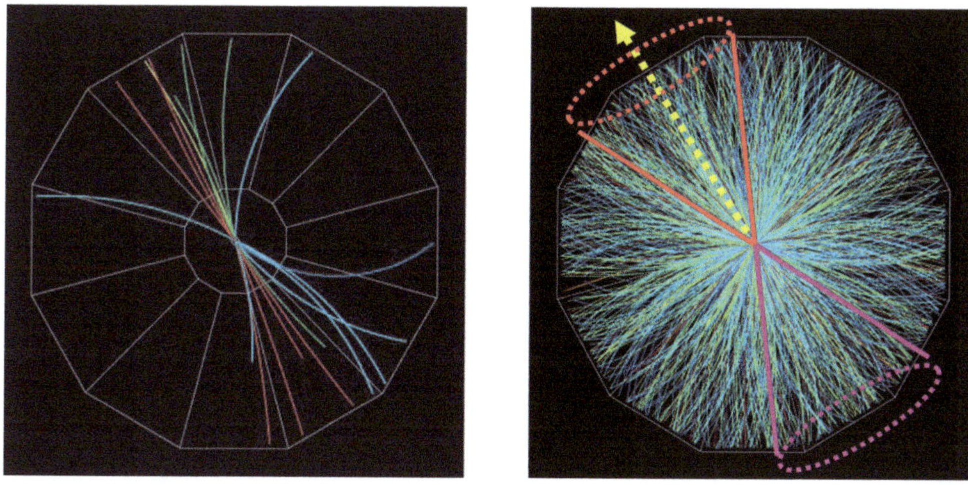

Figure 9.14. Left panel: an experimental reconstruction of an event in pp collision. The two-jet structure is clearly seen. Right panel: the same for an event in Au+Au collisions. Notice that due to the large number of particles in the final state, the jet structure is obscured. The figures are reproduced with permission from the STAR group's website (www.star.bnl.gov).

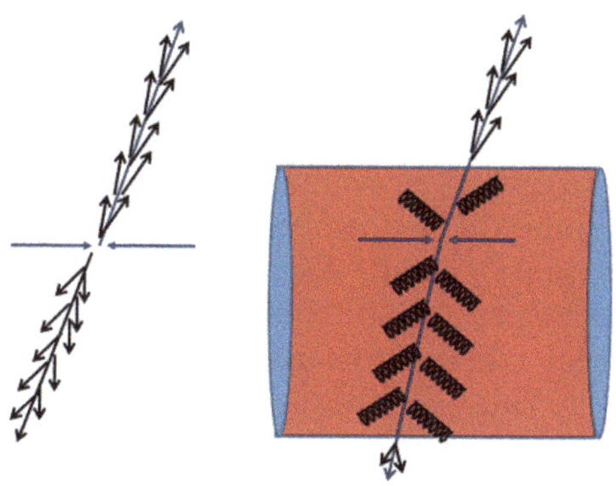

Figure 9.15. Qualitative understanding of jet quenching. Left panel: two scattered quarks fragment into two jets in a vacuum. Right panel: in a medium, one of the scattered quarks loses energy inside the medium and gets suppressed.

characteristic collimated hadrons from fragmentation of outgoing partons are called jets (see figure 9.15). Naturally, the most common structure seen is the two-jet event. Three-jet events are also seen and result from reactions such as $q\bar{q} \rightarrow q\bar{q}g$. From theoretical considerations, Xin-Nian Wang and Miklos Gyulassy predicted the jet quenching phenomenon [65]. They argued that a partonic jet, if traveling through a medium, will lose its energy by gluon emission. A schematic representation is given in figure 9.15. The energy degraded parton will ultimately fragment into a lower number of particles than it would have in the absence of the medium.

After the prediction of the jet quenching phenomenon, both the PHENIX and STAR collaborations discovered the phenomenon in Au+Au collisions at the RHIC

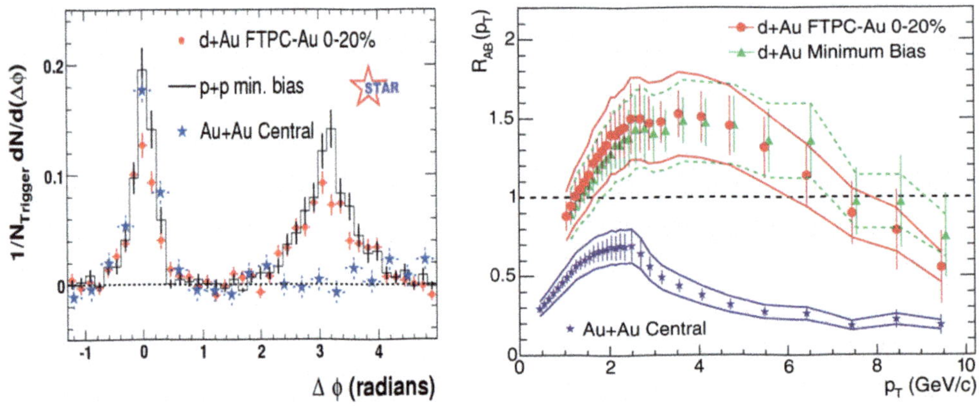

Figure 9.16. Left panel: STAR experimental results on two-particle correlation in the azimuthal plane, in $\sqrt{s}_{\rm NN} = 200$ GeV, p+p, d+Au and Au+Au collisions. Note the two-peak structure in p+p and d+Au collisions. In Au+Au collisions, however, the away side peak has vanished. The figure is reproduced with permission from [69]. Right panel: STAR measurements for the nuclear modification factor in central Au+Au and minimum bias d+Au collisions at the RHIC energy $\sqrt{s}_{\rm NN} = 200$ GeV. The figure is reproduced from [69] with permission.

[66–69]. In the left panel of figure 9.16, di-hadron correlation in the azimuthal plane, in p+p, d+Au and Au+Au collisions, is shown. The data are obtained in the following manner. A high p_T trigger particle is fixed and in coincidence with the trigger particle, associated particles are measured as a function of the azimuthal angle. $\Delta_\phi = \phi_{\rm associate} - \phi_{\rm trigger}$ is the difference of the azimuthal angle between the trigger and associated particles. In p+p and d+Au collisions, di-hadron correlations show a double peak structure, which can be understood in terms of two-jet events. The peak at $\Delta\phi = 0$ is called the near side peak (nearer to the trigger particle) and the peak at $\Delta\phi = \pi$ is called the away side peak (away from the trigger particle). In Au+Au collisions, however, the away side peak is strongly suppressed. Strong suppression of the away side peak is the experimental evidence of jet quenching. The understanding is as follows (see figure 9.15): a di-jet is produced near the surface of the medium. One of the jets escapes into the vacuum and fragments. The other enters the medium and loses its energy in the medium before fragmentation.

Another manifestation of jet quenching is the high p_T suppression, i.e. smaller number of high $p_{\rm T}$ particles in A+A collisions than in p+p collisions, scaled by the collision number. High $p_{\rm T}$ suppression is usually expressed in terms of the nuclear modification factor (R_{AA}),

$$R_{AA} = \frac{{\rm d}N^{AA}/{\rm d}y{\rm d}^2p_{\rm T}}{T_{AA}{\rm d}\sigma^{pp}/{\rm d}y{\rm d}^2p_{\rm T}}, \tag{9.56}$$

where T_{AA} in equation (9.56) is the nuclear thickness function, calculable in a Glauber model. If AA collision is a superposition of pp collisions, the expected ratio is unity. In the right panel of figure 9.16 STAR measurements for R_{AA}, for charged hadrons, in $\sqrt{s}_{NN} = 200$ GeV central Au+Au collisions and in minimum bias d+Au collisions are shown. In Au+Au collisions, hadrons are largely suppressed at high p_T. In d+Au collisions, on the other hand, hadrons are not suppressed. Rather, in the intermediate p_T range, they are enhanced. Enhancement in d+Au collision is understood as the 'Cronin effect' [70]. The Cronin effect was discovered in proton–nucleus collisions in the 1970s [70]. It was observed that in proton–nucleus collisions, at intermediate p_T range, the transverse momentum spectrum is hardened compared to proton–proton collisions. The hardening disappears at very large p_T. A corresponding depletion was seen at low transverse momenta, accompanied by a softening of the spectrum. The effect can be understood in terms of multiple scatterings of partons from the proton off partons from the nucleus. As a result of such scatterings, the partons acquire a transverse momentum kick, shifting their momenta from lower to higher values, causing the observed enhancement at high p_T and depletion at low p_T. At very high p_T, due to a higher twist effect, multiple scatterings are suppressed by powers of p_T. While charged particles do show high p_T suppression in Au+Au collisions, photons are not suppressed. This provides strong evidence that high p_T suppression is not an initial state effect, but a final state effect. A high density colored medium is created in the collision and causes the suppression.

Qualitative and quantitative understanding of jet quenching can be obtained in QCD. In leading order pQCD, the production cross section for the hadron h in $A + B \to h + X$ reactions can be written as

$$E\frac{\mathrm{d}\sigma}{\mathrm{d}^3p_h} = K \sum_{ab\to cd} \int \mathrm{d}x_a \int \mathrm{d}x_b f_{a/A}\left(x_a, Q^2\right) f_{b/B}\left(x_b, Q^2\right) \times \delta(s + t + u)\frac{1}{\pi z_c}\frac{\mathrm{d}\sigma}{\mathrm{d}t}(ab \to cd) D_{h/c}\left(z_c, \mu^2\right). \tag{9.57}$$

In equation (9.57), $f_{a/A}(x,Q^2)$ is the structure function of the parton a in the hadron A, $f_{b/B}(x, Q^2)$ is the structure function of the parton b in the hadron B. The Q^2 is the factorization scale. $\frac{\mathrm{d}\sigma}{\mathrm{d}t}$ is the partonic cross section for the reaction $a + b \to c + d$. $D_{h/c}\left(z_c, \mu^2\right)$ is the fragmentation function for the parton c into hadron h, μ^2 being the fragmentation scale. $z_c = \frac{E_h}{E_c}$ is the fraction of the partonic energy carried by the hadron h. K in equation (9.57) takes into account higher order effects. They can be partially included by generalizing the parton distribution to account for the intrinsic k_T smearing,

$$\tilde{f}_{a/A}\left(x_a, k_T, Q^2\right) \approx f_{a/A}\left(x_a, Q^2\right) g\left(k_T\right) \qquad g\left(k_T\right) = \frac{e^{-k_T^2/\langle k_T\rangle^2}}{\pi\langle k_T\rangle^2} \tag{9.58}$$

where the width $\langle k_T^2\rangle$ of the Gaussian is related to initial state vacuum radiation. Parton energy loss in the medium effectively modifies the 'vacuum' fragmentation function. In the first approximation, this effect can be taken into account by replacing the vacuum fragmentation functions by effective 'in medium' fragmentation functions,

$$z_c D'_{h/c}\left(z_c, \mu^2\right) = z'_c D_{h/c}\left(z'_c, \mu^2\right) + z_g D_{h/g}\left(z_g, \mu^2\right) \tag{9.59}$$

$$z'_c = \frac{p_h}{p_c - \Delta E_c\left(p_c, \phi\right)} \qquad z_g = \frac{p_h}{\Delta E_c\left(p_c, \phi\right)}. \tag{9.60}$$

The first term is the fragmentation function of the parton c after losing energy ΔE (p_c, ϕ) due to medium induced gluon radiation. The second term is the feedback due to the fragmentation from the radiated gluons. The fragmentation function $D_{h/c}(z)$ is a rapidly falling function of z and increase in z will lead to reduced production for the hadron h. Parton energy loss ΔE is important in quantifying the high p_T suppression. ΔE can be decomposed into collisional and radiative contributions. While the former originate from the energy transfer from the moving parton to the medium particles, the latter are caused by radiation from the moving particle.

9.5.1 Collisional energy loss

Consider the propagation of a test parton Q through a partonic medium. It suffers energy loss in elastic collisions with the constituents (quarks and gluons) of the medium. The energy loss per unit length can be obtained by integrating the differential cross section, weighted by the energy loss $\omega = E_{\text{in}} - E_{\text{out}}$,

$$-\frac{\mathrm{d}E}{\mathrm{d}z} = \sum_{p=q,g} \int \mathrm{d}^3k \rho_p(k) \int \mathrm{d}t J\omega \frac{\mathrm{d}\sigma^{Qp\to Qp}}{\mathrm{d}t} \tag{9.61}$$

where ρ_q and ρ_g are the quark and gluon density of the medium, J is the flux factor and t is the usual Mandelstam variable. For $t \ll s$,

$$\frac{\mathrm{d}\sigma^{Qp\to Qp}}{\mathrm{d}t} \approx C_p \frac{2\pi\alpha_s^2}{t^2} \tag{9.62}$$

with $C_q = \frac{N_c^2 - 1}{2N_c^2}$, $C_g = 1$. In the limit of massless quarks, several approximations can be made. The cross section is singular and Bjorken [71] first evaluated collisional energy loss with two reasonable upper and lower limits,

$$-\frac{\mathrm{d}E}{\mathrm{d}z} = \frac{8}{3}\pi\alpha_s^2 T^2\left(1 + \frac{N_f}{6}\right)\log\frac{q_{\max}}{q_{\min}}. \tag{9.63}$$

Bjorken's estimates for the upper and lower limits were, $q_{\max} = \sqrt{4TE}$ and $q_{\min} = \Lambda = 0.5 - 1.0\,\text{GeV}$. The arbitrary lower limit can be eliminated if the plasma screening effect is accounted for [72]. The first field theoretic treatment of collisional energy loss was given by Braaten and Thoma [73, 74]. For a quark of mass M, average energy loss per interaction was defined as

$$\Delta E = \frac{1}{\Gamma}\int_M^\infty \mathrm{d}E'(E - E')\frac{\mathrm{d}\Gamma(E, E')}{\mathrm{d}E'} \tag{9.64}$$

where $\frac{\mathrm{d}\Gamma(E,E')}{\mathrm{d}E'}$ is the differential interaction rate. Average distance between interactions being $\Delta z = \frac{v}{\Gamma}$, energy loss per unit length can be obtained as

$$\frac{\Delta E}{\mathrm{d}z} = \frac{1}{v}\int_M^\infty \mathrm{d}E'(E - E')\frac{\mathrm{d}\Gamma(E, E')}{\mathrm{d}E'}. \tag{9.65}$$

The interaction rate Γ is related to self-energy. Introducing a separation scale q^* between soft and hard scattering, they calculated collisional energy loss for heavy quarks. For $q < q^*$ screening is included by using the effective perturbation theory for high temperature QCD (hard thermal loop) developed by Braaten and Pisarski [48]. Tree level diagrams are sufficient for $q > q^*$. The method was used by Thoma [75] to calculate energy loss for light quarks. When the two contributions are added, the arbitrary cut off q^* drops out, leaving only the undetermined upper limit of momentum transfer $q_{\max} = \tilde{q}$. The result is [75]

$$-\frac{\mathrm{d}E}{\mathrm{d}x} = \frac{8}{3}\pi\alpha_s^2\, T^2\left(1 + \frac{N_f}{6}\right)\log\left(2^{N_f/2(6+N_f)}0.920\frac{\sqrt{\tilde{q}T}}{m_g}\right) \tag{9.66}$$

where $\tilde{q}$ is the upper cut off for the momentum transfer. m_g is the effective gluon mass defined as $m_g = \mu/3$, with Debye screening mass $\mu = \sqrt{1 + N_f/6}\,gT$.

9.5.2 Radiative energy loss

Gluon radiation is the dominating mechanism for parton energy loss in a dense medium. A hard parton passing through the medium interacts with various scattering centers and splits into an outgoing parton and a gluon (figure 9.17).

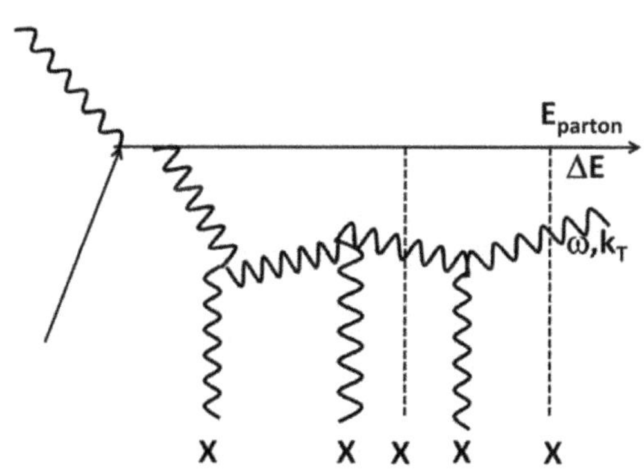

Figure 9.17. A typical diagram showing gluon radiation from a parton passing through a medium. X represents the scattering centers.

Discussions on energy loss in the medium are generally characterized by the following variables: (i) the mean free path $\lambda = 1/(\rho\sigma)$, ρ the medium density and σ the particle medium cross section, (ii) the opacity $N = L/\lambda$ or the number of scattering centers in a medium of thickness L, (iii) the Debye mass $m_D(T) \sim gT$, g the coupling parameter, (iv) the transport coefficient $\hat{q} = m_D^2/\lambda$, which controls the radiative energy loss and (v) the diffusion constant $D = \mu_{\text{mob}}T$ characterizing the dynamics of heavy, non-relativistic particles, μ_{mob} being the mobility of the particles, i.e. the ratio of drift velocity and applied force.

One generally calculates the radiation spectrum $\omega\frac{\mathrm{d}I}{\mathrm{d}\omega}$ or $\omega\frac{\mathrm{d}I}{\mathrm{d}\omega \mathrm{d}^2k_T}$ and obtains the energy loss as

$$\Delta E = \int^{E}\int^{k_{T\max}} \omega\frac{\mathrm{d}I}{\mathrm{d}\omega \mathrm{d}^2k_T}\mathrm{d}^2k_T\mathrm{d}\omega \qquad \Delta E = \int^{E}\omega\frac{\mathrm{d}I}{\mathrm{d}\omega}\mathrm{d}\omega. \tag{9.67}$$

The calculation procedure for radiative energy loss is rather complex. Several methods have been proposed, e.g. the BDMPS [76, 77], GLV [78, 79], Arnold, Moore and Yaffe [83–85] and higher twist [86, 87] methods. In the following, some qualitative results for parton energy loss will be discussed. More detailed information may be found in [88–92].

Energy loss of a charged particle in a medium is a well-studied process in QED. If λ is the mean free path of the charged particle, depending upon the thickness of the medium (L), one can distinguish two regimes: (i) the Bethe–Heitler regime ($L \ll \lambda$), when scattering centers act independently, and (ii) the LPM regime ($L \gg \lambda$), when multiple scattering centers act coherently and radiation is suppressed compared to the Bethe–Heitler regime.

Discussions on radiative energy loss are facilitated by considering the formation time or length. The formation time/length can be interpreted as the length at which a photon/gluon of energy ω emitted at angle θ acquires a phase of order 1 in the frame moving with the particle. It can be obtained as

$$1 = \omega t - k_{\parallel}l = l\left(\omega - \sqrt{\omega^2 - \omega^2\theta^2}\right) \approx l\omega\theta^2 \tag{9.68}$$

giving

$$l_{\text{form}} \approx \frac{1}{\omega\theta^2} = \frac{\omega}{k_T^2} \qquad k_T = \omega\theta. \tag{9.69}$$

The important point is that even if the particle scatters several times before traveling the distance l_{form}, it can emit only one photon/gluon of energy ω with transverse momentum k_T. It is then possible to distinguish between three regimes.

(i) *The Bethe–Heitler regime.* For $l_{\text{form}} \ll \lambda$, each scattering center acts as an independent source of radiation. The induced gluon spectrum is thus proportional to the (Bethe–Heitler) gluon spectrum induced by a single scattering. In this regime, radiation spectrum per unit length, in the limit $E \to \infty$ (neglecting logarithmic factors), behaves as

$$\omega \frac{\mathrm{d}I}{\mathrm{d}\omega \mathrm{d}z}\bigg|_{\mathrm{BH}} \approx \frac{\alpha_\mathrm{s}}{\pi} N_\mathrm{c} \frac{1}{\lambda}. \tag{9.70}$$

The energy loss per unit distance $\frac{\mathrm{d}E}{\mathrm{d}z}$ can be obtained by integrating the gluon spectrum over $0 \leqslant \omega \leqslant E$. In the Bethe–Heitler regime,

$$-\frac{\mathrm{d}E}{\mathrm{d}z}\bigg|_{\mathrm{BH}} \approx \frac{\alpha_\mathrm{s}}{\pi} N_\mathrm{c} \frac{1}{\lambda} E. \tag{9.71}$$

(ii) *The LPM regime*. For $\lambda \ll l_{\mathrm{form}} \ll L$ a group of scattering centers acts as a single scattering source. Let l_{coh} be the coherence length associated with the formation time of a gluon radiated by a group of scattering centers acting coherently to produce the radiation,

$$l_{\mathrm{coh}} \approx \frac{\omega}{\left\langle k_T^2 \right\rangle_{l_{\mathrm{coh}}}} \tag{9.72}$$

with

$$\langle k_T^2 \rangle_{l_{\mathrm{coh}}} \approx \frac{l_{\mathrm{coh}}}{\lambda} \mu^2 = N_{\mathrm{coh}} \mu^2 \tag{9.73}$$

with μ the typical momentum transfer in a single scattering and N_{coh} the number of coherent scatterers. One can estimate

$$l_{\mathrm{coh}} \approx \sqrt{\frac{\lambda}{\mu^2}\omega}; \; N_{\mathrm{coh}} \approx \sqrt{\frac{\omega}{\lambda \mu^2}} = \sqrt{\frac{\omega}{E_{\mathrm{LPM}}}} \tag{9.74}$$

where the LPM energy scale $E_{\mathrm{LPM}} = \lambda \mu^2$ is introduced. Radiation spectrum per unit length can be obtained as

$$\omega \frac{\mathrm{d}I}{\mathrm{d}\omega \mathrm{d}z}\bigg|_{\mathrm{LPM}} \approx \frac{1}{N_{\mathrm{coh}}} \omega \frac{\mathrm{d}I}{\mathrm{d}\omega \mathrm{d}z}\bigg|_{\mathrm{BH}} = \frac{\alpha_\mathrm{s}}{\pi} N_\mathrm{c} \frac{1}{\lambda} \sqrt{\frac{E_{\mathrm{LPM}}}{\omega}} = \frac{\alpha_\mathrm{s}}{\pi} N_\mathrm{c} \sqrt{\frac{\mu^2}{\lambda} \frac{1}{\omega}}. \tag{9.75}$$

Compared to the Bethe–Heitler regime, radiation is reduced by the factor $\sqrt{E_{\mathrm{LPM}}/\omega}$ due to the coherence effect. The energy loss per unit length is easily obtained,

$$-\frac{\mathrm{d}E}{\mathrm{d}z}\bigg|_{\mathrm{LPM}} = \frac{\alpha_\mathrm{s}}{\pi} N_\mathrm{c} \sqrt{\frac{\mu^2 E}{\lambda}} = \frac{\alpha_\mathrm{s}}{\pi} \frac{N_\mathrm{c}}{\lambda} \sqrt{E_{\mathrm{LPM}} E}. \tag{9.76}$$

(iii) *The factorization regime.* For $l_{\text{coh}} \gg L$, when $\omega > \omega_{\text{fact}} = L^2\mu^2/\lambda$, all the scattering centers in the medium act coherently as a source of radiation. N_{coh} can be estimated as L/λ and the radiation spectrum as

$$\omega \frac{\mathrm{d}I}{\mathrm{d}\omega \mathrm{d}z}\bigg|_{\text{fact}} \approx \frac{1}{N_{\text{coh}}} \omega \frac{\mathrm{d}I}{\mathrm{d}\omega \mathrm{d}z}\bigg|_{\text{BH}} = \frac{\alpha_s}{\pi} N_c \frac{1}{L}. \tag{9.77}$$

Energy loss in this regime can be estimated as

$$-\frac{\mathrm{d}E}{\mathrm{d}z}\bigg|_{\text{fact}} \approx \frac{\alpha_s}{\pi} N_c \sqrt{\frac{\mu^2}{\lambda}\omega_{\text{fact}}} = \frac{\alpha_s}{\pi} N_c \frac{\langle k_T^2 \rangle}{\lambda} L. \tag{9.78}$$

Bibliography

[1] Matsui T and Satz H 1986 *Phys. Lett.* B **178** 416
[2] Brambilla N *et al* (Quarkonium Working Group Collaboration) 2004 arXiv: hep-ph/0412158
[3] Brambilla N *et al* 2011 *Eur. Phys. J.* C **71** 1534
[4] Einhorn M B and Ellis S D 1975 *Phys. Rev.* D **12** 2007
[5] Ellis S D, Einhorn M B and Quigg C 1976 *Phys. Rev. Lett.* **36** 1263
[6] Carlson C E and Suaya R 1976 *Phys. Rev.* D **14** 3115
[7] Artoisenet P, Lansberg J P and Maltoni F 2007 *Phys. Lett.* B **653** 60
[8] Campbell J M, Maltoni F and Tramontano F 2007 *Phys. Rev. Lett.* **98** 252002
[9] Fritzsch H 1977 *Phys. Lett.* B **67** 217
[10] Halzen F 1977 *Phys. Lett.* B **69** 105
[11] Gluck M, Owens J F and Reya E 1978 *Phys. Rev.* D **17** 2324
[12] Barger V D, Keung W-Y and Phillips R J N 1980 *Phys. Lett.* B **91** 253
[13] Amundson J F, Eboli O J P, Gregores E M and Halzen F 1997 *Phys. Lett.* B **390** 323
[14] Bodwin G T, Braaten E and Lepage G P 1995 *Phys. Rev.* D **51** 1125
Bodwin G T, Braaten E and Lepage G P 1997 *Phys. Rev.* D **55** 5853
[15] Bodwin G T 2012 arXiv: 1208.5506
[16] Nayak G C, Qiu J-W and Sterman G F 2005 *Phys. Lett.* B **613** 45
[17] Nayak G C, Qiu J-W and Sterman G F 2005 *Phys. Rev.* D **72** 114012
[18] Kang Z-B, Qiu J-W and Sterman G 2011 *Nucl. Phys.* B **214** 39
[19] Vogt R 1999 *Phys. Rep.* **310** 197
[20] Gerschel C and Huefner J 1999 *Annu. Rev. Nucl. Part. Sci.* **49** 255
[21] Abreu M C *et al* (NA50 Collaboration) 2000 *Phys. Lett.* B **477** 28
[22] Adler S S *et al* (PHENIX Collaboration) 2004 *Phys. Rev. Lett.* **92** 051802
[23] Adler S S *et al* (PHENIX Collaboration) 2006 *Phys. Rev. Lett.* **96** 012304
[24] Abelev B *et al* (ALICE Collaboration) 2012 *Phys. Rev. Lett.* **109** 072301
[25] Adare A *et al* (PHENIX Collaboration) 2007 *Phys. Rev. Lett.* **98** 232301
[26] Adare A *et al* (PHENIX Collaboration) 2008 arXiv: 0801.0220
[27] Thews R L, Schroedter M and Rafelski J 2001 *Phys. Rev.* C **63** 054905
[28] Braun-Munzinger P and Stachel J 2000 *Phys. Lett.* B **490** 196
[29] Rafelski J and Mueller B 1982 *Phys. Rev. Lett.* **82** 1066
[30] Koch P, Muller B and Rafelski J 1986 *Phys. Rep.* **142** 167
[31] Koch P and Rafelski J 1985 *Nucl. Phys.* A **444** 678

[32] Srivastava D K, Mustafa M G and Muller B 1997 *Phys. Rev.* C **56** 1064
[33] Wroblewski A 1985 *Acta Phys. Polon.* B **16** 379
[34] Becattini F, Manninen J and Gazdzicki M 2006 *Phys. Rev.* C **73** 044905
[35] Chen J H (for the STAR Collaboration) 2008 *J. Phys. G: Nucl. Part. Phys.* **35** 104053
[36] Agakishiev G *et al* (STAR Collaboration) 2012 *Phys. Rev. Lett.* **108** 072301
[37] Abelev *et al* (STAR Collaboration) 2007 *Phys. Rev. Lett.* **99** 112301
[38] Hamieh S, Redlich K and Tounsi A 2000 *Phys. Lett.* B **486** 61
[39] Tounsi A and Redlich K 2001 arXiv: hep-ph/0111159
[40] Tounsi A, Mischke A and Redlich K 2003 *Nucl. Phys.* A **715** 565
[41] Bass S A, Muller B and Srivastava D K 2002 *Phys. Rev.* C **66** 061902
Bass S A, Muller B and Srivastava D K 2003 *Phys. Rev. Lett.* **90** 082301
Bass S A, Muller B and Srivastava D K 2004 *Phys. Rev. Lett.* **93** 162301
Renk T, Bass S A and Srivastava D K 2006 *Phys. Lett.* B **632** 632
[42] Chatterjee R, Bhattacharya L and Srivastava D K 2009 arXiv: 0901.3610
[43] Alam J, Sinha B and Raha S 1996 *Phys. Rep.* **273** 243
[44] Alam J, Sarkar S, Roy P, Hatsuda T and Sinha B 2001 *Ann. Phys.* **286** 159
[45] Kapusta J, Lichard P and Seibert D 1991 *Phys. Rev.* D **44** 2774
Kapusta J, Lichard P and Seibert D 1993 *Phys. Rev.* D **47** 4171 (erratum)
[46] Nadeau H, Kapusta J and Lichard P 1992 *Phys. Rev.* C **45** 3034
[47] Baier R, Nakkagawa H, Niegawa A and Redlich K 1992 *Z. Phys.* C **53** 433
[48] Braaten E and Pisarski R D 1990 *Nucl. Phys.* B **337** 569
[49] Aurenche P, Gelis F, Kobes R and Zaraket H 1998 *Phys. Rev.* D **58** 085003
[50] Aurenche P, Gelis F and Zaraket H 2000 *Phys. Rev.* D **61** 116001
[51] Aurenche P, Gelis F and Zaraket H 2000 *Phys. Rev.* D **62** 096012
[52] Arnold P, Moore G D and Yaffe L G 2001 *JHEP* **0112** 009
[53] Xiong L, Shuryak E V and Brown G E 1992 *Phys. Rev.* D **46** 3798
[54] Turbide S, Rapp R and Gale C 2005 *Phys. Rev.* C **69** 014903
[55] Adler S S *et al* (PHENIX Collaboration) 2007 *Phys. Rev. Lett.* **98** 012002
[56] Adare A *et al* (PHENIX Collaboration) 2010 *Phys. Rev. Lett.* **104** 132301
[57] Adler S S *et al* (PHENIX Collaboration) 2005 *Phys. Rev. Lett.* **94** 232301
[58] Srivastava D K 2001 *Pramana* **57** 235
Srivastava D K and Sinha B 2001 *Phys. Rev.* C **64** 034902
[59] Alam J E, Sarkar S, Hatsuda T, Nayak T K and Sinha B 2001 *Phys. Rev.* C **63** 021901
[60] Huovinen P, Ruuskanen P V and Rasanen S S 2002 *Phys. Lett.* B **535** 109
Rasanen S S 2003 *Nucl. Phys.* A **715** 717
[61] Turbide S, Rapp R and Gale C 2004 *Phys. Rev.* C **69** 014903
[62] d'Enterria D G and Peressounko D 2006 *Eur. Phys. J.* C **46** 451
[63] Kajantie K, Kapusta J, McLerran L and Mekijan A 1986 *Phys. Rev.* D **34** 2746
[64] Kajantie K, Kataja M, McLerran L and Ruuskanen P V 1986 *Phys. Rev.* D **34** 881
[65] Wang X-N and Gyulassy M 1992 *Phys. Rev. Lett.* **68** 1480
[66] Adler C *et al* (STAR Collaboration) 2002 *Phys. Rev. Lett.* **89** 202301
[67] Adcox K *et al* (PHENIX Collaboration) 2002 *Phys. Rev. Lett.* **88** 022301
[68] Adler C *et al* (STAR Collaboration) 2003 *Phys. Rev. Lett.* **90** 082302
[69] Adams J *et al* (STAR Collaboration) 2003 *Phys. Rev. Lett.* **91** 072304
[70] Cronin J W, Frisch H J, Shochet M J, Boymond J P, Piroue P A and Sumner R L 1973 *Phys. Rev. Lett.* **31** 1426
[71] Bjorken J D 1982 *Fermilab publication* Pub-82/59-THY

[72] Thoma M H and Gyulassy M 1991 *Nucl. Phys.* B **351** 491

[73] Braaten E and Thoma M H 1991 *Phys. Rev.* D **44** 1298

[74] Braaten E and Thoma M H 1991 *Phys. Rev.* D **44** 2625

[75] Thoma M H 1991 *Lett. Phys.* B **273** 128

[76] Baier R, Dokshitzer Y L, Mueller A H, Peigne S and Schiff D 1997 *Nucl. Phys.* B **484** 265

[77] Baier R, Dokshitzer Y L, Mueller A H, Peigne S and Schiff D 1997 *Nucl. Phys.* B **483** 291

[78] Gyulassy M, Levai P and Vitev I 2001 *Nucl. Phys.* B **594** 371

[79] Gyulassy M, Levai P and Vitev I 2000 *Phys. Rev. Lett.* **85** 5535

[80] Vitev I and Gyulassy M 2002 *Phys. Rev. Lett.* **89** 252301

[81] Wang X-N 2004 *Phys. Rev.* C **70** 031901

[82] Alam J, Roy P and Dutt-Mazumder A K 2006 arXiv: hep-ph/0604131

[83] Arnold P B, Moore G D and Yaffe L G 2002 *JHEP* **0206** 030

[84] Jeon S and Moore G D 2005 *Phys. Rev.* C **71** 034901

[85] Turbide S, Gale C, Jeon S and Moore G D 2005 *Phys. Rev.* C **72** 014906

[86] Guo X-f and Wang X-N 2000 *Phys. Rev. Lett.* **85** 3591

[87] Wang X-N and Guo X-f 2001 *Nucl. Phys.* A **696** 788

[88] Baier R, Schiff D and Zakharov B G 2000 *Annu. Rev. Nucl. Part. Sci.* **50** 37

[89] Gyulassy M, Vitev I, Wang X-N and Zhang B-W 2004 Jet quenching and radiative energy loss in clense nuclear matter *Quark–gluon Plasma* vol 3, ed R C Hwa and X H N Wang (Singapore: World Scientific) pp 123–91

[90] Peigne S and Smilga A V 2009 *Phys. Usp.* **52** 659
Peigne S and Smilga A V 2009 *Usp. Fiz. Nauk* **179** 697

[91] d'Enterria D 2009 arXiv: 0902.2011

[92] d'Enterria D and Betz B 2010 In search of the QCD–gravity correspondence *Lecture Notes in Physics* vol 785, ed S Sarkar, H Satz and B Sinha (Berlin: Springer) pp 285–339

IOP Publishing

A Short Course on Relativistic Heavy Ion Collisions

A K Chaudhuri

Appendix A

A.1 Natural units

Throughout this book, we have used natural units. In relativistic quantum mechanics, algebraic equations are considerably simplified by natural units. In natural units, universal constants,

$$\text{speed of light in vacuum: } c = 2.997\,924\,58 \times 10^{8}\ \text{m s}^{-1}, \tag{A.1a}$$

$$\text{reduced Planck constant: } \hbar = \frac{h}{2\pi} = 1.054\,571\,596 \times 10^{-34}\ \text{Js}, \tag{A.1b}$$

$$\text{gravitational constant: } G = 6.673 \times 10^{-11}\ \text{m}^3\text{kg}^{-1}\,\text{s}^{-2}, \tag{A.1c}$$

$$\text{Boltzmann constant: } k = 1.380\,650\,3 \times 10^{-23}\ \text{JK}^{-1}, \tag{A.1d}$$

$$= 8.617\,352 \times 10^{-5}\ \text{eVK}^{-1} \tag{A.1e}$$

are set to unity,

$$c = \hbar = k = G = 1. \tag{A.2}$$

All the observables can then have only one dimension, say that of energy.

$$[Energy] = [Mass] = [Temperature] = [Length]^{-1} = [Time]^{-1}$$

In relativistic quantum mechanics, equations generally contain a multitude of $\hbar$ and c. Algebraic equations are considerably simplified in natural units. The equations can be restored to normal units by inserting appropriate powers of the fundamental constants. As an example, consider the energy–momentum relation in natural units,

$$E^2 = \mathbf{p}^2 + M^2. \tag{A.3}$$

doi:10.1088/978-0-750-31060-4ch10

It should be calculated as,

$$E^2 = \mathbf{p}c^2 + M^2c^4. \tag{A.4}$$

It is useful to remember the conversion factors,

$$\hbar c = 197.33 \text{ MeV·fm} \tag{A.5}$$

$$1\text{K} = 0.0862 \text{ MeV} \tag{A.6}$$

A.2 Planck units

In 1906 Max Planck introduced three units for mass, length and time, now famous as Planck units. The important point is that they are constructed with the fundamental constants, speed of light in vacuum (c), reduced Planck constant ($\hbar$) and Newton's gravitational constant (G). The three fundamental constants are related to three very basic aspects of our Universe: $\hbar$ signifies the quantum aspect, c the relativity and G the gravitational phenomena. Planck units thus embody the three very basic aspects of the Universe.

$$\text{Planck Length: } L_p = \sqrt{\frac{\hbar G}{c^3}} = 1.6 \times 10^{-35} \text{ cm} \tag{A.7}$$

$$\text{Planck Mass: } M_p = \sqrt{\frac{\hbar c}{G}} = 1.3 \times 10^{19} \text{ GeV} \tag{A.8}$$

$$\text{Planck Time: } T_p = \sqrt{\frac{\hbar G}{c^5}} = 5.4 \times 10^{-44} \text{ s} \tag{A.9}$$

The explicit form of the Planck length, mass, or time can be obtained from dimensional analysis. For example, for the Planck length, we can write

$$L_p = c^{n_1}G^{n_2}\hbar^{n_3} \tag{A.10}$$

$$= \left[LT^{-1}\right]^{n_1}\left[M^{-1}L^3\,T^{-2}\right]^{n_2}\left[M^1L^2\,T^{-1}\right]^{n_3} \tag{A.11}$$

$$= L^{n_1+3n_2+2n_3}T^{-n_1-2n_2-n_3}M^{-n_2+n_3}. \tag{A.12}$$

For L_p to have the dimension of length, it is then required that

$$n_1 + 3n_2 + 2n_3 = 1 \tag{A.13}$$

$$-n_1 - 2n_2 - n_3 = 0 \tag{A.14}$$

$$-n_2 + n_3 = 0. \tag{A.15}$$

One easily obtains for the Planck length, $L_p = \sqrt{\frac{\hbar G}{c^3}}$.

One can also write down the expression for Planck temperature.

Planck temperature: $\frac{M_p c^2}{K} = 1.42 \times 10^{32}$ K.

A.3 Wick's theorem

In quantum mechanics, one essentially calculates the vacuum expectation value of an operator $\hat{A}$. The expectation value of any operator can be written as the product of creation and annihilation operators. Wick's theorem gives a simple prescription for the evaluation of a product of any number of creation and annihilation operators. Proof of Wick's theorem can be found in any standard textbook on quantum electrodynamics. Here, we simply state the theorem. Wick's theorem can be stated as follows.

(i) The vacuum expectation value of the product of any number of creation and annihilation operators is equal to the sum of the products of all possible expectation values of these operators taken in pairs (contraction). In each pair, the factors must be placed in the same order as in the original product, which may require interchanging of the operators. For fermions, each interchange of operators will bring one negative sign.

(ii) The vacuum expectation value of $\hat{a}^\dagger\hat{a}^\dagger$, $\hat{a}\hat{a}$ is zero.

If $\hat{A}_i$ is either a creation ($a^\dagger$) or annihilation (a) operator for bosons, Wick's theorem can be stated more formally as

$$\langle \hat{A}_1\hat{A}_2\ldots\hat{A}_{m-1}\hat{A}_m\rangle = \sum_P \langle \hat{A}_1\hat{A}_2\rangle\langle \hat{A}_3\hat{A}_4\rangle\ldots\langle \hat{A}_{m-1}\hat{A}_m\rangle \tag{A.16}$$

where the sum runs over all distinct permutations of m-indices. The above equations can also be written as

$$\langle \hat{A}_1\hat{A}_2\ldots\hat{A}_{m-1}\hat{A}_m\rangle = \langle \hat{A}_1\hat{A}_2\rangle\langle \hat{A}_3\hat{A}_4\ldots\hat{A}_m\rangle + \langle \hat{A}_1\hat{A}_3\rangle\langle \hat{A}_2\hat{A}_4\ldots\hat{A}_m\rangle$$
$$+ \ldots.\ \langle \hat{A}_1\hat{A}_m\rangle\langle \hat{A}_2\hat{A}_3\ldots\hat{A}_{m-1}\rangle, \tag{A.17}$$

and then applying this relation recursively to all of the multiple operator averages until only pairs of operators remain. If the number of operators is odd, Wick's theorem gives the simple result,

$$\langle \hat{A}_1\hat{A}_2\ldots\hat{A}_{2m}\hat{A}_{2m+1}\rangle = 0. \tag{A.18}$$

Wick's theorem is augmented by the commutation relations for creation and annihilation operators, which for bosonic systems are

$$\left[a_i^\dagger, a_j^\dagger\right] = [a_i, a_j] = 0 \tag{A.19a}$$

$$\left[a_i, a_j^\dagger\right] = \delta_{ij}. \tag{A.19b}$$

For fermions, the commutators should be changed to anticommutators[1],

$$\left\{a_i^\dagger, a_j^\dagger\right\} = \{a_i, a_j\} = 0 \tag{A.20a}$$

$$\left\{a_i, a_j^\dagger\right\} = \delta_{ij}. \tag{A.20b}$$

For illustrative purposes, we evaluate the product of four-operators. For bosons, it can be evaluated as

$$\langle \hat{A}_1\hat{A}_2\hat{A}_3\hat{A}_4\rangle = \langle \hat{A}_1\hat{A}_2\rangle\langle \hat{A}_3\hat{A}_4\rangle + \langle \hat{A}_1\hat{A}_3\rangle\langle \hat{A}_2\hat{A}_4\rangle + \langle \hat{A}_1\hat{A}_4\rangle\langle \hat{A}_2\hat{A}_3\rangle \tag{A.21}$$

or for fermions, it can be evaluated as

$$\langle \hat{A}_1\hat{A}_2\hat{A}_3\hat{A}_4\rangle = \langle \hat{A}_1\hat{A}_2\rangle\langle \hat{A}_3\hat{A}_4\rangle - \langle \hat{A}_1\hat{A}_3\rangle\langle \hat{A}_2\hat{A}_4\rangle + \langle \hat{A}_1\hat{A}_4\rangle\langle \hat{A}_2\hat{A}_3\rangle. \tag{A.22}$$

Note that the sign of the second term is changed. For the second term, we have to interchange: 1234 → (−)1324. The third term requires two consecutive changes, 1234 → (−)1243 → (−)(−)1423.

Let us consider a specific example where,

$$\hat{A}_1 = a_{\mathbf{p}_1}^\dagger \qquad \hat{A}_2 = a_{\mathbf{p}_2}^\dagger \qquad \hat{A}_3 = a_{\mathbf{p}_2} \qquad \hat{A}_4 = a_{\mathbf{p}_1}. \tag{A.23}$$

The product, $\left\langle a_{\mathbf{p}_1}^\dagger a_{\mathbf{p}_2}^\dagger a_{\mathbf{p}_2} a_{\mathbf{p}_1}\right\rangle$ can be interpreted as destruction of a particle of momentum $\mathbf{p}_1$, destruction of another particle of momentum $\mathbf{p}_2$ followed by creation of a particle of momentum $\mathbf{p}_2$, followed by creation of another particle of momentum $\mathbf{p}_1$. Using Wick's theorem, the expectation value can be written as

$$\left\langle a_{p1}^\dagger a_{p2}^\dagger a_{p2} a_{p1}\right\rangle = \left\langle a_{p1}^\dagger a_{p2}^\dagger\right\rangle\left\langle a_{p2} a_{p1}\right\rangle \pm \left\langle a_{p1}^\dagger a_{p2}\right\rangle\left\langle a_{p2}^\dagger a_{p1}\right\rangle + \left\langle a_{p1}^\dagger a_{p1}\right\rangle\left\langle a_{p2}^\dagger a_{p2}\right\rangle \tag{A.24}$$

$$= \pm\left\langle a_{p1}^\dagger a_{p2}\right\rangle\left\langle a_{p2}^\dagger a_{p1}\right\rangle + \left\langle a_{p1}^\dagger a_{p1}\right\rangle\left\langle a_{p2}^\dagger a_{p2}\right\rangle. \tag{A.25}$$

The ± signs are, respectively, for bosonic and fermionic systems.

A.4 The density matrix

Consider a quantum system with Hamiltonian H. Let the eigenfunctions be ϕ_n,

$$H\left|\phi_n\right\rangle = \epsilon_n\left|\phi_n\right\rangle. \tag{A.26}$$

[1] The commutator of two operators A and B is defined as

$$[A, B] = AB - BA.$$

The anticommutator of two operators A and B is defined as

$$\{A, B\} = AB + BA.$$

One defines a pure state $|\psi\rangle$ as the linear superposition of the eigenstates,

$$|\psi\rangle = \sum_n c_n |\phi_n\rangle. \tag{A.27}$$

Pure states can be represented by a 'ray' in Hilbert[2] space. The expectation value of an operator A in the state ψ is evaluated as

$$\langle A\rangle = \langle\psi|A|\psi\rangle = \sum_{n,m} c_n c_m^* \langle\phi_m|A|\phi_n\rangle. \tag{A.28}$$

If we define a matrix, ρ, as

$$\rho_{nm} = c_n c_m^*, \tag{A.29}$$

the expectation value can be written as

$$\langle A\rangle = \sum_{n,m} \rho_{nm} A_{mn} = \mathrm{Tr}[\rho A] \tag{A.30}$$

where TrA is the summation of the diagonal elements of the matrix A. ρ is called the density matrix of the system in state ψ. It can be equivalently expressed as

$$\rho = |\psi\rangle\langle\psi|. \tag{A.31}$$

The definition of a density matrix can easily be extended to non-pure or mixed states. In a mixed state, the wave function is an incoherent superposition of a number of pure states $|\psi^{(k)}\rangle$ with a statistical weight factor $p^{(k)}$ such that $\sum_k p^{(k)} = 1$. In each pure state ψ_k, the operator A has an expectation value $\langle A\rangle_k = \langle\psi^{(k)}|A|\psi^{(k)}\rangle$. The expectation value in the mixed state is then obtained as

$$\langle A\rangle = \sum_k p_k \langle\psi^{(k)}|A|\psi^{(k)}\rangle = \sum_k p_k \sum_{m,n} c_n^{(k)} c_m^{*(k)} = \sum_k p_k \sum_{m,n} c_n^{(k)} c_m^{*(k)} \langle\phi_m|A|\phi_n\rangle$$
$$= \sum_{m,n} \langle\phi_m|A|\phi_n\rangle \sum_k p_k c_n^{(k)} c_m^{*(k)}. \tag{A.32}$$

If we define the density matrix ρ as

$$\rho = \sum_k p_k |\psi^{(k)}\rangle\langle\psi^{(k)}| \tag{A.33}$$

or equivalently,

$$\rho_{nm} = \sum_k p_k c_n^{(k)} c_m^{*(k)} \tag{A.34}$$

the expectation value in the mixed state is again evaluated as

$$\langle A\rangle = \sum_{m,n} A_{mn}\rho_{nm} = \mathrm{Tr}[A\rho]. \tag{A.35}$$

[2] Hilbert space is a generalization of Euclidean space in finite or infinite dimensions. It is defined as a real or complex vector space with a definition of inner product.

Evaluating expectation value is the same for pure or mixed states. The only difference is the way in which the elements of the density matrix ρ are obtained.

A density matrix has the following properties.

(i) The condition that $\langle A \rangle$ is real for every Hermitian operator A, requires ρ to be Hermitian,

$$\rho_{mn} = \rho_{nm}^{*}. \tag{A.36}$$

(ii) The condition that the unit operator $\mathcal{I}$ has the mean value 1, requires trace of ρ to be unity,

$$\langle \mathcal{I} \rangle = \mathrm{Tr}[\rho \mathcal{I}] = \mathrm{Tr}\rho = 1. \tag{A.37}$$

(iii) ρ is positive definite, i.e. every diagonal element of ρ in any matrix representation must be nonnegative,

$$\rho_{kk} \geqslant 0. \tag{A.38}$$

(iv) The Hermitian matrix ρ can be reduced to diagonal form by a unitarity transformation,

$$\rho_{kl}\delta_{kl} = \sum_{m,n} T_{km}\rho_{mn}T_{nl}^{-1}. \tag{A.39}$$

Denoting the diagonal density matrix by ρ_k, we have

$$\sum_k \rho_k^2 \leqslant \left(\sum_k \rho_k\right)^2 = (\mathrm{Tr}\rho)^2 = 1 \tag{A.40}$$

or

$$\mathrm{Tr}\rho^2 \leqslant 1. \tag{A.41}$$

$\mathrm{Tr}\rho^2 = 1$ only for pure states. For a pure state, all but one diagonal element of ρ_k is zero. The non-zero element has a value of 1. In a pure state, the density matrix is idempotent operator $\rho^2 = \rho$. For a mixed state, diagonal elements give the probability of occupying a quantum state $|k\rangle$. The off-diagonal elements ρ_{nm}, $n \neq m$ are called coherence. Since ρ is Hermitian and can be diagonalized, it is always possible to choose a basis where coherences are zero.

The time evolution of the density matrix can be easily obtained from the time dependent Schrödinger equation,

$$\frac{\partial|\psi\rangle}{\partial t} = -\frac{\mathrm{i}}{\hbar}H|\psi\rangle \qquad \frac{\partial\langle\psi|}{\partial t} = +\frac{\mathrm{i}}{\hbar}\langle\psi|H \tag{A.42}$$

$$\frac{\partial \rho}{\partial t} = \frac{\partial}{\partial t}|\psi\rangle\langle\psi| = \frac{\partial|\psi\rangle}{\partial t}\langle\psi| + |\psi\rangle\frac{\partial\langle\psi|}{\partial t}$$
$$= -\frac{\mathrm{i}}{\hbar}H|\psi\rangle\langle\psi| + \frac{\mathrm{i}}{\hbar}|\psi\rangle\langle\psi|H = -\frac{\mathrm{i}}{\hbar}[H\rho - \rho H]$$
$$= \frac{\mathrm{i}}{\hbar}[H, \rho]. \tag{A.43}$$

The above equation is called the von-Neumann equation.

Formally, the density matrix is an alternative representation of the state of a quantum system where wave function is not directly used to calculate the expectation values. Indeed, one prefers equation (A.35) to be the definition of the density matrix rather than equation (A.34). It represents a minimum set of input data which serves to calculate the mean value of any operator A for a system prepared according to given specifications. The information from which this set of data is derived is equivalent to a knowledge of the mean values of as many independent operators A_i as there are independent parameters in the matrix ρ_{mn}. In fact the initial information on the state of the system is often conveniently expressed as a set of (A^i) from which the ρ_{mn} are determined by solving a system of equation (A.35), one for each (A^i). Equation (A.35), which gives $(\langle A\rangle)$ for a generic A in terms of the matrix ρ, may be looked upon as a device to calculate a generic $(\langle A\rangle)$ from advance knowledge of a special set of $(\langle A^i\rangle)$.

A.5 The SU(N) group

A group G is a set of elements $\{a,b,c,d\ldots\}$ for which a 'multiplication' operation is defined with the following properties:

(i) *Closure:* if $a, b \in G$ then $a \cdot b \in G$.

(ii) *Identity:* G contains an 'identity' element $I \in G$ such that $a \cdot I = I \cdot a = a$.

(iii) *Inverse:* For every $a \in G$ there is an inverse element $a^{-1} \in G$ such that $a \cdot a^{-1} = a^{-1} \cdot a = I$.

(iv) *Associativity:* the multiplication operation is associative, i.e. $a \cdot (b \cdot c) = (a \cdot b) \cdot c$.

Groups can be finite (group elements in finite number), infinite (infinite number of group elements), discrete or continuous (group elements vary continuously). Symmetry transformations such as parity operations form a finite, discrete group, with two elements, P and I, with $P^2 = I$. Groups of all integers with addition as the 'group multiplication', and zero as the identity element is a discrete group, with an infinite number of elements. A group is called continuous when group elements are defined in terms of parameters that take continuous values. Rotation about different axes forms a continuous group.

The abelian and non-abelian groups. When a multiplication operation is commutative,

$$a \cdot b = b \cdot a \tag{A.44}$$

the group is called an abelian group[3], otherwise the group is non-abelian.

The Lie group. The Lie group is named after the Norwegian mathematician Marius Sophus Lie. He invented the theory of continuous symmetry. A continuous group where group elements are analytic functions of the parameters is called a Lie group. The Lie group has a special structure: all group elements can be written as

$$g = \exp\left(\sum_i \mathrm{i} R_i T_i\right) \tag{A.45}$$

where T_i are Hermitian matrices and R_i are real numbers. The T_i are called the generators of the Lie group.

Special unitary group SU(N) is a Lie group, isomorphic to that of all $N \times N$ special unitary matrices,

$$\det U = 1 \tag{A.46a}$$

$$U^\dagger U = 1. \tag{A.46b}$$

In general $N \times N$ complex matrices have $2N^2$ arbitrary real parameters. The condition $U^\dagger U = 1$ imposes N^2 conditions and $\det U = 1$ an additional one. Hence SU(N) has $N^2 - 1$ arbitrary parameters. Correspondingly SU(N) has $N^2 - 1$ generators, T_α, obeying,

$$\left[T_\alpha, T_\beta\right] = \mathrm{i} \sum_{\gamma=1}^{\gamma=N} f_{\alpha\beta\gamma} T_\gamma. \tag{A.47}$$

$f_{\alpha\beta\gamma}$ are the 'antisymmetric' structure constants (change sign for interchange of consecutive indices, $f_{\alpha\beta\gamma} = -f_{\alpha\gamma\beta} = f_{\gamma\alpha\beta}$). One immediately notes that SU(N) is non-abelian, and generators or group elements do not commute (in an abelian group, structure constants are zero and generator and group elements commute).

Representation of group. An abstract group G may be represented by a set of matrices $D(G)$. The set $D(G)$ itself constitutes a group such that the correspondence in the multiplication property is preserved: $D(a \cdot b) = D(a) \cdot D(b) \in D(G)$. If the mapping $G \to D(G)$ is one-to-one, then the mapping is isomorphic and 'faithful', otherwise it is 'homomorphic'. If $D(G)$ is a set of $M \times M$ matrices, the dimension of the representation is M. If the matrices can be reduced to block diagonal form

[3] The abelian group is named after the Norwegian mathematician Niels Henrik Abel. He, independently of the French mathematician, Evariste Galois, invented group theory. Apart from group theory, Abel is famous for his 'impossibility theorem', that general algebraic equations of order five or more cannot be solved analytically. Abel was largely unrecognized during his lifetime. In 1829, he died at the early age of 26 years.

(with at least two blocks), the representation is 'reducible', otherwise it is an 'irreducible' representation. SU(N) has two irreducible representations:

Fundamental representation of SU(N). If group elements are represented by $N \times N$ matrices, the representation is called a fundamental representation.

Adjoint representation of SU(N). This is the representation of the generators. Adjoint representation of SU(N) is generated by the structure constants f_{abc}. They can be used to define $(N^2 - 1) \times (N^2 - 1)$ matrices,

$$F^a_{bc} = f_{abc}. \tag{A.48}$$

Matrices F_a also satisfy the Lie algebra, $[F_a, F_b] = if_{abc}F_c$.

The Casimir operator. Casimir operators commute with all the group generators. The SU(N) group has $N - 1$ Casimir operators. In the fundamental representation, the Casimir operator is

$$C_F = \frac{N^2 - 1}{2N}\mathbf{1}_N, \tag{A.49}$$

$\mathbf{1}_N$ being the $N \times N$ unit matrix. In the adjoint representation, the Casimir operator is

$$C_A = N\mathbf{1}_{N^2-1}. \tag{A.50}$$

Naturally SU(3) has eight generators, $T_\alpha = \frac{1}{2}\lambda_\alpha$, $(\alpha = 1, 2, \ldots 8)$. λ_α are (3×3) Gell-Mann matrices, and act on the (color) basis states

$$x = \begin{pmatrix} x_1 \\ x_2 \\ x_3 \end{pmatrix}. \tag{A.51}$$

For the sake of completeness, we have noted the structure constants for SU(3) in table A.1. The eight Gell-Mann matrices are also listed,

$$\begin{aligned}
\lambda_1 &= \begin{pmatrix} 0 & 1 & 0 \\ 1 & 0 & 0 \\ 0 & 0 & 0 \end{pmatrix} & \lambda_2 &= \begin{pmatrix} 0 & -i & 0 \\ i & 0 & 0 \\ 0 & 0 & 0 \end{pmatrix}, \\
\lambda_3 &= \begin{pmatrix} 1 & 0 & 0 \\ 0 & -1 & 0 \\ 0 & 0 & 0 \end{pmatrix} & \lambda_4 &= \begin{pmatrix} 0 & 0 & 1 \\ 0 & 0 & 0 \\ 1 & 0 & 0 \end{pmatrix}, \\
\lambda_5 &= \begin{pmatrix} 0 & 0 & -i \\ 0 & 0 & 0 \\ i & 0 & 0 \end{pmatrix} & \lambda_6 &= \begin{pmatrix} 0 & 0 & 0 \\ 0 & 0 & 1 \\ 0 & 1 & 0 \end{pmatrix}, \\
\lambda_7 &= \begin{pmatrix} 0 & 0 & 0 \\ 0 & 0 & -i \\ 0 & i & 0 \end{pmatrix} & \lambda_8 &= \frac{1}{\sqrt{3}}\begin{pmatrix} 1 & 0 & 0 \\ 0 & 1 & 0 \\ 0 & 0 & -2 \end{pmatrix}.
\end{aligned} \tag{A.52}$$

Table A.1. Structure constant for SU(3).

ijk	123	147	156	246	257	345	367	458	678
f_{ijk}	1	$\frac{1}{2}$	$-\frac{1}{2}$	$\frac{1}{2}$	$\frac{1}{2}$	$\frac{1}{2}$	$-\frac{1}{2}$	$\frac{\sqrt{3}}{2}$	$\frac{\sqrt{3}}{2}$

One does note that Gell-Mann matrices are generalizations of Pauli matrices:

$$\sigma_1 = \begin{pmatrix} 0 & 1 \\ 1 & 0 \end{pmatrix} \qquad \sigma_2 = \begin{pmatrix} 0 & -\mathrm{i} \\ \mathrm{i} & 0 \end{pmatrix} \qquad \sigma_3 = \begin{pmatrix} 1 & 0 \\ 0 & -1 \end{pmatrix}. \tag{A.53}$$

Mathematically, quark fields transform as the fundamental representation. A generic SU(3) matrix can be written as

$$S = \mathrm{e}^{-\frac{\mathrm{i}}{2}\omega_a \lambda_a}. \tag{A.54}$$

An infinitesimal element of the group is represented by the transformation,

$$x' = Sx \tag{A.55a}$$

$$S = \mathrm{e}^{-\frac{\mathrm{i}}{2}\omega_a \lambda_a} \approx 1 - \frac{\mathrm{i}}{2}\omega_a \lambda_a \tag{A.55b}$$

where $\omega_a (\alpha = 1, 2, \ldots 8)$ are arbitrary infinitesimal real numbers.

A.6 The Riemann zeta function

The Riemann zeta function is an extremely important special function of mathematics and is frequently encountered in physics. It is a function of complex variable $s = (x + \mathrm{i}y)$ and is expressed as the infinite series

$$\zeta(s) = \sum_{n=1}^{\infty} \frac{1}{n^s} = \frac{1}{1^s} + \frac{1}{2^s} + \frac{1}{3^s} + \frac{1}{4^s} + \cdots \tag{A.56}$$

The function is intimately related to prime numbers and can also be written as

$$\zeta(s) = \prod_p \left(1 - p^{-s}\right)^{-1} \tag{A.57}$$

where $\prod_p$ is the product over all primes p.

The Riemann zeta function or, more precisely, the Riemann hypothesis played and continues to play an important part in the development of mathematical theory. Riemann zeta functions have trivial and non-trivial zeros. They have zeros at the negative even integers. The Riemann hypothesis states that non-trivial zeros of the zeta function have a real part $\frac{1}{2}$, i.e. non-trivial zeros lie on the line $\frac{1}{2} + \mathrm{i}t$, t being a real number. The hypothesis is one of the most challenging problems in mathematics, and has not been proved so far. Once, Hilbert was asked what would be on his mind if he were resurrected in 1000 years. He answered that he would inquire whether the Riemann hypothesis had been proved.

One can compute

$$\begin{aligned}\zeta(0) &= -1/2\\ \zeta(1) &= \infty\\ \zeta(2) &= \pi^2/6 \approx 1.645\\ \zeta(3) &= 1.202\\ \zeta(4) &= \pi^4/90 = 1.0823.\end{aligned}$$

$\zeta(0) = -\frac{1}{2}$ can be obtained from a globally convergent series for the Riemann zeta function (which provides the analytic continuation of the entire complex plane except at $s = 1$) and is given by

$$\zeta(s) = \frac{1}{1 - 2^{1-s}} \sum_{n=0}^{\infty} \frac{1}{2^{n+1}} \sum_{k=0}^{\infty} (-1)^k \binom{n}{k} (k+1)^{-s} \qquad s \neq 1. \tag{A.58}$$

The series gives

$$\begin{aligned}\zeta(0) &= \frac{1}{1-2} \sum_{n=0}^{\infty} \frac{1}{2^{n+1}} \sum_{k=0}^{\infty} (-1)^k \binom{n}{k} = -\sum_{n=0}^{\infty} \frac{1}{2^{n+1}} \delta_{n,0}\\ &= -\frac{1}{2}.\end{aligned} \tag{A.59}$$

One also notes an important relation between the Riemann zeta function and the Dirichlet eta function,

$$\eta(s) = \sum_{n=1}^{\infty} \frac{(-1)^{n-1}}{n^s} = \left(1 - 2^{1-s}\right)\zeta(s), \tag{A.60}$$

giving

$$\begin{aligned}\eta(0) &= (-1)\zeta(0) = 1/2\\ \eta(1) &= \ln 2\\ \eta(2) &= \frac{1}{2}\zeta(2) = \frac{1}{2}\frac{\pi^2}{6}\\ \eta(3) &= \frac{3}{4}\zeta(3) = \frac{3}{4}1.202\\ \eta(4) &= \frac{7}{8}\zeta(4) = \frac{7}{8}\frac{\pi^4}{90}.\end{aligned}$$

A.7 The Legendre transformation

Let $f = f(x, \zeta)$ be a function of two variables, x and ζ. Let

$$p = \frac{\partial f}{\partial x}. \tag{A.61}$$

We want to obtain a function $g(p, \zeta)$ which is uniquely defined by the function $f(x, \zeta)$. The function is the Legendre transformation of f,

$$g(p, \zeta) = f(x, \zeta) - px. \tag{A.62}$$

The transformation can be easily generalized to more than two variables. For example, if

$$f\left(x_1, x_2, \ldots, x_n, \zeta\right) \tag{A.63}$$

with

$$p_i = \frac{\partial f}{\partial x_i} \qquad i = 1, 2, \ldots n, \tag{A.64}$$

then the Legendre transformation of f is

$$g\left(p_1, p_2, \ldots, p_n, \zeta\right) = f(x_1, x_2, \ldots, x_n) - \sum_i p_i x_i. \tag{A.65}$$

The most well-known Legendre transformation is between the Lagrangian $\mathcal{L}(q, \dot{q})$ and Hamiltonian $H(p, q)$. The Hamiltonian is the Legendre transformation of the Lagrangian,

$$H(p, q) = \mathcal{L}(q, \dot{q}) - p.\dot{q} \tag{A.66}$$

with

$$p = \frac{\partial \mathcal{L}(q, \dot{q})}{\partial \dot{q}}. \tag{A.67}$$

A.8 The Dirac equation in a spherical cavity

By late 1960, it was firmly established that all the known hadrons can be built from three building blocks, the colored quarks. Non-observation of free quarks necessitated some kind of model to confine quarks within a hadron. Towards this end, Bogolioubov proposed a model of hadrons where quarks are confined by a scalar potential of infinite depth. Later, the model has developed as the MIT bag model. In the following, we briefly sketch the Bogolioubov model.

The Dirac equation for a particle of mass m under the influence of scalar potential $U(x)$ can be written as

$$\left[i\gamma^\mu \partial_\mu - (m + U(x))\right]\psi(x) = 0 \tag{A.68}$$

where ψ is a four-component vector, the Dirac spinor. For completeness, the usual forms of 4×4 γ matrices are given below,

$$\gamma^0 = \begin{pmatrix} I & 0 \\ 0 & -I \end{pmatrix} \qquad \gamma = \begin{pmatrix} 0 & \boldsymbol{\sigma} \\ -\boldsymbol{\sigma} & -0 \end{pmatrix} \tag{A.69}$$

where I is a 2×2 unit matrix and σ are the Pauli matrices,

$$\sigma_x = \begin{pmatrix} 0 & 1 \\ 1 & 0 \end{pmatrix} \qquad \sigma_y = \begin{pmatrix} 0 & -\mathrm{i} \\ \mathrm{i} & -0 \end{pmatrix} \qquad \sigma_z = \begin{pmatrix} 1 & 0 \\ 0 & -1 \end{pmatrix}. \tag{A.70}$$

One also defines a γ_5 matrix,

$$\gamma_5 = \gamma_5^{\dagger} = \mathrm{i}\gamma^0\gamma^1\gamma^2\gamma^3 = \begin{pmatrix} 0 & I \\ -I & -0 \end{pmatrix}. \tag{A.71}$$

γ matrices obey the anticommutation relations,

$$\{\gamma^{\mu}, \gamma^{\nu}\} = 2g^{\mu\nu} \qquad \{\gamma^{\mu}, \gamma^{5}\} = 0. \tag{A.72}$$

Writing $\psi(x) = \psi(\mathbf{x})e^{-iEt}$, the Dirac equation becomes.

$$[\gamma \cdot \mathbf{p} + (m + U(x))]\psi(\mathbf{x}) = E\gamma^0\psi(\mathbf{x}). \tag{A.73}$$

Writing the wave function $\psi(\mathbf{x}) = \begin{pmatrix} \psi_+ \\ \psi_- \end{pmatrix}$, the Dirac equation can be decomposed into two coupled equations,

$$\boldsymbol{\sigma} \cdot \mathbf{p}\psi_+ - [m + U(\mathbf{x})]\psi_- = E\psi_- \tag{A.74a}$$

$$\boldsymbol{\sigma} \cdot \mathbf{p}\psi_- + [m + U(\mathbf{x})]\psi_+ = E\psi_+. \tag{A.74b}$$

Unlike the non-relativistic case, even for a central potential, orbital angular momentum is not a good quantum number. The conserved quantities are energy E, j^2, j_3 and $K = \gamma^0(\Sigma \cdot \mathbf{l} + 1)$ where,

$$\mathbf{j} = \mathbf{l} + \frac{1}{2}\Sigma \qquad \Sigma = \begin{pmatrix} \boldsymbol{\sigma} & 0 \\ 0 & \boldsymbol{\sigma} \end{pmatrix}. \tag{A.75}$$

The eigenvalue equations can be written as

$$j^2\psi = j(j + 1)\psi \tag{A.76a}$$

$$j_3\psi = m\psi \tag{A.76b}$$

$$K\psi = -\kappa\psi. \tag{A.76c}$$

From equation (A.76c) it follows that,

$$(\boldsymbol{\sigma} \cdot \mathbf{l} + 1)\psi_+ = -\kappa\psi_+ \tag{A.77a}$$

$$(\boldsymbol{\sigma} \cdot \mathbf{l} + 1)\psi_- = +\kappa\psi_-. \tag{A.77b}$$

It can be shown that $\kappa = \pm\left(j + \frac{1}{2}\right)$ and the Dirac wave function ψ can be labeled by the quantum numbers, m and κ. Let us write the Dirac spinor as

$$\psi = \begin{pmatrix} \psi_+ \\ \psi_- \end{pmatrix} = \begin{pmatrix} g(r)\chi_\kappa^m \\ if(r)\chi_{-\kappa}^m \end{pmatrix}. \tag{A.78}$$

Consider the term $(\boldsymbol{\sigma} \cdot \mathbf{p})\psi_+$ in equation (A.74a). Using the identity

$$\begin{aligned} \boldsymbol{\sigma} \cdot \mathbf{p} &= \frac{(\boldsymbol{\sigma} \cdot \mathbf{r})}{r^2}(\boldsymbol{\sigma} \cdot \mathbf{r})(\boldsymbol{\sigma} \cdot \mathbf{p}) \\ &= \frac{(\boldsymbol{\sigma} \cdot \hat{\mathbf{r}})}{r}[\mathbf{r} \cdot \mathbf{p} + i\boldsymbol{\sigma} \cdot (\mathbf{r} \times \mathbf{p})] \end{aligned} \tag{A.79}$$

it can be written as

$$\begin{aligned} \boldsymbol{\sigma} \cdot \mathbf{p} g(r)\chi_\kappa^m &= \frac{\boldsymbol{\sigma} \cdot \hat{\mathbf{r}}}{r}[(\mathbf{r} \cdot \mathbf{p}) + \mathrm{i}\boldsymbol{\sigma} \cdot \mathbf{l}]g(r)\chi_\kappa^m \\ &= -\mathrm{i}\left[\frac{\mathrm{d}g}{\mathrm{d}r} + (\kappa + 1)\frac{g(r)}{r}\right](\boldsymbol{\sigma} \cdot \hat{\mathbf{r}})\chi_\kappa^m \\ &= \mathrm{i}\left[\frac{\mathrm{d}g}{\mathrm{d}r} + (\kappa + 1)\frac{g(r)}{r}\right]\chi_{-\kappa}^m. \end{aligned} \tag{A.80}$$

In the last line, we have used the relation

$$(\boldsymbol{\sigma} \cdot \hat{\mathbf{r}})\chi_\kappa^m = -\chi_{-\kappa}^m. \tag{A.81}$$

The relation can be proven exactly. We just note that the relation makes sense since $(\sigma \cdot \hat{\mathbf{r}})^2 = 1$ and $(\sigma \cdot \hat{\mathbf{r}})$ is a pseudo-scalar. Inserting equation (A.80) into equation (A.74a) we obtain,

$$\frac{\mathrm{d}g}{\mathrm{d}r} + \frac{\kappa + 1}{r}g - (m + U(r))f = Ef \tag{A.82a}$$

and similarly,

$$\frac{\mathrm{d}f}{\mathrm{d}r} + \frac{1 - \kappa}{r}f - (m + U(r))g = -Eg. \tag{A.82b}$$

Equations (A.82a) and (A.82b) are the Dirac equations for an m mass fermion in a scalar potential $U(r)$. The coupled equations can be solved with appropriate boundary conditions. Solutions are particularly simple for a massless quark, in a spherical cavity of radius R, when

$$U(r) = 0 \qquad r < R \tag{A.83a}$$

$$U(r) = \infty \qquad r \geqslant R. \tag{A.83b}$$

In the ground state $\kappa = -1$ and inside the cavity, equations (A.82*a*) and (A.82*b*) can be rewritten as

$$\frac{dg}{d(Er)} = f \tag{A.84a}$$

$$\frac{df}{d(Er)} + \frac{2}{Er}f = -g. \tag{A.84b}$$

Using $z = Er$ and inserting equation (A.84*a*) in equation (A.84*b*), one obtains

$$\frac{d^2g}{dz^2} + \frac{2}{z}\frac{dg}{dz} + g = z^2\frac{d^2g}{dz^2} + 2z\frac{dg}{dz} + z^2g = 0. \tag{A.85}$$

Now the solution of the differential equation,

$$Z^2\frac{d^2w}{dz^2} + 2z\frac{dw}{dz} + \left[z^2 - n(n+1)\right]w = 0, \tag{A.86}$$

is a spherical Bessel function of the first kind,

$$j_n(z) = \sqrt{\frac{\pi}{2z}}J_{n+\frac{1}{2}}(z). \tag{A.87}$$

One can immediately write the solution of equation (A.85),

$$g(r) = \mathcal{N}j_0(Er) = \mathcal{N}\frac{\sin Er}{Er} \tag{A.88}$$

with some normalization constant $\mathcal{N}$. Similarly, inserting equation (A.84*b*) in to equation (A.84*a*) one obtains

$$z^2\frac{d^2f}{dz^2} + 2z\frac{df}{dz} + \left(z^2 - 2\right)f = 0, \tag{A.89}$$

and may write the solution as

$$f(r) = \mathcal{N}j_1(Er) = \mathcal{N}\left[\frac{\sin Er}{(Er)^2} - \frac{\cos Er}{(Er)}\right]. \tag{A.90}$$

The confinement condition that $\bar{\psi}\psi = 0$ at the surface gives

$$\begin{aligned} \bar{\psi}\psi|_{r=R} &= \left[j_0(ER)\right]^2 - \left[j_1(ER)\right]^2 = 0 \\ \text{or} \quad j_0(ER) &= j_1(ER). \end{aligned} \tag{A.91}$$

From the tabulated values of the spherical Bessel functions, the equation is satisfied for

$$E = \frac{2.04}{R}. \tag{A.92}$$

A.9 Grassmann numbers

Grassmann numbers are mathematical constructions such that they are anti-commuting. A collection of Grassmann variables θ_i are independent elements of an algebra which contains real numbers that anticommute with each other, but commute with ordinary numbers x,

$$\theta_i\theta_j = -\theta_j\theta_i \tag{A.93}$$

$$\theta_i x = x\theta i \tag{A.94}$$

$$\theta_i^2 = 0. \tag{A.95}$$

One also notes that the operations of integration and differentiation are identical in Grassmann algebra,

$$\int 1\mathrm{d}\theta = 0 \tag{A.96}$$

$$\int \theta\mathrm{d}\theta = 1. \tag{A.97}$$

Grassmann numbers can always be represented as matrices. In general, a Grassmann algebra on n generators can be represented by $2^n \times 2^n$ square matrices.

Let us consider the integral

$$I = \int \mathrm{d}\eta^\dagger \int \mathrm{d}\eta \mathrm{e}^{-\eta^\dagger a\eta} \tag{A.98}$$

where a is an ordinary number. The exponential can be expanded as

$$\mathrm{e}^{-\eta^\dagger a\eta} = 1 - \eta^\dagger a\eta \tag{A.99}$$

and the integral can be evaluated as

$$I = \int \mathrm{d}\eta^\dagger \int \mathrm{d}\eta 1 - a \int \mathrm{d}\eta^\dagger \int \mathrm{d}\eta\eta^\dagger\eta \tag{A.100}$$

$$= a \int \mathrm{d}\eta^\dagger \Big(\int \mathrm{d}\eta\eta \Big) \eta^\dagger = a \int \mathrm{d}\eta^\dagger \eta^\dagger = a = \mathrm{e}^{\ln a}. \tag{A.101}$$

For two Grassmann variables,

$$\int \mathrm{d}\eta_1^\dagger \int \mathrm{d}\eta_1 \mathrm{e}^{-\eta_1^\dagger a_1 \eta_1} \int \mathrm{d}\eta_2^\dagger \int \mathrm{d}\eta_2 \mathrm{e}^{-\eta_2^\dagger a_2 \eta_2} = a_1 a_2 = \mathrm{e}^{\ln a_1 a_2}. \tag{A.102}$$

The quantity $a_1 a_2$ is the determinant of the diagonal matrix with a_1 and a_2 on the diagonal. In a general basis then we write in matrix notation

$$\prod_i \int \mathrm{d}\eta_i^\dagger \int \mathrm{d}\eta_i \mathrm{e}^{-\eta_i^\dagger \mathbf{A} \eta_i} = \det A = \exp[\mathrm{Tr} \ln A]. \tag{A.103}$$

A.10 The Wigner function

In classical statistical mechanics, we have encountered the distribution function $f(x, p)$ which gives the probability of finding particles with momenta p at the position x. The Wigner function is the quantum analog of the classical distribution function. In classical mechanics, one deals with trajectories, while quantum mechanics, on the other hand, inherently deals with probability. Connections between the two (if any) can be obtained by comparing ensembles of trajectories in phase space for the classical case and density distributions, e.g. Wigner functions for the quantum case.

In quantum mechanics, given a wave function ψ, the expectation value of an operator A is given as

$$\langle A \rangle = \langle \psi | A | \psi \rangle. \tag{A.104}$$

Let us first consider Weyl transformation for A,

$$\tilde{A}(x, p) = \int \mathrm{d}y \mathrm{e}^{-\mathrm{i}py} \left\langle x + \frac{y}{2} \right| A \left| x - \frac{y}{2} \right\rangle \tag{A.105}$$

where the operator has been expressed in the coordinate basis. If the operator is expressed in the momentum basis, Weyl transformation is obtained as

$$\tilde{A}(x, p) = \int \mathrm{d}u \mathrm{e}^{-\mathrm{i}xu} \left\langle p + \frac{u}{2} \right| A \left| p - \frac{u}{2} \right\rangle. \tag{A.106}$$

The Weyl transformation converts an operator into a function of x and p. An important property of Weyl transformation is,

$$\mathrm{Tr}[AB] = \frac{1}{h} \int \int \tilde{A}\tilde{B} \mathrm{d}x \mathrm{d}p. \tag{A.107}$$

In appendix A.4 we introduced the notion of the density matrix. In terms of the density matrix, the expectation value of an operator is

$$\langle A \rangle = \mathrm{Tr}[\rho A]. \tag{A.108}$$

It can now be written as

$$\langle A \rangle = \mathrm{Tr}\left[\tilde{\rho}\tilde{A}\right] = \frac{1}{h} \int \tilde{\rho}\tilde{A} \mathrm{d}x \mathrm{d}p. \tag{A.109}$$

The Wigner function is defined as

$$\begin{aligned} W(x, p) &= \frac{\tilde{\rho}}{h} = \frac{1}{h} \int \mathrm{e}^{-\mathrm{i}py} \left\langle x + \frac{y}{2} \right| \rho \left| x - \frac{y}{2} \right\rangle \mathrm{d}y \\ &= \frac{1}{h} \int \mathrm{e}^{-\mathrm{i}py} \psi\left(x + \frac{y}{2}\right) \psi^*\left(x - \frac{y}{2}\right) \mathrm{d}y \end{aligned} \tag{A.110}$$

In terms of the Wigner function, the expectation value of the operator A is

$$\langle A\rangle = \int \tilde{A}(x,p)W(x,p)\mathrm{d}x\mathrm{d}p. \tag{A.111}$$

The expression closely resembles the expression for the average value in classical mechanics in terms of the distribution function $f(x,p)$,

$$\langle A\rangle = \int A(x,p)f(x,p)\mathrm{d}x\mathrm{d}p. \tag{A.112}$$

It is easy to prove that the Wigner function, when integrated over p, gives the probability distribution for x, and when integrated over x, gives the probability distribution for p.

www.ingramcontent.com/pod-product-compliance
Ingram Content Group UK Ltd.
Pitfield, Milton Keynes, MK11 3LW, UK
UKHW051901100926
8556IPUK00005B/721

9 780750 31061